Prosody, Phonology and Phonetics

The series will publish studies in the general area of Speech Prosody with a particular (but non-exclusive) focus on the importance of phonetics and phonology in this field. The topic of speech prosody is today a far larger area of research than is often realised. The number of papers on the topic presented at large international conferences such as Interspeech and ICPhS is considerable and regularly increasing. The proposed book series would be the natural place to publish extended versions of papers presented at the Speech Prosody Conferences, in particular the papers presented in Special Sessions at the conference. This could potentially involve the publication of 3 or 4 volumes every two years ensuring a stable future for the book series. If such publications are produced fairly rapidly, they will in turn provide a strong incentive for the organisation of other special sessions at future Speech Prosody conferences.

More information about this series at http://www.springer.com/series/11951

Benjamin Weiss · Jürgen Trouvain ·
Melissa Barkat-Defradas ·
John J. Ohala
Editors

Voice Attractiveness

Studies on Sexy, Likable, and Charismatic Speakers

Springer

Editors
Benjamin Weiss
Technische Universität Berlin
Berlin, Germany

Jürgen Trouvain
Saarland University
Saarbrücken, Germany

Melissa Barkat-Defradas
ISEM
Montpellier, France

John J. Ohala
International Computer Science Institute
Berkeley, CA, USA

ISSN 2197-8700 ISSN 2197-8719 (electronic)
Prosody, Phonology and Phonetics
ISBN 978-981-15-6626-4 ISBN 978-981-15-6627-1 (eBook)
https://doi.org/10.1007/978-981-15-6627-1

This Springer imprint is published by the registered company Springer Nature Singapore Pte Ltd.
The registered company address is: 152 Beach Road, #21-01/04 Gateway East, Singapore 189721, Singapore

John Ohala passed away in August 2020, shortly before the publication of this volume. We dedicate this book to the memory of a great researcher.

Preface

At the Interspeech conference 2015, in Dresden, John (Ohala) asked Jürgen (Trouvain) what he thinks about organizing a special session on attractive voices, maybe for the next conference in this series. A former visiting researcher in Berkeley, Melissa (Barkat-Defradas), had already expressed some ideas on such an event on this topic. John has a long-standing interest in evolutionary aspects of speech and voice, Melissa works in an interdisciplinary research team on all kinds of aspects of evolution, and Jürgen has some background in paralinguistic characteristics of speech. At the same conference in Dresden, Jürgen introduced Benjamin (Weiss) to John with Benjamin as the optimal complement to this team since he has published several papers on social likability of voices.

It was then at Interspeech in Stockholm 2017, that we were able to organize the planned special session on voice attractiveness. We considered this event as the perfect setting for presenting research dealing with many aspects: perceived vocal preferences of men, women, and synthesized voices in well-defined social situations, acoustic correlates of voice attractiveness/pleasantness/charisma, interrelations between vocal features and individual physical and physiological characteristics, consequences for sexual selection, predictive value of voice for personality and for other psychological traits, experimental definition of esthetic standards for the vocal signal, cultural variation of voice attractiveness/pleasantness and standards, and also the link between vocal pathology and vocal characteristics. In Stockholm we agreed on a follow-up publication where the authors have more space than in a conference paper with its strict limitations. Moreover, also those colleagues could be reached that were not participants of this conference.

The special session was a success in our view. In total, we had nine accepted contributions. Authors from six papers of this session are also aboard in this volume. In addition to these, there are ten further contributions for this publication, having a total of seventeen papers when we add the introductory chapter. It is our belief that both collections, the nine conference papers, and the seventeen articles in this volume, can provide a useful overview on the state-of-the-art research on voice attractiveness, voice likability, and vocal charisma. We also hope that these studies represent a fruitful fundament for further thoughts and investigations of an exciting field of speech and voice research.

As many book projects of this size, the editing process took longer than expected. This delay is mainly but note entirely due to health reasons of some of the editors. We would like to thank all authors for their patience and the publishing house for the provided support.

Berlin, Germany	Benjamin Weiss
Saarbrücken, Germany	Jürgen Trouvain
Montpellier, France	Melissa Barkat-Defradas
Berkeley, USA	John J. Ohala
April 2020	

Contents

Editors and Contributors

About the Editors

Benjamin Weiss received his Ph.D. in 2008, in phonetics from Humboldt-University, Berlin. Since then he has extensively studied acoustic correlates of pleasant and likable voices, taking into account also speaking styles and conversational behavior in order to build quantitative models. He was visiting fellow at the University of Western Sydney and the University of Technology Sydney. In 2019, he completed his habilitation on human dialog and speech-based (multimodal) HCI. Since September 2020, he is an Associate Professor at the School of Intelligence, Hanyang University, Seoul.

Jürgen Trouvain received his Ph.D. in Phonetics in 2004, from Saarland University (Germany), where he works as a Senior Researcher and Lecturer at the Department of Language Science and Technology. His research fields include nonverbal vocalizations such as breathing and laughing, as well as non-native speech and phonetic learner corpora. He has acted as an organizer for several international conferences and workshops.

Melissa Barkat-Defradas obtained her Ph.D. in Forensic Linguistics at the University of Lyon, in 2000, and received the Young Researcher Award for her work in Automatic Language Identification. After a research fellowship at UC Berkeley, she joined the French National Centre for Scientific Research. She is now a full-time Researcher at The Institute of Evolutionary Sciences of Montpellier (France), where she actively contributes to developing interdisciplinary research by bridging the gap between experimental phonetics and evolutionary biology. She is particularly interested in the selective forces that may explain the emergence of articulated language in humans.

John J. Ohala is an Emeritus Professor of Linguistics at the University of California, Berkeley, and a Research Scientist at the International Computer Science Institute, Berkeley. He has had a major impact on the field of speech communication. His research interests focus on experimental phonology and phonetics and ethological aspects of communication, including speech perception, sound change, phonetic and phonological universals, psycholinguistic studies in phonology, and sound symbolism. He proposed an innovative ethological hypothesis, which unifies—via "the frequency code"—such diverse behavioral phenomena as the cross-language use of voice pitch for questions and statements, the systematic use of consonants, vowels, and tones in sound symbolical vocabulary, the "smile," and sexual dimorphism of the vocal anatomy in adult humans.

Contributors

Anita Auszmann Department of Head and Neck Surgery and Linguistics, University of California, Los Angeles, CA, USA

Molly Babel Department of Linguistics, University of British Columbia, BC, Canada

Melissa Barkat-Defradas Institut des Sciences de l'Evolution de Montpellier, University of Montpellier, Centre National de la Recherche Scientifique, Institut pour la Recherche et le Développement, Ecole Pratique des Hautes Etudes – Place Eugène Bataillon, Montpellier, France

Timo Baumann Universität Hamburg, Language Technology Group, Hamburg, Germany

Pascal Belin Institut de Neurosciences de La Timone, CNRS et Aix-Marseille Université Département de Psychologie, Université de Montréal, Montreal, Canada

Geke Boomsma Utrecht institute of Linguistics, Utrecht University, Utrecht, The Netherlands

Hans Rutger Bosker Max Planck Institute for Psycholinguistics, Nijmegen, The Netherlands; Psychology of Language Department, Donders Institute for Brain Cognition and Behaviour, Radboud University, Nijmegen, The Netherlands

Alexander Brem Innovation and Technology Management, Friedrich-Alexander-Universität Erlangen-Nürnberg, Erlangen, Germany

Felix Burkhardt audEERING GmbH, Berlin, Germany

Leigh Clark School of Information, & Communication Studies, University College Dublin, Dublin, Ireland; Computational Foundry, Swansea University, Swansea, UK

Benjamin R. Cowan School of Information, & Communication Studies, University College Dublin, Dublin, Ireland

Bruce R. Gerratt Department of Head and Neck Surgery, University of California, Los Angeles, CA, USA

Julia Hirschberg Columbia University, NYC, New York, NY, USA

Jody Kreiman Department of Head and Neck Surgery and Linguistics, University of California, Los Angeles, CA, USA

Grant McGuire Department of Linguistics, University of California Santa Cruz, Santa Cruz, CA, USA

Jan Michalsky University of Oldenburg, Oldenburg, Germany

Oliver Niebuhr Mads Clausen Institute, Centre for Electrical Engineering, University of Southern, Odense, Denmark

Yasunari Obuchi School of Media Science, Tokyo University of Technology, Hachioji, Tokyo, Japan

Abdulmalik Ofemile English Department, FCT College of Education, Zuba, Abuja, Nigeria

Hugo Quené Utrecht institute of Linguistics, Utrecht University, Utrecht, The Netherlands

Michel Raymond Institut des Sciences de l'Evolution de Montpellier, University of Montpellier, Centre National de la Recherche Scientifique, Institut pour la Recherche et le Développement, Ecole Pratique des Hautes Etudes – Place Eugène Bataillon, Montpellier, France

Andrew Rosenberg Google LLC, NYC, New York, NY, USA

Heike Schoormann University of Oldenburg, Oldenburg, Germany

Rosario Signorello Laboratoire de Phonétique et Phonologie, CNRS & Sorbonne Nouvelle, Paris, France

Alexandre Suire Institut des Sciences de l'Evolution de Montpellier, University of Montpellier, Centre National de la Recherche Scientifique, Institut pour la Recherche et le Développement, Ecole Pratique des Hautes Etudes – Place Eugène Bataillon, Montpellier, France

Ilaria Torre Department of Electronic and Electrical Engineering, Trinity College Dublin, Dublin, Ireland

Jürgen Trouvain Saarland University, Saarbrücken, Germany

Romée van Erning Utrecht Institute of Linguistics, Utrecht University, Utrecht, The Netherlands

Benjamin Weiss Technische Universität Berlin, Berlin, Germany

Laurence White School of Education, Communication and Language Sciences, Newcastle University, Newcastle, UK

Chloe Willis Department of Linguistics, University of California Santa Barbara, Santa Barbara, CA, USA

Part I
General Considerations

Chapter 1
Voice Attractiveness: Concepts, Methods, and Data

Jürgen Trouvain, Benjamin Weiss, and Melissa Barkat-Defradas

Abstract This book comprises contributions on vocal aspects of attractiveness, social likability, and charisma. Despite some apparent distinct characteristics of these three concepts, there are not only similarities, but even interdependencies to be considered. This chapter introduces and regards the concepts studied, methods applied, and material selected in the contributions. Based on this structured summary, we argue to increase interdisciplinary and even holistic efforts in order to better understand the concepts for voice and speech in humans and machines.

Keywords Attractiveness · Charisma · Likability · Sexual selection · Interdisciplinary · Holistic view · Structured summary · Speech production · Speech perception

1.1 Introduction

Probably, everybody has an idea of the meaning or meanings of *attractive* and *attractiveness* on the one side, and of voice and speaker on the other. It is also likely that everybody has their own ideas, which voices sound attractive—either in general or in specific contexts. But these ideas show by no means homogeneous structures and similar definitions.

A book on voice attractiveness attracts researchers, be it as authors and/or readers, who look at this topic from different angles as the subtitle of this book indicates. A *sexy* speaker is not the same as a *likable* speaker, and a *charismatic* speaker is different

---

J. Trouvain
Saarland University, Campus C7.2, 66123 Saarbrücken, Germany
e-mail: trouvain@coli.uni-saarland.de

B. Weiss (✉)
Technische Universität Berlin, Ernst-Reuter-Platz 7, 10405 Berlin, Germany
e-mail: benjamin.weiss@tu-berlin.de

M. Barkat-Defradas
University of Montpellier, Place Eugène Bataillon cc065, 34090 Montpellier cedex 05, France
e-mail: melissa.barkat-defradas@umontpellier.fr

B. Weiss et al. (eds.), *Voice Attractiveness*, Prosody, Phonology and Phonetics,
https://doi.org/10.1007/978-981-15-6627-1_1

again. These differences of how attractiveness is considered are also reflected in the chapters of this book. Likewise, the definition of speaker and voice is heterogeneously used, too. For this reason, we first attempt to shed some light onto the diversity of concepts we face in the upcoming chapters.

There is a broad range of different methods used in the studies of this volume. Many perform experimental research to investigate aspects of production, acoustics, and perception of attractive speech. There are some studies with a focus on modeling of data with respect to attractiveness, whereas other studies review how speech technology can be applied taking the (missing) attractiveness of voices into account. The data types that were used in the studies of this volume also show a large span. They range from manipulations of monosyllabic stimuli over single words and sentences in controlled settings up to many minutes of spontaneous conversational speech. The recap of the diversity of methods and data in this collection is followed by some concluding remarks on the emerging field of voice attractiveness, a research field that attracts researcher from many disciplines.

1.2 Concepts

1.2.1 Voice, Speaker, and Speech

The contributions of this collection consider the *voice* and *voice attractiveness* in different ways. Voice is not only seen in a narrow sense where it refers only to glottal activity. Voice in a wider sense additionally includes supra-glottal activities such as tongue raising, pharyngeal constriction, nasality or lip spreading (Laver, 1980), so that for instance formants as acoustic correlates of supra-laryngeal resonances are taken into account. For several studies, prosody plays an important role, reflected by fundamental frequency (F0), intensity, pauses and duration from a suprasegmental point of view. Further, timing parameters refer to entrainment in dialogs.

Naively, one would not assume that a voice that is considered as "normal", "stereotypical" or "average" would correlate to attractiveness. Nevertheless, three papers of this volume look more closely to the acoustic parameters of the "mean" voice and its perception of attractiveness—partially with somewhat surprising results.

Kreiman et al. (this volume) show that listeners differ regarding the question of what it means for a voice to sound "normal". There seem to be individual, rather consistent, strategies to label how normal or not normal a voice sounds. In their study, listeners assessed a wide range of one second samples of female speakers. From several acoustic parameters, the most relevant for explaining some amount of variance in the labels are fundamental frequency and its variation, as well as the first two formants, but not others that are typically associated with voice quality. However, the authors could not find a simple or generally valid answer, the situation is rather complex because several factors like the listener, the context, the purpose of the judgment, and of course the individual voice have to take into account.

The topic of recalling a voice from memory, an everyday task for everybody of us, is analyzed in Babel et al. (this volume). They show in a set of experiments with monosyllabic words as stimulus material that subjective stereotypicality and attractiveness affect the performance to remember a voice. Overall, they found support for the statement that less stereotypical voices and less attractive voices were better memorized.

Belin (this volume) reports of findings of experiments where identical short syllables of multiple voices of the same sex were averaged. The more voices were averaged the "speakers" of the averaged voice samples were perceived as more and more attractive. (similar to a visual effect concerning face attractiveness). Obviously, the main responsible factors for this effect are the reduced "distance-to-mean" for differences between F0 and the first formant, and an increased "texture smoothness" reflected by a raised harmonics-to-noise ratio.

There are also studies with stimuli to be rated that are longer than just one syllable or just one second. These studies concentrate more on speech prosody. Quené et al. (this volume), for instance, control for tempo and F0 in stimuli sentences, and Bosker (this volume) analyzed amplitude modulation in authentic speech samples. The review of charismatic speech of Rosenberg and Hirschberg (this volume) centers at prosody in all possible aspects, whereas, for instance, Weiss et al. (this volume) investigate acoustic parameters that reflect prosody (F0, intensity, rate), segmental properties (formants, spectral features) but also the voice in a narrow sense (shimmer, jitter, harmonics-to-noise ratio). These examples show that the vocal part in voice attractiveness can be referred to very different aspects of voice and speech when performing research in this field.

1.2.2 Sexual Selection and Voice Attractiveness

A sexy speaker can be seen as somebody who underlines her or his perceived sexual attractiveness—often unconsciously—with her or his voice and speech behavior. Though the voice is the privileged medium for interpersonal communication, it is not solely useful for conveying semantic information to other people. As a matter of fact, voice should also be regarded as a powerful social object, whose role is crucial in the context of human relationships. Indeed, by using oral communication, speakers are not only able to share their ideas and emotions, but they are also able to signal some reliable sociobiological features to their interlocutors such as sex, age, health, and social status, among others. There is a large body of scientific literature, for instance Scherer (1978), which describe the links between voice characteristics and personality traits, or the works by Laver and Trudgill (1979) and Bezooijen (1995), who studied voice as a social and cultural marker, or either still, Banse and Scherer (1996) whose work investigate how voice is used to express one's emotional state.

All of these authors, to name a few, have demonstrated that voice goes far beyond its primary linguistic function. Yet, interestingly, researches in Humanities mostly tackled the topic of vocal function independently of any evolutionary considerations. However, as early as 1890, Darwin addressed the issue within the frame of sexual selection by drawing intriguing parallels between animal vocalizations and the human voice:

> The sexes of many animals incessantly call for each other during the breeding-season; and in not a few cases, the male endeavors thus to charm or excite the female. This, indeed, seems to have been the primeval use and means of development of the voice [...]. When male animals utter sounds in order to please the females, they would naturally employ those which are sweet to the ears of the species; and it appears that the same sounds are often pleasing to widely different animals, owing to the similarity of their nervous systems, as we ourselves perceive in the singing of birds and even in the chirping of certain tree-frogs giving us pleasure. (Darwin, 1890, pp. 90–96).

Darwin's original idea according to which vocalizations allow the transmitter to attract females' attention and express his reproductive intentions make it legitimate to address the issue of human voice attractiveness in the specific context of human mating. As a matter of fact, as it is developed in the first contribution of Suire, Raymond, and Barkat–Defradas (this volume), it is reasonable to think that sexual selection—the mechanism which promotes biological and social traits that confer a reproductive benefit—has also intervened in the shaping of human vocal dimorphism; the attractiveness of a voice being a proxy, or a reinforcing signal, for other physical characteristics. By providing an overview of the research that lies at the crossroad of the human voice and evolutionary biology, the authors aim at demonstrating that sexual selection provides an interesting theoretical framework to understand the functional role of the human voice from an evolutionary perspective. Indeed, several studies have demonstrated the existence of a vocal attractiveness stereotype, which suggests that voice is an honest signal[1] of phenotypic quality in the same way as other physical features like, for example, the waist-to-hip ratio.[2]

Such an assumption raises the question of what makes a voice attractive? In their survey of the literature, Rosenberg and Hirschberg (this volume) examine the concept of vocal attractiveness itself. The authors consider the concept as highly context-dependent and discriminate between several types of attraction (i.e., political charisma, business leadership, nonsexual attraction and, last but not least, romantic desirability) each one of them being associated with specific articulatory, acoustic, and prosodic traits. They also show that though voice attractiveness is a complicated and exceptionally subjective phenomenon, evidence suggests some shared cross-cultural patterns that must have been shaped in the course of evolution by the selective pressure induced by the preferences of one sex for the vocal attributes of the other. The topic of vocal preferences has given rise to a large body of literature on the evolution of vocal preferences, which generally speaking, reveals that low-pitched

[1]Signals are traits that have evolved specifically because they change the behavior of receivers in ways that benefit the signaler. For example, peacock resplendent tail feathers are honest since they truly signal reproductive fitness of their bearer to the receiver.

[2]The waist-to-hip ratio (WHR) is the dimensionless ratio of the circumference of the waist to that of the hip. WHR correlates with health and fertility (with different optimal values in males and females).

masculine voices are universally preferred by women, such voices being perceived as related to a high quality phenotype. Conversely, men tend to prefer high-pitched feminine voices that are perceptually associated with youth and fertility at least in English. For more details of evolutionary mechanisms of attractive voices like mate choice see the systematic review of vocal preferences in humans by Barkat–Defradas, Raymond and Suire (this volume). Quené et al. (this volume) also confirm the expected pattern that men with lower-pitched voices tend to be rated as more attractive by (heterosexual) female listeners. They also reveal the importance of fast tempo in voice attractiveness evaluation. Indeed, their results based on manipulated speech show that the female raters judged masculine voices as less attractive if the F0 was artificially raised and the tempo decreased.

In their speed dating study, Michalsky and Schoormann (this volume) investigated the effects of perceived attractiveness and conversational quality on entrainment. In analyzing speed dating dialogs, prosodic disentrainment, in terms of pitch differences, is related to facial attractiveness for interlocutors of opposing sex. However, this result is inhibited by high conversational quality for females, and low conversational quality for males.

1.2.3 Likability and Social Attractiveness

A likable speaker is seen as somebody who underlines her or his perceived social attractiveness or pleasantness with her or his voice and speech behavior. There are several potential aspects that may constitute likability. For example, from the two of the most stable interpersonal concepts for unacquainted persons, benevolence (or warmth, communion) and competence (or agency, capability) (Abele, Cuddy, Judd, & Yzerbyt, 2008; Schaller, 2008; Fiske, Cuddy, & Glick, 2006), the first dimension (benevolence) is often assumed to resemble likability (DePaulo, Kenny, Hoover, Webb, & Oliver, 1987; Fiske et al., 2006; Argyle, 1988). However, liking-aversion may conceptually comprise the second dimension of competence as well (McCroskey & McCain, 1974), even in speech (Putnam & Street, 1984). Actually, there is much evidence from questionnaire analysis in a speech during dimension reduction that evaluative questionnaire items, such as "likable", can be apparent in both dimensions, benevolence and competence, or neither (Cuddy, Fiske, & Glick, 2008; Brown, Strong, & Rencher, 1973, 1985; Hart & Brown, 1974; Street & Brady, 1982; Weirich, 2010; Weiss & Möller, 2011). Given these empirical results, it can be argued that the so-called benevolence is just one possible but a very likely attribution to a person, which affects a speaker's social attractiveness, especially in a first impression.

Concerning voice acoustics, there are only few correlates of likability that show at least some robustness to changes in material, most notably increased pitch variability and tempo, while the results of average pitch reveal to be more complex, at least in German (Weiss et al., this volume).

While such results aim at correlates of averaged ratings on a scale, paired comparisons allow for a much finer measure of preference in likability. This method is, unfortunately, much more effort. Therefore, a crowd-based procedure is presented to collect such data efficiently, and it was used to train a model for predicting preferences of pairs of stimuli (Baumann, this volume).

In order to better take into regard the individual aspects of attractiveness, a method is presented that extracts overall voice attractiveness and listeners' preferences from paired comparisons, so that voices' likability can be estimated by the inner product of the two vectors of attractiveness and preferences (Obuchi, this volume).

1.2.4 Charisma and Leadership

A charismatic speaker is seen as somebody who underlines her or his perceived leadership, persuasive power, enthusiasm, and passion with her or his voice and speech behavior. Charisma is, just like likability, a social evaluation. However, likability typically refers to a dialogic situation, or in passive listening test, to the anticipation of a dialog—without any predefined difference in social status. In contrast to this, charisma is typically about an individual affecting a group of people, and thus implies some kind of social superiority. Charismatic people stand out, formally by social status or rank, or situationally by other's acknowledgment of their specialty. Therefore, the typical domains to study charisma in voice are speeches or talks of famous people, such as politicians and managers. A passionate and motivating speech by such people represents an often used, and sometimes even requested and anticipated, method of leadership. A discursive overview of what a charismatic voice actually is, can be found in Signorello (this volume).

The focus on public speeches and talks when dealing with charisma, complicates, on the one hand, differentiating between effects of a speech's presentation from those that originate in the fame, attributions, and social status. On the other hand, instead of relying on ratings in the laboratory, there a plenty of potentially valid indicators of charisma of those famous people including type of applause, (social) media reaction, and election results. For example, during a party conference of the German social democrats in 1995, the chairman was replaced by his vice-chairman—atypically early at this specific date—after an inspiring and enthusiastic speech of that vice-chairman. Given rather similar contents, sometimes even identical formulations, this outcome of the election was analyzed not regarding rhetorics, but speaking style instead (Paeschke & Sendlmeier, 1997). Such occurrences not only show that charisma is blended with power and leadership, but also exemplify the relevance of voice and speech for charisma. In this volume, the relevance of prosody and attire is studied for speeches of leading senior managers (Brem & Niebuhr, this volume). And in Bosker (this volume), a closer look on the modulation spectrum, which is related to speech rhythm, is taken for speeches from the US presidential campaign candidates Hillary Clinton and Donald Trump.

1.3 Methods

From a methodological perspective, we can divide studies on voice attractiveness in three fields. Investigations of the possible effects of different kinds of attractiveness and their vocal correlates are covered by *experimental research*. In addition to this research direction, *modeling* of processes how individual voices in audio samples attract listeners represents a further field of study. Finally, *technological applications* should be viewed as an own field of research in voice attractiveness.

1.3.1 Experimental Research

Human attractiveness is typically considered as a subjective concept. Therefore, experimental research is dominated by collecting explicit and implicit human ratings and decisions. The simplest methodological approach is to present stimuli and explicitly ask for ratings; on a scale if sequentially presented, or as a preference in the case of comparing stimuli. Such listening and ratings are, for example, conducted by Babel et al. (this volume). They collected a variety of subjective characteristics, among them perceptual similarity, applying a comparison of pairs of stimuli on a single scale, and perceptual attractiveness, collecting ratings in a sequential procedure for each stimulus individually. The latter method is also frequently used in the studies evaluated by Belin (this volume). Quené et al. (this volume) explicitly argue in favor of the sequential approach with absolute ratings instead of a forced preference choice of a direct comparison, as they want to avoid drawing attention to the signal manipulations they have conducted. There are various variants applied, often taken advantage of graphical computer interfaces, for example, to sort and assign short stimuli of a set to labels (Kreiman et al., this volume).

Instead of explicitly asking for measures of attractiveness, implicit measures can be attempted to collect, in order to avoid a social bias of the subjects. Such approaches comprise observations of social decisions, for example, counting the number of direct interactions in gaming or game-like tasks (Krause, Back, Egloff, & Schmukle, 2014). Other observations refer to the number of friends, or offspring (or explicitly asking to disclose the number of sexual partners). Such long-term or retrospective observations and surveys are, however, difficult to relate to specific traits, such as vocal characteristics.

1.3.2 Modeling

Quantitative modeling of subjective human ratings, such a sexual or social attractiveness, serves in principle two purposes. One is to describe the relations, e.g., correlations, found with parameters of interest in a given data set. Such a model could be a starting point for a prediction model, but does not provide explanatory power as in a scientific theory. For the case of voice attractiveness, typical model parameters are acoustic or articulatory measures. Another purpose is to actually explain interdependencies between parameters and ratings in a quantitative way. However, in the latter case, the parameters chosen and the kind of relationship have to be confirmed by methodological means ensuring a causal relationship. Synthesizing or resynthesizing speech represents the most popular approach to control for the variables in question. It also aims at providing proof for a causal relationship. As the knowledge base is enhanced by empirical studies incrementally, each study might fulfill both purposes to some degree. For example, the linear models of social attractiveness of Weiss et al. (this volume) build on hypotheses drawn from several scientific methods in order to add evidence for acoustic-perceptual relations, but its main result is a simple data description.

Baumann (this volume), present a methodological approach, that does comprises not only the acoustic modeling part, but also a method to efficiently collect preference ratings for stimulus pairs. Such pairwise preferences for German spoken Wikipedia articles were acoustically correlated directly, and modeled as relative preferences by means of a recurrent neural network.

In a related approach, Obushi (this volume) collected pairwise preferences for a Japanese greeting phrase. The ratings are multidimensionally analyzed, taking into account the listeners' differences as well, and modeled by multiple acoustics features applying machine learning.

1.3.3 Technological Applications

Voice attractiveness can play an essential role in human-machine interaction (HMI) as two contributions in this volume show. There is a tendency that "people tend to attribute personality traits to computers and robots as if they were human agents" (Nass, Moon, Fogg, Reeves, & Dryer, 1995). That means that the human-sounding voices of talking and conversational computers can also be considered as personalized machines. In addition, machines can act for humans, for instance, when a speech synthesizer is used as a speech prosthesis for people who cannot clearly and fluently articulate anymore. From a view of listening to talking machines, we all know that it is most of the time rather boring and less interesting when faced with an artificial voice and synthesized speech, be it when street names are announced in car navigation or when interacting with a dialog system. For conversational agents, e.g., intelligent personal assistants, it is a particular challenge to show skills that are required for smooth dialogs that span aspects of timing up to common grounding. Thus, voice selection and voice modeling should be an integral part of the design in HMI tools. The paper collected in this volume are not empirical studies with existent systems but are reviews in which important thoughts are developed before experiments that test the usability of certain aspects of voice attractiveness are performed.

Torre and White (this volume) focus on the characteristics of a robot's voice in human-robot interaction. They are particularly interested in how vocal elements can contribute to the impression of trustworthiness. They review studies in which a robot's voice was analyzed or manipulated, always with a particular view on trustworthiness. Naturalness and "machine-likeness", cognitive load, incongruity with the robot's behavior in general and the robot's appearance such as its size, gender, accent, and interaction context. Furthermore, they argue that the design of robot voices should come with an unambiguous appearance and function, because unrealistic expectations of robot performance in human users should be avoided.

The human evaluation in regard to different kinds of attractiveness represent immanent social and cognitive processes. Such evaluations are, however, not limited to other living persons. Instead, interactive systems, especially those using speech, are known to evoke similar processes (Reeves & Nass, 1996; Nass & Brave, 2005). And with the emergence of speech interaction with computers in the form of personal smartphone assistants, smart home devices, virtual persons, and human-like (social) robots, the users' appraisal of the verbal and nonverbal behavior of such interactive computers are receiving much attention.

One observation specific to anthropomorphic computers is the so-called "uncanny valley" effect. It describes an overall increase in familiarity (or attractiveness or likability) with increasing human-likeliness (or level of details) of the systems features and movements that is disrupted by a sudden decrease in familiarity close to perfect human-likeliness (Mori, 2012). This awkward or eerie feeling for a close to human, but obviously not natural synthesis is typically explained by a shift in reference from artificial to human and can be circumvented by reducing the level of human-likeliness or choosing an artificial metaphor (e.g., a puppet or cartoon) instead of a human. This effect is mostly studied for visual perceptions of the body and face of a robot or virtual person and their animated movements. However, in Clark (this volume), results for the evaluation of three linguistic strategies, politeness, relational work, and vague language are discussed in their usage for speech interfaces and their potential mismatch with the expectations in human users, and thus their potential to cause an uncanny valley effect.

One important sub-concept of social attractiveness is trust (McAleer, Todorov, & Berlin, 2014; Weiss, Wechsung, Kühnel, & Möller, 2015). In Torre and White (this volume) the effects of robot voices' gender, naturalness, prosody, and accent on trust perception in users are presented and systematized. Overall, there are effects, but they depend on the context and user group. For example, a regional accent showed an increased credibility to a standard accent when being knowledgeable, but the opposite in the case of being unknowledgeable.

1.4 Data

The material used in studies on voice attractiveness varies widely, from monosyllabic stimuli recorded in the lab to large extracts of authentic speech material that was not produced for research. This stylistic diversity is also reflected in the contributions for this volume. Thus, it seems fair to separate three kinds of sources, controlled experimental data, naturalistic lab data, and natural field data “from the wild”.

1.4.1 Controlled Experimental Data

One major source of the material stems from lab experiments, where new recordings are conducted for a specific purpose with already defined acoustic and perceptual analytic methods to be applied on. Such recordings are usually very short, for example (sustained) vowels, syllables or words. They can also not be considered as socially authentic, i.e., they do not aim to resemble real-life social communication situations. Due to its short duration, such material lacks major prosodic aspects, e.g., intonation contour or emphasis variation, as well as any natural situational grounding, affecting, e.g., speaking rate. Controlling for such aspects, however, allows to focus on topics like voice quality and person identification/similarity, while explicitly controlling for the just mentioned effects.

Examples of experimental data are Belin (this volume), who uses averaged short syllables of multiple voices, for which attractiveness ratings are collected. Kreiman et al., (this volume) analyzes steady state vowels (one second duration) regarding “normal” voice quality, whereas Babel et al., and Obuchi (both this volume) used single (monosyllabic, respectively multisyllabic) words for perception tests.

On some occasions, full sentences, or even a paragraph, are read by speakers in a lab with similar aims. The practical implications include potential laborious manual work to extract specific segments for analysis, and to take into account richer linguistic context, while the read speech style in a controlled environment allows to analyze not only segmental and micro-prosodic, but also macro-prosodic parameters. Therefore, it is not a coincidence to find a mixture of material types from experimental data in the cited literature for our topics that refer to social attributions and traits from speech (Suire et al.; Rosenberg & Hirschberg, both this volume). While some decisions on the material duration are made because of the costs inflicted by the prospective methods (see Sect. 1.3), other reasons to select material originate in the aspects under research.

The syllables used by Belin (this volume) were recorded in the lab, and subsequently post-processed to study the effect of acoustic averaging over speakers. Such a manipulation of speech recordings is another kind of experimental data. Manipulations comprise post-processing of the acoustic speech signal, as well as outright synthesis. Manipulated audio files can be in principle of any duration, but are considered here still as experimental data due to its similarity in careful and specific creation in a laboratory, but also due to the aim of controlling influencing factors—this time by means of inducing a controlled number of manipulations. There are different reasons for such manipulations, most importantly to verify analysis results with even more controlled material, producing stimuli for experiments which are hard or impossible to record, or to obtain speech signal qualities for the domain of computer speech.

The papers in the part on technological applications are good examples, as they all refer to studies in which manipulated or synthesized material, typically shorter utterances in a dialog, are used, or they argue to conduct those (Torre & White; Clark et al., both this volume).

1.4.2 Naturalistic Data Recorded in the Lab

While strictly controlled speech material from the laboratory is a foundation of basic research, there is always the aim to use naturalistic data in order to estimate the strength of effects for real-life situations and to study situational and dialogic aspects that cannot be simulated with—what we call—experimental data. Typically, this means to elicit naturalistic situations and thus also spontaneous material in the lab, often with the help of some supporting material. In contrast to the aforementioned controlled experiments, the lab recordings of naturalistic data are not controlled to the same degree. Here, experimenters aim to control a good acoustic quality, to initiate conversations, and possibly to instruct conversational tasks. That means that the linguistic and phonetic content is not (strictly) controlled for. However, very specific instructions and support material is often provided to support the subjects to elicit the situation, e.g., a game or task, but databases have been created with far less information provided (Schweitzer, Lewandowski, Duran, & Dogil, 2015).

For obtaining attractiveness ratings, Quené et al., (this volume) used sentences from spontaneous interview speech as stimuli that were manipulated. They also used visual data. The situation of speed dating was applied by Michalsky and Schoormann (this volume) to allow for studying the effects of prosodic entrainment in dialog. Simulated telephone conversations on pizza ordering from the Nautilus database, but post-edit to exclude the callee were used by Weiss et al. (this volume).

1.4.3 Data from the Wild

The last category of the material refers to recordings from real situations. Obtaining such data seems to be the easiest one on the first glance. However, it is often practically impossible to ensure sufficient quality and sufficient amount of material given the available resources, especially if there are requirements on the linguistic conditions to be included. In addition, there is often more information on the speakers required, which might be difficult to collect while or after recording, for example, additional physiological measures. Finally, there might be ethical reasons to avoid taking data from the wild.

In this collection, this kind of data was selected to solely study charismatic speakers. Bosker (this volume) selected speech fragments of c. 25 s from mass media recordings of US presidential debates. Brem and Niebuhr (this volume) used audio-visual data (video clips of charismatic management leaders). For natural data, this kind of material is the least uncontrolled, as the speakers are not only professional, but also very aware of the fact of being recorded. Therefore, such field data might not always be considered as truly "wild", but of course, it is as natural as it can be when studying speeches of charismatic leaders.

Sometimes, it is not easy to assign data to one of the categories. For example, read Wikipedia articles used by Baumann (this volume) is comparable on the surface with other naturalistic speech paragraphs read in the lab, except for the varying recording quality. But still, the origin of this material is natural, as the speakers truly recorded themselves with the intention to be listened to by people interested in the Wikipedia articles.

1.5 Conclusions

The word "attractiveness" stems from Latin "ad trahere" and means "dragging or pulling to something". For our topic, people are dragged or pulled to the voice and vocal behavior of somebody else. This relationship unfolds in various dimensions: from sexuality and biology over social likability up to charisma and leadership. It is this diversity of voice attractiveness that we intended to cover in this book. It is our hope to raise awareness with this book for this diversity and the broad range of the various scientific fields involved.

What we see in the contributions to this volume is on the one hand a clear and intended separation of the above-mentioned concepts on the sexual, the likable, and the charismatic speaker. On the other hand, we recognize the interdependencies between the three concepts. The classical example is that a person perceived as beautiful is also regarded as a socially more attractive (Zuckermann & Driver, 1989).

In our view, we deal here with a contrast between simultaneous distinctive concepts that have not only mutual influences and mutual conditionality. We see a need for a unifying theory with respect to the concepts, but also the different methods and data used in the various scientific disciplines. Several contributions in this book provide useful suggestions for such a theory, which can be viewed as a starting point for a more systematic foundation to overcome the current limitations of knowledge.

As an example can serve the frequency code by Ohala (1984): Similarities between languages, cultures, and even species in the use and effect of F0 was argued to originate in biologically grounded separation between "smaller" and "larger" (vocal) individuals. This does not only reflect the sexual dimorphism in terms of sexual selection, but also social aspects of signaling and estimating relational power, submissiveness, even helplessness, and thus supports social roles and interaction. The universal systematic in F0 observed by Ohala concerns charisma, attractiveness, and likability alike. Following this road to connect biological and articulatory bases for acoustic and perceptual effects can be seen as one of the most important elements of a unifying theory.

Interestingly, we observe that *trust* occurs in many contributions and it seems to have an overarching character. Trust, obviously, represents a link between the concepts of the sexual, the social, and the charismatic attractiveness, as it represents a positive attitude towards another. Trust may be considered as an immediate result of attractiveness, whatever the kind of attractiveness and social relation might be. Therefore, it is an important characteristic of human relationships, but also an important feature for Human-Computer Interaction.

References

Abele, A. E., Cuddy, A. J. C., Judd, C. M., & Yzerbyt, V. Y. (2008). Fundamental dimensions of social judgment. Editorial to the Special Issue. *European Journal of Social Psychology, 38*(7), 1063–1065.

Argyle, M. (1988). *Bodily Communication*. New York: Methuen.

Banse, R., & Scherer, K. (1996). Acoustic profiles in vocal emotion expression. *Journal of Personality and Social Psychology, 70*(3), 614–636.

Bezooijen, R. V. (1995). Sociocultural aspects of pitch differences between Japanese and Dutch women. *Language and Speech, 38*, 253–265.

Brown, B. L., Strong, W. J., & Rencher, A. C. (1973). Perceptions of personality from speech: effects of manipulations of acoustical parameter. *Journal of the Acoustical Society of America, 54*(1), 29–35.

Brown, B. L., Giles, H., & Thakerar, J. N. (1985). Speaker evaluation as a function of speech rate, accent, and context. *Language and Communication, 5*(3), 207–220.

Cuddy, A. J., Fiske, S. T., & Glick, P. (2008). Warmth and competence as universal dimensions of social perception: The stereotype content model and the BIAS map. *Advances in Experimental Social Psychology, 40*, 62–149.

Darwin, C. (1890). *The Expression of the Emotions in Man and Animals*. London: John Murray.

DePaulo, B. M., Kenny, D. A., Hoover, C. W., Webb, W., & Oliver, P. V. (1987). Accuracy of person perception: Do people know what kinds of impressions they convey? *Journal of Personality and Social Psychology, 52*(2), 303–315.

Fiske, S. T., Cuddy, A. J., & Glick, P. (2006). Universal dimensions of social cognition: Warmth and competence. *Trends in Cognitive Sciences, 11*(2), 77–83.

Hart, R. J., & Brown, B. L. (1974). Personality information contained in the verbal qualities and in content aspects of speech. *Speech Monographs, 41*, 271–380.

Krause, S., Back, M. D., Egloff, B., & Schmukle, S. C. (2014). Implicit interpersonal attraction in small groups automatically activated evaluations predict actual behavior toward social partners. *Social Psychological and Personality Science, 20*, 671–679.

Laver, J., & Trudgill, P. (1979). Phonetic and linguistic markers in speech. In K. R. Scherer & H. Giles (Eds.), *Social Markers in Speech* (pp. 1–32). Cambridge: Cambridge University Press.

Laver, J. (1980). *The Phonetic Description of Voice Quality*. Cambridge: Cambridge University Press.

McAleer, P., Todorov, A. & Berlin, P. (2014). How do you say 'Hello'? Personality impressions from brief novel voices. *PLOS ONE 9*(3).

McCroskey, J., & McCain, T. (1974). *The Measurement of Interpersonal Attraction. Speech Monographs, 41*, 261–266.

Mori, M. (2012). The uncanny valley. *IEEE Robotics and Automation 19*(2). Originally 1970, Translated by MacDorman, K.F. & Kageki, N., pp. 98–100.

Nass, C., & Brave, S. (2005). *Wired for Speech. How Voice Activates and Advances the Human-Computer Relationship*. MIT Press.

Nass, C., Moon, Y., Fogg, B., Reeves, B., & Dryer, D. (1995). Can computer personalities be human personalities? *International Journal of Human-Computer Studies, 43*, 223–239.

Ohala, J. (1984). An ethological perspective on common cross-language utilization of F0 of voice. *Phonetica, 41*, 1–16.
Paeschke, A., & Sendlmeier, W. F. (1997). Die Reden von Rudolf Scharping und Oskar Lafontaine auf dem Parteitag der SPD im November 1995 in Mannheim –Ein sprechwissenschaftlicher und phonetischer Vergleich von Vortragsstilen. *Zeitschrift für Angewandte Linguistik, 27*, 5–39.
Putnam, W. B., & Street, R. L. J. (1984). The conception and perception of noncontent speech performance: Implications for speech-accommodation theory. *International Journal of the Sociology of Language, 46*, 97–114.
Reeves, B., & Nass, C. (1996). *The Media Equation: How People Treat Computers, Television, and New Media Like Real People and Places*. Cambridge: Cambridge University Press.
Schaller, M. (2008). Evolutionary basis of first impressions. In N. Ambady & J. J. Skowronski (Eds.), *First Impressions* (pp. 15–34). New York: Guilford Press.
Scherer, K. R. (1978). Personality inference from voice quality: The loud voice of extroversion. *European Journal of Social Psychology, 8*(4), 467–487.
Schweitzer, A., Lewandowski, N., Duran, D., & Dogil, G. (2015). Attention, please!—Expanding the GECO database. In *Proceedings of the 18th International Congress of Phonetic Sciences*, Glasgow, paper 620.
Street, Jr. R. L., & Brady, R. M. (1982). Speech rate acceptance ranges as a function of evaluative domain, listener speech rate and communication context. *Communication Monographs 49*(4), 290–308.
Weirich, M. (2010). *Die attraktive Stimme: Vocal Stereotypes. Eine phonetische Analyse anhand akustischer und auditiver Parameter*. Saarbrücken: Verlag Dr. Müller.
Weiss, B., Wechsung, I., Kühnel, C., & Möller, S. (2015). Evaluating embodied conversational agents in multimodal interfaces. *Computational Cognitive Science, 1*(6), 1–21.
Weiss, B., & Möller, S. (2011). Wahrnehmungsdimensionen von Stimme und Sprechweise. 22. Konferenz Elektronische Sprachsignalverarbeitung, Aachen, pp. 261–268.
Zuckermann, M., & Driver, R. E. (1989). What sounds beautiful is good: The vocal attractiveness stereotype. *Journal of Nonverbal Behaviour, 13*, 67–82.

Chapter 2
Prosodic Aspects of the Attractive Voice

Andrew Rosenberg and Julia Hirschberg

Abstract A speaker's voice impacts listeners' perceptions of its owner, leading to inference of gender, age, personality, and even height and weight. In this chapter, we describe research into the qualities of speech that are deemed "attractive" by a listener. There are a number of ways that a person can be found attractive. We will review the research into what makes speakers attractive in the political and business domains, and what vocal properties lead to perceptions of trust. We then turn our attention to research into "likeability" and romantic attraction. While the lexical content of a speaker's speech is important to their attractiveness, we focus this survey on prosodic qualities, those acoustic properties that describe "how" the words are said rather than "what" the words are. Of course, attractiveness is subjective; what is attractive to one listener may not be to another. Properties of the listener and other contextual qualities can have a significant impact on the voices which are found to be attractive. The most comprehensive research in this topic includes analyses of both the speaker and the listener, since attraction is frequently a mutual phenomenon; when people are attracted to someone, they want to be found attractive in return. We will also summarize work that has investigated attraction dynamics in two-party conversations.

Keywords Likeability · Charisma · Political attractiveness · Business attractiveness · Romantic attraction · Speech prosody · Vocal attractiveness

A. Rosenberg (✉)
Google LLC, NYC, New York, NY, USA
e-mail: rosenberg@google.com

J. Hirschberg
Columbia University, NYC, New York, NY, USA
e-mail: julia@cs.columbia.edu

B. Weiss et al. (eds.), *Voice Attractiveness*, Prosody, Phonology and Phonetics,
https://doi.org/10.1007/978-981-15-6627-1_2

2.1 Understanding Vocal Attractiveness

Attraction is central to human social bonding. It is an expression of whom we choose to be close to and whom we choose to avoid. There are as many types of attraction as there are types of interaction. In this chapter we will survey the prosodic qualities of different types of attractive voices.

A person's speech communicates a wide variety of information about the speaker. Not only information that they are trying to communicate, but information about the speaker themselves is important in this regard. This information enables listeners to assess the gender and age of a speaker, their emotional state, and aspects of both their personality, and physicality, all while listening to a person speak. These qualities may be more or less attractive to a listener based on their inherent preferences and other situational factors. For example, in the case of political attractiveness, there are times when anger in a speaker can resonate with a listener and will be perceived positively, while in other contexts anger is deemed inappropriate and, therefore, unattractive.

We divide this survey into five sections, based upon different types of attractiveness. In Sect. 2.2 we discuss political attractiveness. Political figures attract and retain followers through their speeches, interviews, and other public performance. Understanding what allows a speaker to gain political authority has been a source of investigation in political science and sociology for many years. Of late, more computational approaches have been brought to bear in assessing what kind of speech is perceived as charismatic. Also related to this is the kind of charisma that is found in business leaders (cf. Sect. 2.3). The business community takes communication and leadership very seriously. A significant amount of work has examined the speech of entrepreneurs and established (and sometimes beloved) executives in hopes of understanding what draws investors and employees to a business leader. Central to both of these types of attractiveness is trust. In Sect. 2.4 we will survey research that strives to identify what makes a voice sound trustworthy. Researchers also tend to distinguish two more social types of attraction: likeability (Sect. 2.5) and romantic attraction (Sect. 2.6). These types of attractiveness are not identical, but neither are they orthogonal. Types of attraction may overlap with one another. Leaders who are politically attractive may also be perceived as likeable. In addition, physical attraction can impact the degree to which people are trusted. The types of voices that signal qualities of business success may be attractive to some people as friends or romantic partners, but may be unattractive to others.

In all of these analyses of vocal attractiveness, spoken communication is an important avenue to establishing the central social bond. People appear to have relatively consistent preferences regarding vocal attractiveness. Many of these vocal qualities are associated with other speaker properties that are considered attractive; for example, male body size in the case of romantic attractiveness, or enthusiasm and dynamism in the case of political and business leaders, are correlated with attractiveness.

Of course, attractiveness is not an objective phenomenon. Qualities of the listener also contribute to their perceptions of attraction. These can include sexual preference in romantic attraction or political bias in assessing political attractiveness. Similarly, some voices and messages resonate more or less with a listener on the basis of any number of factors—memories, contextual relevance, broader business or political context, or other idiosyncrasies.

Another quality that adds a layer of complexity to understanding the attractive voice is the interplay between inherent and performance qualities of the voice. In general, studies are looking to assess what makes a voice inherently attractive, but the same voice may be used in ways that are more or less attractive. Most studies avoid direct assessment of this distinction. Some will look at the same speaker in different venues or types of speech (cf. Sect. 2.2.1). Other work, particularly in studying romantic attractiveness (cf. Sect. 2.6), will contextualize speech in two-party conversations and consider qualities and assessments of the two speakers. Distinguishing the influence of the voice itself and the way it is used in a particular stimulus remains an open question in these studies. Overall assessments of attractiveness in each of these domains is a combination of both inherent qualities of the voice and how it is being used in the specific utterance that is being assessed.

Moreover, attraction is often a dynamic process in which conversational partners are simultaneously being attracted (or repelled) by an interlocutor while demonstrating their own preference for their partner to be attracted to (or repelled by) them. This contemporaneous perception and performance can make analysis challenging. For example, male voices which are spoken lower in the speakers' pitch range and with a relatively large formant dispersion tend to be found attractive by heterosexual women. But men who are attracted and are signaling their attraction to a conversational partner demonstrate the same qualities. So should we consider this voice to be attractive or flirtatious?

While there are relatively few clear, consistent, and universal answers to what makes speech attractive even in a specific context, to a specific group, there are some broad conclusions in the literature centered around identifying prosodic properties of an attractive voice. This chapter is an attempt to summarize the current understanding, highlight gaps and inconsistencies, and provide some directions for future inquiry.

2.2 Political Attractiveness and Charisma

Charisma is defined as the ability to persuade and command authority by virtue of personal qualities rather than through formal institutional (political, organizational, or military) structures (Weber, 1947). Viewed from this perspective, charisma is a challenge for institutional stability because it represents a path to leadership that eschews standard institutional pathways to power. Alternately, charisma is an important driver of revolutionary change specifically because it does not require specific structures to grant power; rather, it is a quality attributed to a person by her or his followers.

There is a wealth of political science and sociology research on charismatic leaders and movements, including importantly (Weber, 1947; Boss, 1976; Marcus, 1961). In this section, we will survey research that has used empirical techniques to investigate charismatic political speech. In Sect. 2.2.1, we will survey studies that have looked at spoken correlates of charismatic speech. We will summarize work that has sought to define charisma empirically in Sect. 2.2.2.

2.2.1 Vocal Correlates of Charisma

Rosenberg and Hirschberg (2005, 2009) describe the first set of studies that attempt to measure the vocal and lexical correlates of charisma in American English. This study presented 45 speech segments to eight subjects. Materials were chosen to balance speakers, topics, and genres. A small set of speakers were chosen from those whose public speech covered a similar set of topics, and for whom speech tokens could be found in a wide variety of genres, or speaking styles. Since the experiment was designed during the winter and spring of 2004, there was abundant speech material available for the nine candidates running at that time for the Democratic Party's nomination for President. Speakers were limited to Democrats in this study to confine the range of opinions presented in the tokens, as it has been suggested in the literature (Boss, 1976; Dowis, 2000; Weber, 1947) that a listener's agreement with a speaker bears upon their judgment of that speaker's charisma. Segments were selected from a variety of topics in order to test the influence of topic on subject judgments of charisma. Five speech tokens were chosen from each speaker, one on each of the following topics: health care, postwar Iraq, Pres. Bush's tax plan, the candidate's reason for running, and a content-neutral topic (e.g., greetings). For these five tokens, genre was also varied among the following types: interview, debate, stump speech, campaign ad.

Subjects were presented with each of the stimuli twice, with a 2 s silence between presentations. They were asked to respond to 26 statements about the speaker including "The speaker is charismatic." The order of presentation of stimuli and statements was randomized for each subject.

Using the subject responses, a mean score measuring the degree to which the speech in each token was calculated in order to examine the extent to which the subject believed that the speaker was charismatic. Colloquially this was referred to this as "how charismatic" the utterance was—despite charisma being a quality of the speaker rather than the speech itself. With this mean charisma score for each token, it was possible to analyze acoustic–prosodic qualities of the speech to identify correlates with charisma. These qualities were identified by measuring pitch, intensity, speaking rate, and duration features of the tokens in the experiment and then measuring the degree of correlation between these features and subject ratings of the charismatic statement. Results of these analyses showed significant positive correlations between charisma ratings and the duration of the speech, whether measured in words, seconds, or number of phrases. These results also showed positive correlations between enthusiastic and passionate ratings and mean and maximum F0, intensity, and speaking rate. More colloquially this means, higher pitched, louder, and faster speech is considered to be more passionate and more enthusiastic (with caveats that the perceptual properties of pitch and loudness are not identical to the acoustic measurements of mean and maximum F0 and intensity). Additionally, a positive correlation between standard deviation of F0 and ratings of enthusiastic and passionate speech was observed in male speakers.

In a later study, Rosenberg and Hirschberg (2009) extended this analysis to include ToBI labeling (Beckman & Hirschberg, 2005) of the segments. In this study, phrase boundary prosody was classified into three types: rising pitch (L-H%; H-H%), falling pitch (L-L%; L-), and plateau or flat pitch (H-L%; H-). Results showed that the rate of rising tokens negatively correlates with charisma. Rising intonation is used in questions, and can be associated with uncertainty. Neither of these qualities is consistent with "persuasiveness," a component of charisma. Consistent with this, the L*+H pitch accent type, also associated with uncertainty, had a negative correlation with charisma. The L*+H pitch accent is realized with low pitch on a prominent syllable nucleus which rises, typically reaching a peak after the nucleus boundary. In addition, prosody associated with "new" information (H* pitch accents) was positively correlated with charisma, while prosody associated with "given" information (downstepped contours: H* !H* L-L%) was negatively correlated. H* pitch accents are high tone pitch peaks that are more or less time-synchronized with intensity peaks occurring within syllable nuclei. Downstepped high pitch accents, !H*, are H* pitch accents that occur after a previous high tone, and have a lower pitch height during their high tone. The "downstepped" contour is a shorthand to describe a high tone, followed by one or more downstepped high tones with a L-L% phrase ending.

Other notable efforts in measuring vocal correlates to charisma have investigated political speech in other languages and countries. From this work we can look for evidence of linguistic and or cultural biases in the perception or production of charisma. Disentangling these factors (linguistic vs. cultural; perception vs. production) is virtually impossible given the size of these studies and additional confounds (speaker/listener demographics and other biases, political, social, and temporal context to name a few) that all analyses in this space are subject.

Cullen and Harte (2018) analyzed a relatively large set (945 utterances) of longitudinal speech material from a single speaker, over seven years (2007–2012). This material, compiled as the Irish Political Speech Database, has a number of useful qualities. By focusing on a single speaker, many political biasing elements are controlled for. By including many recording contexts (talk shows, parliamentary speeches) differences in genre can be accounted for. The longitudinal aspect also allows polling data to be associated with the politician's speech, facilitating investigation of how popularity or standing impact communication. This work also included automatic classification of charisma based on acoustic–prosodic features. The authors found that prosodic features, based on pitch, intensity, and duration, outperformed spectral features. The specific performance of this classifier is somewhat immaterial—the broad applicability of a single speaker model for a paralinguistic task is *extremely* limited. But the relative value of the acoustic signal is revealing—charisma is found here to be a function of suprasegmental qualities more than voice quality (as captured by spectral features).

Biadsy, Rosenberg, Carlson, Hirschberg, and Strangert (2008) significantly extended the studies described in Rosenberg and Hirschberg (2005, 2009. The original American English stimuli were additionally rated by native Swedish and Palestinian Arabic speakers, and a subsequent study presenting Palestinian Arabic speech to speakers of the American English and Palestinian Arabic was conducted. Comparative analysis of the original study with these four new studies allowed the identification of some vocal correlates of charisma that appear to be robust to differences in the language of the speaker or listener. Others appeared to be sensitive to the language of the listener, regardless of the language of the speaker, and still others are specific to the speaker/listening configuration. For example, across all experiments, mean pitch, pitch range, mean and standard deviation of intensity, and stimulus duration all positively correlated with charisma ratings regardless of the language spoken and the native language of the rater. Conversely, the presence of disfluencies negatively correlated with charisma in all experiments, though this correlation was weakest for Swedish judgments of American English.

The studies also found that raters tended to pattern similarly in response to many aspects of the stimuli regardless of their native language. For instance, when assessing English stimuli, minimum F0 was positively correlated with charisma. However, when assessing Palestinian Arabic utterances, this feature was negatively correlated for Palestinian subjects, and not significant for American subjects. Also both groups judging Arabic data rated speech to be more charismatic that exhibits larger standard deviations in F0 but none of the groups judging English showed the same effect.

Finally, a third group of correlates appeared to be specific to the language of both speaker and listener. For example, the speaking rate was positively correlated with charisma judgments only for American and Swedish ratings of English: the faster the speech, the more charismatic the speaker was deemed to be. However, when Palestinians judged Arabic speakers, speaking rate approached a negative correlation with charisma, with no correlation between speaking rate and charisma when Palestinians judged American English or Americans judged Palestinian Arabic.

This is not the only work that has looked at cross-cultural biases in perceptions and production of charisma. Though not every investigation found clear differences on the basis of culture or nationality. For example, Cullen et al. (2014) also found that native Irish raters and Amazon Mechanical Turk workers, who are largely American, were quite consistent in their assessment of Irish Political speech with respect to charisma.

Pejčić (2014) investigated persuasiveness in Serbian and British political speech, which appears clearly related to charisma. This study presented five samples of Serbian political speeches and five samples of British speeches to 113 Serbian subjects asking them to respond to a subset of the 26 statements used in Rosenberg and Hirschberg (2009) on a 5-point Likert scale. Acoustic analysis was performed on the tokens from both languages considered as a common population, and also on each language in isolation. When pooling both languages, relatively few statistically significant correlates with persuasiveness were observed. These were the standard deviations for F0 peaks in narrow-focused rising nuclear tones, their percentage in Tone Units' F0 range and the maximum F0 of their Tone Units. Anecdotal observations suggest roughly that larger F0 excursions were positively associated with persuasion in Serbian speech, but negatively associated with British speech, at least when rated by Serbian speakers.

In addition to these studies, there are a number of descriptive investigations of the speaking style of politicians, particularly concerning the recognition of charisma. Pèrez (2016) contrasted the speech of the Venezuelan politicians, Hugo Châvez and Josė Luis Rodrìguez Zapatero, characterizing Châvez as using a "revolutionary" style, consistent with charismatic authority, whereas Zapatero uses a more "traditional" style, consistent with institutional authority. Ryant and Liberman (2016) proposed a number of visualization techniques to investigate and compare prosodic qualities of speech, using U.S. Presidents Barack Obama and George W. Bush as examples.

2.2.2 *Defining Charisma*

Careful reading will reveal that the studies described in Sect. 2.2.1 side-step any definition of "charisma." Specifically, subjects in Rosenberg and Hirschberg (2005) were simply asked to respond to the statement "The speaker is charismatic," which does very little to identify the personal or vocal qualities that lead to this perception.

Researchers in other fields have posited a number of factors that contribute to perceptions of charisma. Boss (1976) sees charismatic leaders emerging from an *important crisis* met by an *inspiring message* delivered by a messenger with a *gift of grace*. Marcus takes a more specific view identifying charisma as a product of the faith of a potential leader's *listener-followers* (Marcus, 1961). While these are useful perspectives on political attractiveness and authority, they provide little direction when we try to empirically quantify charisma and charismatic speech.

In Rosenberg and Hirschberg (2005), subjects were asked to respond to the statement "the speaker is charismatic." But the subjects also responded to 25 other statements about the speaker and his or her speech. Most of these were of the form "The speaker is X," where X was one of the following: *charismatic, angry, spontaneous, passionate, desperate, confident, accusatory, boring, threatening, informative, intense, enthusiastic, persuasive, charming, powerful, ordinary, tough, friendly, knowledgeable, trustworthy, intelligent, believable, convincing, reasonable*. These attributes represent a subset of those often associated in the literature with charisma. "The speaker's message is clear" and "I agree with the speaker" were also included as statements to be rated. Using these ratings, along with the ratings of charisma, it was possible to determine which *other* qualities were highly correlated with charisma, to help in developing a "functional" definition of this term. Rather than offering a formal definition of charisma as a sociopolitical concept or a vocal characteristic, these results indicate how the subjects themselves understood charisma and how they were using the term. Specific results can be found in Table 2.1. These results confirmed some of the conventional wisdom of what we mean when we say charismatic—specifically, a charismatic speaker is **charming**—and what we believe charisma to

Table 2.1 Statements showing the most consistent subject responses with the statement "The speaker is charismatic"

Statement	κ
The speaker is enthusiastic	0.606
The speaker is charming	0.602
The speaker is persuasive	0.561
The speaker is boring	−0.513
The speaker is passionate	0.512
The speaker is convincing	0.503

be used for—a charismatic speaker is **convincing** and **persuasive**. However, they also provide support for claims found in Dowis (2000) and Boss (1976) that charismatic speakers should be passionate and enthusiastic and, by extension, not boring. It was also interesting to see that responses to the *desperate, threatening, accusatory,* and *angry* qualities showed no positive or negative ($|\kappa| < 0.15$) agreement with the charismatic statement. Apparently, a charismatic speaker *can* demonstrate these qualities, but, at least among the subjects in this study, they neither promote nor inhibit perceptions of charisma.

A similar approach to defining charisma was undertaken in Signorello, D'Errico, Poggi, and Demolin (2012). This study administered a free-form web survey, asking 58 French participants to provide adjectives that are consistent or inconsistent with the term "charisma" as they understood it. Retaining only adjectives that were reported by more than one subject, the authors identified 40 terms that were positively associated with charisma and 27 that were negatively associated. To facilitate understanding, the authors grouped these into five categories (1) Pathos, (2) Ethos Benevolence, (3) Ethos Competence, (4) Ethos Dominance, and (5) Emotional Induction Effects. Table 2.2 is reproduced from Signorello et al. (2012). Note that *charming, persuasive, enthusiastic,* and '*boring*' appear in both Signorello et al. (2012) and Rosenberg and Hirschberg (2009) despite the studies using French and American participants, respectively.

One divergent finding did appear however: while Rosenberg and Hirschberg (2009) found no correlation between *threatening* and *anger*, Signorello et al. (2012) identified through factor analysis an *Authoritarian-Threatening* factor which in their study **is** a factor, including the terms *determined, authoritarian, leader, confident* as well as the more aggressive terms *Who Scares, cold. dishonest* and *menacing*.

While not directly related to defining charisma, but related to political speech, an interesting idea presented in Cullen and Harte (2018) addresses vocal attractiveness more broadly. The Irish Political Speech Database is labeled for six attributes: *charisma, boring, enthusiastic, inspiring, likeable*. From these six, Cullen and Harte (2018), define Overall Speaker Appeal (OSA) as the average of these six ratings (including negative boredom ratings). The correlation of these attributes may limit

Table 2.2 The 67 positive and negative adjectives related to charisma. Reproduced from

Dimension	Positive adjectives	Negative adjectives
Pathos	Passionate, empathetic, enthusiastic, reassuring	Cold, indifferent
Ethos benevolence	Extroverted, positive, spontaneous, trustworthy, honest, fair, friendly, easygoing, makes the others feel important	Untrustworthy, dishonest, egocentric, individualistic, introverted
Ethos competence	Visionary, organized, smart, sagacious, creative, competent, wise, enterprising, determined, resolute, who propose, seductive, exuberant, sincere, clear, communicative	Inefficient, inadequate, uncertain, faithless, unclear, menacing
Ethos dominance	Dynamic, calm, active, courageous, confident, vigorous, strong, leader, authoritarian, captivating, who persuade, who convince	Apathetic, timorous, weak, conformist, unimportant, who scare
Emotional induction effects	Charming, attractive, pleasant, sexy, bewitching, eloquent, influential	Boring

the efficiency of this measure, but the attempt to summarize these signals into a single measure is potentially valuable, even if the specific formulation might benefit from modification.

2.3 Business Attractiveness

Business organizations are an area in which leadership and authority have clear impacts. There are many organizational structures that are used in business activities, but all instill participants with distinct, decision-making authority. Within these structures, charismatic authority can be manifested the way (Weber, 1947) formulated it—as an alternative to established, institutional authority. This would be revealed by a situation where employees look to a co-worker who is not in a management or reporting structure for direction rather than their direct manager. A more common way to think about charismatic leadership in a business context is when charismatic authority is aligned with institutional authority. This allows us to think about "how charismatic" is one manager, one CEO, or one founder over another.

While there is always an element of "trust" in a leader–follower relationship, this is somewhat more quantifiable in business relationships. Investors are entrusting their capital to the efforts of a founder when they invest in a business. While the specific leadership of a founder may be more essential to a start-up, opinions about the CEO can have an impact on institutional investing in well-established corporations.

We previously noted some of the complications in defining charisma. The use of a limited number of speakers who have a cultural consensus of being charismatic is one way to get around a broader definition. One thread of work undertaken by Oliver Niebuhr and colleagues has been to study Steve Jobs, former CEO and co-founder of Apple Inc., as an exemplar of a charismatic business leader. Niebuhr, Brem, Novák-Tót, and Voße (2016b) posit a profile of charismatic speech based on a reading of previous political studies (cf. Sect. 2.2). This is summarized as having high and varied pitch, high and varied intensity, a fast speaking rate, few disfluencies, a large number of emphatic words, but with varied realizations and high rhythmic variation. By automatic analysis of two landmark speeches (launching the iPhone 4 and iPad 2) they find that Steve Jobs does in fact fit this profile.

This research direction is continued in a number of works via a contrastive analysis of Steve Jobs and Mark Zuckerberg, founder and CEO of Facebook (Mixdorff, Niebuhr, & Hönemann, 2018; Niebuhr, Voße, & Brem, 2016a, 2018b). The approach here is based on the common perceptions of Steve Jobs as a charismatic speaker and Mark Zuckerberg as a less charismatic speaker, though both were CEOs of major corporations at the time their speech was collected for analysis.

Niebuhr et al. (2016a) find that Jobs has shorter phrases, fewer and shorter hesitations, and a more dynamic use of pitch and rhythm than Zuckerberg. While Jobs speaks quickly (compared to "normal" speech), Zuckerberg's speaking rate is even higher. This contributes to strong phonetic reductions in his speech which may negatively impact perceptions of charisma. Applying the Fujisaki model of intonation Fujisaki and Hirose (1984), Mixdorff et al. (2018) enable a more specific analysis of how the two CEOs manipulate pitch in their speeches. In general, this analysis brings insight into the earlier (and overly simplistic) findings that high pitch leads to perceptions of charisma. These two speakers differ more in how they reset their pitch ranges across phrases and the strength of their excursions. This work is then expanded upon in Niebuhr et al. (2018b) where the timing and shape of pitch accents are examined. Moreover, the authors find that a large vowel space, limited place of assimilation, and a clear differentiation between voiced and unvoiced stops all differentiate Jobs from Zuckerberg. These factors all contribute to fast, dynamic speech that is clearly pronounced.

While analysis of specific business leaders enables clear contrastive discussion, there is more work that looks at business speech in entrepreneurship more generally. Weninger, Krajewski, Batliner, and Schuller (2012) extracted speeches from 143 male business leaders that were shared on YouTube. They collected ratings of charisma and attempted to automatically predict the human ratings with acoustic and linguistic features. The raters were 10 psychology Ph.D. students, 5 male and 5 female.[1] This work investigated a large number (1,582) of acoustic–prosodic features, in addition to lexical features derived from automatic speech recognition transcripts of the speeches. This work finds that charisma can be automatically detected with 61.9% accuracy, significantly over chance level, based on acoustic-prosodic and lexical features.

[1] No statistically significant gender effects in the ratings of charisma were discovered.

While the previous studies looked at established business leaders (Niebuhr, Brem, & Tegtmeier, 2017) investigated start-up state entrepreneurs, since "a decisive part of their strategy and daily work is to persuade others." Leaders of these early stage businesses need to convince both investors, suppliers, and customers of the legitimacy of their nascent technology, developing products and services, and of the likely market demand. In this study, 45 participants gave the same elevator pitch, 15 practiced with no feedback, 15 received visual feedback, and 15 received feedback based on the Steve-Jobs-as-charismatic-exemplar acoustic model described above. They found that speakers who received acoustic feedback about their speech were rated 41% more charismatic following training, significantly more than those who received no feedback (24% more charismatic) or those who received visual feedback (12% more charismatic).

Extending this investigation of entrepreneurial speech into spectral qualities contributing to voice quality, Niebuhr et al. (2018a) found that a fuller and less breathy voice also led to higher speaker charisma ratings. This may be consistent with findings that suggest that clear or easily understood speech is an important element to charisma.

Much of the study of business attractiveness has been focused on analysis of speech spoken by men. On one hand, this can limit variability to facilitate analysis. On the other, it perpetuates patriarchal norms, implicitly treating charisma—and specifically business leadership—as a quality only associated with male speech. This thus limits our ability to understand charisma in female speakers. Novák-Tót, Niebuhr, & Chen, (2017) investigated the bias in the perception of speeches delivered by American female executives Oprah Winfrey and Ginni Rometti and male executive Steve Jobs. No information as to the gender of the raters was provided. They found that female speech that is judged to be as charismatic as male speech demonstrates more and stronger acoustic cues to charisma. This suggests that this gender bias may be compensated for by making a greater effort by the female speakers. Significantly more work is necessary with regard to the charisma of female leaders both in business and politics alike.

2.4 Vocal Correlates of Trust

Trust and attractiveness are closely related. Some studies have found that people trust romantically attractive strangers more than unattractive ones, e.g., Wilson and Eckel (2006). While others have found that the relationship is not so simple. Sofer, Dotsch, Wigboldus and Todorov (2015) found that more "typical" faces elicited more trust, rather than the most attractive faces. In this work, "typical" faces were constructed as an averaged composite of 92 faces, while the "attractive" face was an averaged composite of the 12 most attractive in the used data set. However, in an investigation of responses to dating profiles, McGloin and Denes (2018) found that attractive men were considered trustworthy, but attractive women were not. It is worth noting that in both of these studies, the presented face was exhibiting a “neutral” expression. Smiling or grimacing would likely impact impressions of attractiveness, pleasantness, trustworthiness, and likeability in unanticipated ways. When we think about attractiveness more broadly, as we have done in this chapter, trust is a necessary component to political, business, and nonsexual attractiveness. In the political and business roles, attractiveness can endow abilities to the person. They can obtain political power via elections or they can obtain commercial power through investment. Trusting the person is necessary when granting these abilities and responsibilities to the person.

In an analysis of deceptive and truthful, trusted, and mistrusted speech in the Columbia Cross-Cultural Deception (CXD) corpus, Levitan, Maredia, and Hirschberg, (2018) found significant differences in trusted and mistrusted speech. The CXD corpus is a study of deceptive versus nondeceptive speech from native speakers of Standard American English (SAE) and Mandarin Chinese (MC), all speaking in English. The participants were balanced between male and female speakers and native speakers of English and Chinese. It contains interviews between 340 subjects in 122 h of speech. A variation of a fake resume paradigm was used to collect the data. All subjects were previously unacquainted, and pairs of subjects played a “lying game” with each other. Each subject filled out a 24-item biographical questionnaire and was instructed to create false answers for a random half of the questions. They also reported demographic information including gender and native language, and completed the NEO-FFI personality inventory. The speech was recorded in a double-walled sound booth, where the two subjects were separated by a curtain to ensure no visual contact. For the first half of the game, one subject assumed the role of the interviewer, while the other answered the biographical questions, lying for half and telling the truth for the other; questions chosen in each category were balanced across the corpus. For the second half of the game, the subjects’ roles were reversed, and the interviewer became the interviewee. During the experiment, the interviewer was encouraged to ask follow-up questions to aid them in determining the truth of the interviewee’s answers. Interviewers recorded their judgments for each of the 24 questions, providing information about human perception of deception. Subjects were incentivized monetarily: for every response to the 24 questions that the interviewer judged correctly, the interviewer received an extra $1, while every incorrect judgment cost them $1. Every false answer the interviewee persuaded the interviewer was true gained the interviewee $1, while every false answer the interviewer judged false lost the interviewee $1. The interviewees annotated each of their statements during the interview by pressing a “truth” or “false” key on a computer keyboard. We aligned these annotations with transcriptions of the interviews obtained by speech recognition with crowdsourced corrections and automatically aligned the transcripts with the speech recordings.

Overall, the researchers found that the mistrusted speech in their corpus (interviewee responses that were not believed by interviewers) was significantly more intense (louder) and spoken in a higher pitch range, while the speech that interviewers tend to trust was spoken more rapidly. However, they also found differences between male and female and English and Mandarin Chinese native speakers in these features. While male speakers did tend not be trusted when they spoke in a high pitch range, this was not true of female speakers (note that all features were z-score normalized, so these findings were not influenced by a speaker's "normal" range or loudness or speaking rate). Both genders were trusted more when they spoke more rapidly. Female speakers, however, were trusted more when their voice quality exhibited more jitter and shimmer—instabilities in their pitch and intensity associated with perceived "roughness" or "breathiness." There were also differences in trustworthiness in speakers' native language backgrounds, although all speakers spoke in English. In general, native speakers of Standard American English were more trusted when they exhibited high jitter and shimmer while this was not a significant factor for native speakers of Mandarin Chinese, who were more trusted when they spoke more rapidly. These Chinese speakers were less likely to be trusted when they spoke in a high pitch range and when their overall mean pitch was high; they were also mistrusted when their maximum intensity was high and when their Harmonics-to-Noise (HNR) ratio (another measure of voice quality disorders) was high.

The researchers also examined the gender and the native language of the interviewers that correlated with their judgments whether interviewers are lying or telling the truth. Overall, all interviewers mistrusted speech with a high pitch range and a high maximum intensity and trusted speech spoken rapidly. However, there were major differences between genders. Male interviewers distrusted speech with high mean pitch and maximum intensity and trusted fast speaking rate while females only mistrusted high jitter and shimmer. Comparing native English speakers to native Mandarin speakers, the researchers found fewer differences: both mistrusted high-intensity speech and trusted faster speaking rate, but only native English speakers mistrusted high pitch range.

2.5 Likeability or Nonsexual Social Attractiveness

The distinction between finding a voice "pleasant" to listen to, and finding the speaker to be socially attractive as in "I like this person" is difficult to distinguish in research protocols. These two facets may overlap, they may even be identical for some listeners, but there may be differences that are elided in the research in this space.

There are several factors that have been found to contribute to likeability in speech. Strangert and Gustafson (2008) found that the speaker should be proficient. That is, the speech should include limited disfluencies and a reasonably high speaking rate. For clear speech, Weiss and Burkhardt (2010) found that warm/relaxed speech correlated significantly with likeability.[2] This included less pressed, more breathy voice quality and lower spectral center of gravity.

Weiss and Burkhardt (2012) performed a focused analysis of 30 speakers rated as highly likeable and 30 that were highly not-likeable, drawn from the material used in the 2012 Interspeech paralinguistics challenge (which is discussed in detail

[2]Note the difference in likeability correlating with *relaxed* speech, while charismatic speech (cf. Sect. 2.2.1 correlates with passion and enthusiasm.

below). The presence of positive factors of likeability was found in all speakers. These included minimal disfluencies and no discernible accent. However, unlikable speakers show higher pitch, lower articulation rate, and lower pronunciation precision. This suggests that these factors can make a speaker "unlikeable," although perhaps the mere absence of negative attributes is sufficient for an unknown speaker of a relatively short amount of speech to be viewed as "likeable."

Regarding the "no discernible accent" finding of Weiss and Burkhardt (2012), there appears to be a more nuanced relationship between social factors like likeability and trust and a speaker's accent. For example, Tavernier (2007) examined perceptions of Flemish speaker's responses to English speech. They found the highest social attractiveness and trust ratings to come from RP (Native British) speech, with the lowest ratings coming from Flemish-accented English, despite the raters being Flemish speakers themselves. Looking at American English, Preston (1999) found broad differences in social assessments on the basis of the internal regional accent of American speakers, including a finding that northern speakers are considered to be less friendly than southern speakers by students in Michigan.

Baumann (2017) collected pairwise likability ratings from more than 220 speakers and over 160 raters. This work found very limited acoustic correlations with rater preferences. Only measures related to the acoustic fidelity of the recording showed significant correlations, while prosodic qualities showed trends that did not reach statistical significance. However, the authors did find an interesting relationship between gender and likeability. Both male and female raters responded to male speech similarly. However, female speech was rated as much more likeable by female raters than by male raters.

As in the study of charisma, qualities of the *listener* do not receive as much research attention as qualities of the *speaker*. This is particularly true in the case of likeability. Social attractiveness necessarily involves two parties and is a subjective quality. We do not all want to be friends with the same people. The attitude and behaviors of the listener can impact the speaker and reveal the dynamics of establishing, maintaining, or undermining social attractiveness.

Schweitzer, Lewandowski, and Duran (2017) directly addressed this facet of likeability. This work examined dialogs between pairs of German female speakers who both rated their dialog partners following their conversation. This work treats likeability as social and participatory. By investigating only dialogs between two female participants, this study avoids the biasing on the part of speaker or listener based on gender. While it was not explicitly measured, there is an assumption in this work that the participants were all heterosexual, therefore, the potential for overlap between likeability (social attractiveness) and sexual attractiveness is diminished. It is worth mentioning that in work that investigates social and sexual attractiveness, the sexual preferences of the participants are particularly relevant. As such, it is necessary to collect or verify information about the sexual preferences of subject participants.

The experiment consisted of 46 two-party dialogs between 13 participants. Dialogs were collected in situations where the speakers could see each other, and where they were visually separated. Each dialog was spontaneous and unconstrained, and lasted approximately 25 min. After the conversation both participants responded to a questionnaire about how likeable, competent, friendly, and self-confident they found their conversational partner.

The authors found limited confirmation of pitch and voice quality correlates to likeability in this study. Specifically, they found no effect of absolute pitch or pitch range. Neither were effects of shimmer, jitter, or HNR observed. However, they did find a number of entrainment or "convergence" based effects. These relate to how the acoustic–prosodic and lexical qualities of two (or more) speakers either become more or less similar over the course of a dialog. The authors found that lexical entrainment, when interlocutors use the same words, is a reliable predictor of likeability. In multimodal conversations, where the participants could see each other, they found convergence of peak F0 height made a speaker appear *less* likeable.

The Interspeech Paralinguistics Challenge is an annual shared task with results presented at the Interspeech Conference each fall. The organizers distribute speech data sets labeled for some paralinguistic quality which are partitioned into train, development, and evaluations sets. Previous tasks have included classification of emotion, sleepiness, and intoxication among many others. The 2012 challenge included a task to classify the likeability of a speaker on the basis of a short utterance. Sentences were drawn from the aGender corpus (Burkhardt, Eckert, Johannsen, & Stegmann, (2010), and originally collected for the prediction of age and gender. The longest utterance for each speaker was selected. This resulted in 800 speakers balanced between male and female and divided into three age ranges (young: 15–24; middle: 25–54; senior: 55–85). These were rated on a 7-point Likert scale of likeability by 32 participants (17M; 15F) aged 20–42 years. Ratings were adjusted based on evaluator reliability and discretized into Likeable and Not-Likeable classes for classification. The organizers of the challenge found no impact of the rater's age or gender on ratings, but the age and gender of the *speaker* did have a significant impact. These challenges have served as a venue for the broader research community to test the limits of automatic analysis of paralinguistics. In many situations, in part because of the short time frame, and limited meta data available for the challenge data sets, a good number of submissions associated with these challenges tend to be applications of feature selection, e.g., Pohjalainen, Kadioglu, and Räsänen (2012), Wu (2012) and classification approaches, e.g., Cummins, Epps, and Kua (2012), Lu and Sha (2012), Brueckner and Schuller (2012), Sanchez, Lawson, Vergyri, and Bratt (2012), Some of these are quite novel to these tasks yet include only limited analyses of the underlying phenomena. One exception can be found when participants develop novel acoustic features for analysis. This was undertaken by Buisman and Postma (2012) in this likability challenge. They found that spectral information extracted via log-gabor-filter-based features were able to predict likeability with higher accuracy than a much larger set of features included in the OpenSmile baseline (Eyben, Wöllmer, & Schuller, 2010).

Additionally, Montaciè and Caraty (2012) developed specific pitch and intonation feature sets based on MOMEL (Hirst, 1987) and INTSINT (Louw and Barnard, 2004). MOMEL is a stylization technique which smooths out microprosody from a pitch contour, while INTSINT discretizes the contour into "key ranges" describing the speaker's pitch range, and "contextual labels" describing the relationship between the pitch at a given target to the previous target. A set of features based on the MOMEL and INTSINT processes were developed to help predict likability and also personality traits (another task of the 2012 Interspeech Paralinguistics Challenge). While the specific correlations between likeability and these novel features are not presented, the use of intonational features was useful for the prediction of likeability where they were not useful for predicting personality traits. This suggests that these features may be particularly well suited to likeability, rather than being generally valuable features for paralinguistic analysis. There are conflicting results about correlations between pitch and likeability. These seem to suggest that either the specific formulation of intonational features is critical, or the relationship is nuanced and significantly influenced by other factors.

2.6 Romantic Attractiveness

Romantic attraction is a complicated phenomenon that involves the synthesis of a wide array of signals to determine romantic interest. The current understanding of this topic involves an interplay of influences too complicated to summarize here. Here we will focus only on the work that has investigated qualities of the voice that lead a listener to find a speaker romantically attractive, or not.

While romantic attractiveness is exceptionally subjective, research has been undertaken to identify voices that are typically found to be more (or less) attractive. In this work, compared to much of the work surveyed elsewhere in the paper, characteristics of the listener are measured, and generally controlled for. However, a significant number of studies in this area conflate the influence of gender and sexual orientation in considering the qualities of the listener. Some studies investigate how males react to female voices or faces and others will study how females respond to male voices. In doing this, there is an assumption that all of the participants are, in fact, attracted romantically or sexually to members of the opposite sex. When these studies do not report the sexual orientation of the subjects, it stands to reason that the question was not asked of the participants. This is a significant methodological problem with this body of work. Through this section we will highlight whether a study has in fact reported the sexual orientation of the subjects or not, and suggest that future studies take this into consideration. We would also suggest that gender questions in recruitment for these studies be broadened to gain an understanding of how transgender, nonbinary, and intersex people assess attractiveness by the voice. None of the surveyed papers address these populations.

In an example of this, Collins and Missing (2003) investigated subject ratings of attractiveness of female voices, and female faces. To account for sexual preference, they used only male raters. However, they do not report whether all participants were heterosexual. In this work, they found strong agreement as to what was an attractive voice, and what was an attractive face, and moreover, attractive voices belonged to attractive faces. They found that voices of younger women are typically higher pitched, as are voices of smaller women, while taller women demonstrate a narrower formant dispersion. The authors' findings suggest that both the visual and auditory signals are communicating complementary information regarding age and body shape. The finding that men find high-pitched women's voices attractive has been identified elsewhere as well, including by Feinberg, DeBruine, Jones, and Perrett (2008b).

On the other hand, Feinberg, Jones, Little, Burt, and Perrett (2005), Collins and Missing (2003), and Hodges-Simeon, Gaulin and Puts (2010) all found that women find men with lower pitched voices to be more attractive. Feinberg, DeBruine, Jones, and Little (2008a) found that both male and female subjects consistently rated the masculinity of male faces and voices and demonstrated preferences for more masculine voices. The claim here is that testosterone information is similarly communicated via the voice and the face. This supports a finding by Saxton et al. (2006) that men with attractive voices also have attractive faces. Interestingly, this result was found in adolescent and adult women, but not in female children. Of these, only Hodges-Simeon et al. (2010) reported the sexual orientation of the participants reported.

Many of these findings are predicated on the idea that attractiveness of a voice is being used as a proxy or a reinforcing signal for other physical characteristics. While there are plausible evolutionary justifications (cf. Puts, Doll, & Hill, 2014) for why some secondary sexual traits are attractive, the value of an attractive voice is less obvious. There is, however, some evidence that attractive voices are correlated with other physical traits that are themselves attractive. For example, Bruckert, Liénard, Lacroix, Kreutzer, and Leboucher (2006) found that male speech with low-frequency formants correlate with age, height, and weight. However, female listeners were only able to reliably estimate the age and weight of a male speaker based on enunciation of vowels. González (2006) found that the pitch of human speech reveals very little about body size when age and gender are controlled for. However, formant dispersion does carry this information. Despite the fact that it is a poor signal, listeners do rely on pitch information to estimate body size. Babel, King, McGuire, and Miller (2011) investigated the vocal correlates of attractiveness particularly as it relates to body size in the perception of opposite-sex voices by both male and female listeners. They found that the ratings of both genders were highly correlated, though males generally rated other males as less attractive than females did. They also found that attractive female voices had high second formants in high vowels, breathy voice quality, reduced pitch variance, and longer durations. However, attractive male voices had shorter durations (consistent with faster speaking rate), higher vowels, lower first formants overall, and higher second formant in /u/s. While this work was motivated by a search for body size correlates, the authors found a much more complicated relationship than expected.

In addition to pitch qualities, speaking rate also matters. Quené, Boomsma, and van Erning (2016) investigated the attractiveness of male voices by heterosexual female listeners as a function of both pitch and speaking rate. They found that faster and lower pitched speech was more attractive. However, tempo only matters if the pitch component is present. Fast but relatively high-pitched speech was not consistently rated as attractive.

In general, there are relatively few published findings about the relationship between voice quality and attractiveness. Babel et al. (2011) found breathy voice to be an indicator of attractiveness in female voices. Barkat-Defradas et al. (2015) found that male voices that are slightly rough (R1 on the GRBAS scale, a measure of dysphonia) are rated as the most attractive by women. The sexual orientation of subjects was not reported in either study.

Given these findings that there are vocal correlates to attractiveness, Fraccaro et al. (2013) investigated whether subjects could intentionally sound more or less attractive. They asked male and females to intentionally raise and lower the pitch of their voice. They found that when male speakers lowered their pitch and female speakers raised theirs, these manipulations did not necessarily lead to increased attractiveness. Additionally, when the male speakers raised their pitch and women lowered theirs, their attractiveness was lowered. This suggests that it is difficult to "fake" an attractive voice. Although we will return to the idea of intention when we discuss entrainment and communication of interest (i.e., flirting).

These trends, that lower pitched (and therefore more masculine) men are considered more attractive, are not independent of other qualities of the subject. Valentová, Roberts, and Havlícek (2013) investigated ratings of attractiveness and masculinity of male voices and faces by homosexual men and heterosexual women. These authors also collected information about the relationship status and sexual restrictiveness. Homosexual male subjects also self-rated themselves on a masculine–feminine scale. (Heterosexual female subjects were not asked to perform this self-rating.) They found no consistent preference for masculine faces by either homosexual men or heterosexual women. Moreover, a preference for masculine voices was only found in coupled heterosexual women and single homosexual men, While a preference for less masculine faces was observed in coupled homosexual men. Homosexual men who considered themselves to be more masculine tended to prefer more masculine voices, but more feminine faces. These findings highlight the complexity of identifying romantically attractive voices. Perceptions of attractiveness are conditioned not only on gender, but also sexual preference, and the gender expression of both the listener and speaker, in addition to other subjective idiosyncrasies. While this (and other) work by Valentova et al. goes further than most in acknowledging and investigating these factors, there remains a wide range of unstudied questions and interactions in this space.

The studies that we have surveyed so far have studied the perceptions of listeners who are not also conversational participants. While there are, of course, situations where this occurs, listening to the radio, an audiobook, a lecture, or other presentation, romantic attraction is more commonly established in two-party conversations. Here attraction is both assessed and performed and the voice is used to both express attraction and promote attractiveness. While this is a more complicated process, a number of efforts have been made to understand how romantic attractiveness works in a conversational setting.

Leongómez et al. (2014) investigated this by examining how adult heterosexual participants spoke when addressing attractive and unattractive potential partners (opposite-sex conversational partners) and potential competitors (same-sex conversational partners). The scenario followed a design similar to video dating and was conducted in both Czech and English. Subjects watched a stimulus video and recorded a response video introducing themselves. In the case of opposite-sex stimuli, the response video was to be played to the person who recorded the initial video. In the case of same-sex stimuli, the response video would be played along with the stimulus video to all opposite-sex subjects. Participants were instructed to explain whether and why they would like to date the potential partner in opposite-sex stimuli, and to explain why they should be chosen over the subject in same-sex stimuli. The stimuli videos were rated for attractiveness by an independent set of raters and comprised the three most and least attractive men and women drawn from a set of 40 participants (20 male and 20 female). They found that male F0 varied most in speech toward attractive women, but female F0 varied more in response to attractive competitors. Also, male minimum pitch was lowered when addressing attractive women. In a follow-up study, the experimenters also found that speech directed *toward* attractive participants was itself considered to be more attractive.

Dating scenarios are especially useful for investigating romantic attractiveness. The previous study used a video-dating paradigm. Another body of work looks at speed dating. In speed dating, participants engage in short (approximately 5 min) face-to-face conversations with potential partners and then fill out a questionnaire about their partner including an opportunity to indicate whether they would like to see the person again. In a speed-dating session, each participant may repeat this experience 10 or more times. In this work, all participants have self-selected to be interested in opposite-sex romantic partners. McFarland, Jurafsky, and Rawlings (2013) recorded speed-dating participants, and analyzed their speech, the content of their conversations, and their responses toward each other. While their analyses are quite comprehensive, we focus on vocal qualities here. Both genders described increased "connection" when they expressed excitement toward their partner. Male participants expressed this excitement through laughter, varied loudness, and reduced pitch variance. Female participants, however, raised and varied their pitch, spoke softer, varied loudness, and took shorter turns. They also found that women felt they "clicked" more with male partners who interrupted them. While this is somewhat unexpected—conventional understanding of interruption is that it is rude—closer inspection of these interruptions suggest that the overlapping speech that leads to a sense of connection was used to demonstrate understanding, through backchanneling and agreement. This is not to say that all interruption is "constructive" or used to demonstrate connection. Interruption can also be rude or dismissive. However, distinguishing the pragmatic effect of interruption can be challenging especially via a reliable automated technique. The study also found that entrainment, the convergence or divergence of vocal qualities between partners, is associated with attractiveness. Specifically, they found that partners who described a connection mimicked each others rate of speech, use of function words, and use of laughter.

Michalsky and Schoormann (2017) also looked at the role of entrainment in attractiveness, again investigated in a speed-dating setting. They focused on measures of pitch convergence. They found that speakers become more similar over time in both register and range, but that this degree of convergence was influenced by how attractive subjects found their conversational partner. In a later study, Michalsky and Schoormann (2018) found that listener reactions of attraction were sensitive to pitch height relative to the speaker's natural pitch range rather than an absolute measure. That is, attractive male voices are not simply low, but they are low in the speaker's pitch range. Conversely, female voices that are considered attractive are high in the woman's pitch range, not just naturally high pitched.

Examining vocal qualities in conversations forces experimenters to attempt to disentangle those aspects that are perceptive (being attractive) from those which are performative (expressing attractiveness). Puts et al. (2011) found that increased pitch and increased formant dispersion in women is found to be attractive and to be perceived as flirtations by other women. Jurafsky, Ranganath, and McFarland (2009) found that women who are labeled as "flirting" by men on speed dates spoke faster and with higher pitch and laugh more. These prosodic qualities overlap completely men who are labeled as "flirting," but men also speak more quietly. When women labeled their male partner as flirting (whether or not they actually were), they laughed more and lowered their intensity. But when men labeled their female partner as flirting, they raised their pitch. These analyses were developed and systematized in Ranganath, Jurafsky, and McFarland (2009). This work attempted to automatically detect flirting in speed-date speech. The most interesting qualities of this work come from identifying which features are used in the perception of flirting but are *not* used in the expression of flirting. For example, men are perceived to flirt when they overlap less and use fewer appreciations, but this is not significant in men who indicated that they were flirting. Similar faster speaking rate has a stronger influence on the perception of flirting than the performance of flirting. For women, laughing, taking fewer longer turns, and asking repair questions are strong indicators of a woman intending to flirt, but are not perceived by their partners as flirtatious.

2.7 Conclusions

In this chapter, we have surveyed the literature on four types of attraction and trust as it relates to a person's speech. We have used the term "charismatic" to describe a speaker who is politically attractive. In general, charismatic speakers are dynamic, passionate, and enthusiastic. These assessments are consistent across a range of listeners. American, Irish, Swedish, and Palestinian subjects have come to similar conclusions. However, the vocal realizations of this passion and dynamism vary by speaker. In general, charismatic political speakers vary their use of pitch, intensity, and speaking rate. Some research suggests that clear comprehensible pronunciation with relatively few disfluencies is also important.

Considering attraction in the business domain, business leaders considered charismatic often demonstrate the same qualities as political leaders. They pronounce words clearly, are rarely disfluent, and demonstrate more varied speech.

In the cases of business and political attractiveness, male and female subjects tended to assess speakers similarly. However, across research in both of these domains, far greater attention has been given to charisma in male speakers. One area that needs further study is what qualities of the female voices lead listeners to find them to be charismatic.

Regarding trust in a speaker, evidence suggests that listeners trust people who speak quickly. Male voices spoken with high pitch led to mistrust and female voices with more breathiness were more trusted. It is worth noting that these qualities are strongly linked to measures of political or business-based charisma.

Considering likeability, listeners tend to prefer voices that clearly enunciate—they have a higher pronunciation precision, but also a higher speaking rate. Other prosodic properties have less of an impact on assessment.

Romantic attraction as it relates to the voice has received quite a bit of research attention. The broad and most consistent finding here suggests than men with low voices and greater formant dispersion are attractive as are women with higher voices and more breathiness. The dynamics of romantic attraction in two-party conversations create an interesting area for research. The voice is involved both as an object of attraction and also a mechanism to demonstrate attraction. When heterosexual male speakers flirt, they lower their pitch, while flirting heterosexual women raise their pitch. Also, when participants are mutually attracted they tend to entrain on a number of prosodic dimensions including speaking rate, the use of laughter, and intensity.

One important caveat in the assessment of romantic attraction is that in many cases the gender of a listener is assumed to be a proxy for sexual preference. This is a methodological problem that can be found in a number of the reviewed studies.

While we have presented these types of attraction as related to each other, they have their own idiosyncrasies both in terms of how they operate socially and in how they are communicated via the voice. These forms of attraction may interact in unpredictable ways. The current research does not consider ways in which the qualities that make a voice attractive in one context may make it more or less attractive in another context or for a distinct social assessment. For example, are voices that are socially likeable more or less like voices that are attractive in business leaders?

In all, our understanding of what makes a voice attractive is fairly limited. There are a number of broad findings, but none of these in isolation is sufficient to either reliably predict attractiveness, or to provide overwhelmingly useful feedback to speakers. This ambiguity of findings can be found in individual studies but is even more clear through this survey. It is possible that it results from the fact that there is more inter-listener variability in both what is attractive and what signals are being relied on to make this decision.

While there is clearly more work to be done on this subject, major areas for further study include (1) investigation of business and political charisma in female speakers, (2) likeability and romantic attraction in nonheterosexual participants, and (3) more thorough consideration of qualities of the listener in identifying not just what is attractive in the speaker's voice, but what particular types of listeners find attractive.

References

Babel, M., King, J., McGuire, G., & Miller, T. (2011). Acoustic determiners of vocal attractiveness go well beyond apparent talker size. In *Laboratory Report: University of British Columbia and University of California, Santa Cruz.*

Barkat-Defradas, M., Fauth, C., Didirkova, I., de La Bretèque, B. A., Hirsch, F., Dodane, C., & Sauvage, J. (2015). "Dysphonia is beautiful: A perceptual and acoustic analysis of vocal roughness. In *18th International Congress of Phonetic Sciences (ICPhS-18).*

Baumann, T. (2017). Large-scale speaker ranking from crowdsourced pairwise listener ratings. In *Proceedings of INTERSPEECH.*

Beckman, M., Hirschberg, J., & Shattuck-Hufnagel, S. (2005). The original ToBI system and the evolution of the ToBI framework. In (pp. 9–54).

Biadsy, F., Rosenberg, A., Carlson, R., Hirschberg, J., & Strangert, E. (2008). A cross-cultural comparison of American, Palestinian, and Swedish perception of charismatic speech. In *Proceedings of Speech Prosody* (Vol. 37).

Boss, G. P. (1976). Essential attributes of the concept of charisma. *Southern Journal of Communication, 41*(3), 300–313.

Bruckert, L., Liènard, J.-S., Lacroix, A., Kreutzer, M., & Leboucher, G. (2006). Women use voice parameters to assess men's characteristics. *Proceedings of the Royal Society of London B: Biological Sciences, 273*(1582), 83–89.

Brueckner, R., & Schuller, B. (2012). Likability classification-a not so deep neural network approach. In *Proceedings of INTERSPEECH.*

Buisman, H., & Postma, E. (2012). The log-Gabor method: Speech classification using spectrogram image analysis. In *Proceedings of INTERSPEECH.*

Burkhardt, F., Eckert, M., Johannsen, W., & Stegmann, J. (2010). A database of age and gender annotated telephone speech. In *Proceedings of LREC.*

Collins, S. A., & Missing, C. (2003). Vocal and visual attractiveness are related in women. *Animal Behaviour, 65*(5), 997–1004.

Cullen, A., & Harte, N. (2018). A longitudinal database of Irish political speech with annotations of speaker ability. *Language Resources and Evaluation, 52*(2), 401–432.

Cullen, A., Hines, A., & Harte, N. (2014). Building a database of political speech: Does culture matter in charisma annotations? In *Proceedings of the 4th International Workshop on Audio/Visual Emotion Challenge* (pp. 27–31). ACM.

Cummins, N., Epps, J., & Kua, J. M. K. (2012). A comparison of classification paradigms for speaker likeability determination. In *Proceedings of INTERSPEECH.*

Dowis, R. (2000). *The lost art of the great speech: how to write it, how to deliver it.* Amacom Books.

Eyben, F., Wöllmer, M., & Schuller, B. (2010). OpenSmile: The Munich versatile and fast open-source audio feature extractor. In *Proceedings of the 18th ACM International Conference on Multimedia* (pp. 1459–1462). ACM.

Feinberg, D. R., Jones, B. C., Little, A. C., Burt, D. M., & Perrett, D. I. (2005). Manipulations of fundamental and formant frequencies influence the attractiveness of human male voices. *Animal Behaviour, 69*(3), 561–568.

Feinberg, D. R., DeBruine, L. M., Jones, B. C., & Little, A. C. (2008a). Correlated preferences for men's facial and vocal masculinity. *Evolution and Human Behavior, 29*(4), 233–241.

Feinberg, D. R., DeBruine, L. M., Jones, B. C., & Perrett, D. I. (2008b). The role of femininity and averageness of voice pitch in aesthetic judgments of women's voices. *Perception, 37*(4), 615–623.

Fraccaro, P. J., O'Connor, J. J., Re, D. E., Jones, B. C., DeBruine, L. M., & Feinberg, D. R. (2013). Faking it: Deliberately altered voice pitch and vocal attractiveness. *Animal Behaviour, 85*(1), 127–136.

Fujisaki, H., & Hirose, K. (1984). Analysis of voice fundamental frequency contours for declarative sentences of Japanese. *Journal of the Acoustical Society of Japan (E), 5*(4), 233–242.

González, J. (2006). Research in acoustics of human speech sounds: Correlates and perception of speaker body size. *Recent Research Developments in Applied Physics, 9*, 1–15.

Hirst, D. (1987). La description linguistique des systèmes prosodiques: une approche cognitive. Ph.D. thesis. Thèse de doctorat d'Etat, Universitè de Provence.

Hodges-Simeon, C. R., Gaulin, S. J., & Puts, D. A. (2010). Different vocal parameters predict perceptions of dominance and attractiveness. *Human Nature*, *21*(4), 406–427.

Jurafsky, D., Ranganath, R., & McFarland, D. (2009). Extracting social meaning: Identifying interactional style in spoken conversation. In *Proceedings of HLT/NAACL* (pp. 638–646). Association for Computational Linguistics.

Leongómez, J. D., Binter, J., Kubicová, L., Stolařová, P., Klapilová, K., Havlíček, J., et al. (2014). Vocal modulation during courtship increases proceptivity even in naive listeners. *Evolution and Human Behavior*, *35*(6), 489–496.

Levitan, S. I., Maredia, A., & Hirschberg, J. (2018). Acoustic-prosodic indicators of deception and trust in interview dialogues. In *Proceedings of INTERSPEECH* (pp. 416–420).

Louw, J., & Barnard, E. (2004). Automatic intonation modeling with INTSINT. In *Proceedings of the Pattern Recognition Association of South Africa* (pp. 107–111).

Lu, D., & Sha, F. (2012). Predicting likability of speakers with Gaussian processes. In *Proceedings of INTERSPEECH*.

Marcus, J. T. (1961). Transcendence and charisma. *The Western Political Quarterly*, *14*(1), 236–241.

McFarland, D. A., Jurafsky, D., & Rawlings, C. (2013). Making the connection: Social bonding in courtship situations. *American Journal of Sociology*, *118*(6), 1596–1649.

McGloin, R., & Denes, A. (2018). Too hot to trust: Examining the relationship between attractiveness, trustworthiness, and desire to date in online dating. *New Media & Society*, *20*(3), 919–936.

Michalsky, J.&Schoormann, H. (2017). Pitch convergence as an effect of perceived attractiveness and likability. In *Proceedings of INTERSPEECH* (pp. 2253–2256).

Michalsky, J., & Schoormann, H. (2018). Opposites attract! Pitch divergence at turn breaks as cause and effect of perceived attractiveness. In *Proceedings of Speech Prosody* (pp. 265–268).

Mixdorff, H., Niebuhr, O., & Hönemann, A. (2018). Model-based prosodic analysis of charismatic speech. In *Proceedings of Speech Prosody* (pp. 1–5).

Montaciè, C., & Caraty, M. -J. (2012). Pitch and intonation contribution to speakers' traits classification. In *Proceedings of INTERSPEECH*.

Niebuhr, O.; Brem, A., Novák-Tót, E., & Voße, J. (2016a). Charisma in business speeches-A contrastive acoustic-prosodic analysis of Steve Jobs and Mark Zuckerberg. In *Proceedings of Speech Prosody*.

Niebuhr, O., Voße, J., & Brem, A. (2016b). What makes a charismatic speaker? A computer-based acoustic-prosodic analysis of Steve Jobs tone of voice. *Computers in Human Behavior*, *64*, 366–382.

Niebuhr, O., Brem, A., & Tegtmeier, S. (2017). Advancing research and practice in entrepreneurship through speech analysis—From descriptive rhetorical terms to phonetically informed acoustic charisma profiles. *Journal of Speech Sciences*, *6*(1), 3–26.

Niebuhr, O., Thumm, J., & Michalsky, J. (2018a). Shapes and timing in charismatic speech-Evidence from sounds and melodies. In *Proceedings of Speech Prosody*.

Niebuhr, O.; Skarnitzl, R., & Tylečková, L. (2018b). The acoustic fingerprint of a charismatic voice-Initial evidence from correlations between long-term spectral features and listener ratings. In *Proceedings of Speech Prosody* (pp. 359–363).

Novák-Tót, E., Niebuhr, O., & Chen, A. (2017). A gender bias in the acoustic-melodic features of charismatic speech? In *Proceedings of INTERSPEECH* (pp. 2248–2252).

Pejčić, A. (2014). Intonational characteristics of persuasiveness in Serbian and English Political debates. *Nouveaux cahiers de linguistique française*, *31*, 141–151.

Pèrez, C. P. (2016). A study of the phono-styles used by two different Spanish-speaking political leaders: Hugo Chávez and Josè L. R. Zapatero. In *Proceedings of Speech Prosody* (pp. 410–414).

Pohjalainen, J., Kadioglu, S., & Räsänen, O. (2012). Feature selection for speaker traits. In *Proceedings of INTERSPEECH*.

Preston, D. R. (1999). A language attitude analysis of regional US Speech: Is northern US English not friendly enough? *Cuadernos de filología inglesa*, *8*, 129–146.

Puts, D. A., Barndt, J. L., Welling, L. L., Dawood, K., & Burriss, R. P. (2011). Intrasexual competition among women: Vocal femininity affects perceptions of attractiveness and flirtatiousness. *Personality and Individual Differences*, *50*(1), 111–115.

Puts, D. A., Doll, L. M., & Hill, A. K. (2014). Sexual selection on human voices. In *Evolutionary perspectives on human sexual psychology and behavior* (pp. 69–86). Springer.

Quenè, H., Boomsma, G., & van Erning, R. (2016). Attractiveness of male speakers: Effects of voice pitch and of speech tempo. In *Proceedings of Speech Prosody* (Vol. 8, pp. 1086–1089).

Ranganath, R., Jurafsky, D., & McFarland, D. (2009). It's not you, it's me: Detecting flirting and its misperception in speed-dates. In *Proceedings of EMNLP* (pp. 334–342). Association for Computational Linguistics.

Rosenberg, A., & Hirschberg, J. (2005). Acoustic/prosodic and lexical correlates of charismatic speech. In *EUROSPEECH*.

Rosenberg, A., & Hirschberg, J. (2009). Charisma perception from text and speech. *Speech Communication*, *51*(7), 640–655.

Ryant, N., & Liberman, M. (2016). Automatic analysis of phonetic speech style dimensions. In *Proceedings of INTERSPEECH* (pp. 77–81).

Sanchez, M. H., Lawson, A., Vergyri, D., & Bratt, H. (2012). Multi-system fusion of extended context prosodic and cepstral features for paralinguistic speaker trait classification. In *Proceedings of INTERSPEECH*.

Saxton, T. K., Caryl, P. G., & Craig Roberts, S. (2006). Vocal and facial attractiveness judgments of children, adolescents and adults: The ontogeny of mate choice. *Ethology*, *112*(12), 1179–1185.

Schweitzer, A., Lewandowski, N., & Duran, D. (2017). Social attractiveness in dialogs. In *Proceedings of Interspeech* (pp. 2243–2247).

Signorello, R., D'Errico, F., Poggi, I., & Demolin, D. (2012). How charisma is perceived from speech: A multidimensional approach. In *2012 ASE/IEEE International Conference on Social Computing*.

Sofer, C., Dotsch, R., Wigboldus, D. H., & Todorov, A. (2015). What is typical is good: The influence of face typicality on perceived trustworthiness. *Psychological Science*, *26*(1), 39–47.

Strangert, E., & Gustafson, J. (2008). What makes a good speaker? Subject ratings, acoustic measurements and perceptual evaluations. In *Proceedings of INTERSPEECH*.

Tavernier, J. (2007). Attitudes towards native and non-native accents of English. Ph.D. thesis. Ghent University.

Valentová, J., Roberts, S. C., & Havlícek, J. (2013). Preferences for facial and vocal masculinity in homosexual men: The role of relationship status, sexual restrictiveness, and self-perceived masculinity. *Perception*, *42*(2), 187–197.

Weber, M. (1947). *The theory of social and economic organization*. New York: Oxford University Press.

Weiss, B. & Burkhardt, F. (2010). Voice attributes affecting likability perception. In *Proceedings of INTERSPEECH*.

Weiss, B., & Burkhardt, F. (2012). Is "not bad" good enough? Aspects of unknown voices' likability. In *Proceedings of INTERSPEECH*.

Weninger, F., Krajewski, J., Batliner, A., & Schuller, B. (2012). The voice of leadership: Models and performances of automatic analysis in online speeches. *IEEE Transactions on Affective Computing*.

Wilson, R. K., & Eckel, C. C. (2006). Judging a book by its cover: Beauty and expectations in the trust game. *Political Research Quarterly*, *59*(2), 189–202.

Wu, D. (2012). Genetic algorithm based feature selection for speaker trait classification. In *Proceedings of INTERSPEECH*.

Chapter 3
The Vocal Attractiveness of Charismatic Leaders

Rosario Signorello

Abstract Social attractiveness in human leaders is defined as charisma, the set of leadership characteristics such as vision, emotions, and dominance used by leaders to share beliefs, persuade listeners, and achieve goals. Charisma is expressed through voice quality manipulations reflecting physiologically-based qualities and culturally-acquired habits to display leadership. These manipulations are adapted by the speakers to the social environment where they intend to be perceived as charismatic. Charisma in political speech is observed here to unveil the biological abilities versus the culturally-mediated strategies in leaders' speech according to different social contexts in which political communication takes place. Manipulations of vocal pitch, loudness, and phonation types are shown to cause both cross-cultural and culture-specific social attractiveness and consequently, are key factors for charisma effectiveness. Charismatic voice is then intentionally and unintentionally controlled by the human leaders to carry the perlocutionary salience of persuasive speech and influence listeners' choice of leadership.

Keywords Vocal charisma · Political speech · Attractiveness of leadership · Biological abilities in vocal persuasion · Cultural descriptors of charisma · Perceived charisma from speech

3.1 Introduction

3.1.1 Charisma Defined as the Social Attractiveness of Group Leaders

In modern literature, the term "charisma" was first popularized by sociologist Max Weber (1920). According to Weber, "charismatic" leaders generally emerge in times of great crisis for a nation, responding to the necessity of strong leadership to over-

R. Signorello (✉)
Laboratoire de Phonétique et Phonologie, CNRS & Sorbonne Nouvelle, Paris, France
e-mail: rosario.signorello@gmail.com

B. Weiss et al. (eds.), *Voice Attractiveness*, Prosody, Phonology and Phonetics,
https://doi.org/10.1007/978-981-15-6627-1_3

come the crisis. This author defines charisma as an "extraordinary quality" of a person who is believed to be endowed with superhuman properties, in such a way as to induce people to acknowledge him as a leader, to the point of making a cult of him. Weber calls this quality "charisma" (from Greek charis, grace), thus considering it a grace, a divine gift that only some enlightened people may possess. Weber does not describe this gift at length, and even considers it beyond human comprehension; yet, the very notion of charisma has been alternatively redefined and challenged.

Some first sketches of charisma may be retrieved from ancient philosophy. According to Heraclitus, only a few individuals are endowed with particular physical and mental skills and virtues, that include, in accordance with Socrates, fast learning capacities, memory, open mind, and vision. These virtues are innate, according to Plato, and make a chief the object of trust, faith, and veneration by other people, which results in the cult of the leader (Cavalli, 1995). Such idea of the charismatic leader was personified in the great dictators of the twentieth century: Hitler, Mussolini, and Stalin.

Previously, research on charisma was mainly conducted in social psychology within the general framework of leadership studies. Some authors consider leadership as an internal trait of individuals (House & Howell, 1992). For example, transformational leaders, which Burns (1978) and Bass (1985) consider to be charismatic, show high values in four of the Big Five factors: extraversion, openness, agreeableness, and conscientiousness (Bono & Judge, 2004).

An opposing view—the contingency perspective, which also includes the contextual approach, contends that leadership and charisma are strongly determined by the context: contextual factors trigger or inhibit particular leadership behaviors, and leadership is interactively constructed by the relationship between leader and followers (Haslam et al., 2011). This contextualist view further develops into the transactional leadership perspective, in which the strength and effectiveness of leadership is determined by a cost-benefit computation, where followers agree to comply with the leader's will to the extent they feel this is functional to their goals. Their behavior is stimulated by rewards and punishments more than trust and identification. This is not the case, however, for transformational leadership, which, introduced by the so-called neo-charismatic school, views a true leader as an authentically charismatic person (Lowe et al., 1996), endowed with vision and capacity for inspiring followers, who works in their interest and aims at their growth (Burns, 1978; Bass, 1985). Neo-charismatic scholars stress the ethical impact of transformational leadership, and warn of the "dark side" of charisma and the inauthentic or pseudo-transformational leaders, who with self-serving aims act in bad faith, consciously or unconsciously. Actually, the charismatic/transformational view integrates sociological and psychological aspects since it sees charisma as a "social process" in which the perception of followers becomes a very central aspect (Shamir, 2000).

The discussion among these diverse perspectives, based on personality or context, transaction or transformation, makes the definition of a charismatic leader and the singling out of charismatic attributes particularly complex. In fact, charisma is a multidimensional construct: it is certainly affected (and constructed) by the values, needs, motivations, and discourses of potential followers, but it also, indubitably depends on the leader's skills, choices, and characteristics. External displays are the perceivable expression of the internal features, and we can distinguish two kinds: one which we call the "charisma of the body" and the other, "charisma of the mind" (Signorello, 2014). Actually, the external features may stem either from the mind or from the body of the leader. Aspects of the charisma of the mind, such as creative and charming ideas or feelings, are displayed by a person's words or actions, while the charisma of the body is displayed by specific aspects of their visual and/or acoustic appearance, determined by their body's multimodal physical traits and behaviors (Shamir, Zakay, Breinin, & Popper, 1993; Bull, 1986; Atkinson, 1984; Rosenberg & Hirschberg, 2009).

The athletic and proud gait of Barack Obama is a way of moving that conveys dignity. But, take Mahatma Gandhi, who was a short, thin shy man, without a loud voice, and who even sometimes stuttered: the features of his charisma did not emanate from his voice or gait, but from the strength of his message, and what revolutionary ideas came from his words and his political action. The first example is a case of the charisma of the body, while the latter is an example of the charisma of the mind: the meaning of a discourse by Gandhi (Bligh & Robinson, 2010). It is through words that his charismatic qualities shine forth. These two forms of expression of charisma—body and mind—may sometimes appear in combination, for example, Barack Obama may be seen as charismatic both for the concepts he proposes and the way he exposes them: he has charisma both of the body and of the mind (Bono & Judge, 2004).

In sum, charismatic persons may have different kinds of charisma which depend on the type of internal charismatic features they possess, the external features that express them, and on their combinations. The aim of the present work is then to highlight the multidimensionality of charisma, and to explore in detail a specific display of political leaders' attractiveness: their voice. The hypothesis of this study is that the charisma of a person can be disentangled into a set of "charismatic features", and that in different persons, particular combinations of these features cluster into peculiar kinds of charisma. So what are the internal features of charisma, and how can we find them out?

3.1.2 Charisma and Voice Behavior: The Charismatic Voice

Group leaders use their voices to communicate their charisma, the set of leadership characteristics, such as vision, emotions, and dominance used by leaders to share beliefs, persuade, and achieve goals. Voice quality reflects leaders' physiologically-based vocal characteristics and culturally-acquired habits and strategies used to shape those characteristics qualitatively. Political speech is studied in order to unveil the biological abilities versus the culturally-mediated strategies of group leaders' charismatic voices. Through voice acoustic analyses and perceptual studies, a cross-culturally similar use of vocal pitch, loudness levels, and ranges in political speech and a culture-specific perceptual effect of overall vocal characteristics like phonation types, prosodic factors, and temporal characteristics were found. Charismatic voices reflect individuals' (a) biological needs to have easy access to resources and (b) cultural needs to show skills that reflect high social status and power.

Voice quality results from speakers' biologically-derived differences in vocal apparatus combined with learned linguistic and cultural habits used to convey their personal identity (Garvin & Ladefoged, 1963; Kreiman & Sidtis, 2011). Voice quality conveys individuals' physical (e.g., size, Ohala, 1994; Pisanski et al., 2014, attractiveness, Zuckerman & Driver, 1989; Collins, 2000), psychological (e.g., personality traits, Scherer, 1972; emotional status, Patel, Scherer, Björkner, & Sundberg, 2011) and social characteristics (e.g., leadership; Surawski & Ossoff, 2006; Tigue et al. 2012; Klofstad, Anderson, & Peters, 2012; dominance, Ohala, 1984). These studies raise the question about whether particular features characterizing political speakers' voices are biologically versus culturally determined, and which type of feature is primary in distinguishing individuals chosen as group leaders from non-leaders.

Besides theoretical discussions on the nature of charisma, some studies investigated how charisma is perceived from voice. Tackling the relationship between the acoustic-prosodic characteristics of a political leader's speech and the perception of his/her charisma, Touati (1993) investigated the prosodic features of rhetoric utterances in French political speech in pre and post-elections discourses. Strangert and Gustafson (2008) examined the relationship between prosodic features and the perception of a speaker as a "good communicator", while Rosenberg and Hirschberg (2009), studied the correlation between acoustic, prosodic, and lexico-syntactic characteristics of political speech and the perception of charisma.

The overview above, introduced our conceptual definition of charisma focused on its psychological multidimensionality that affects social attractiveness, as well as a few theoretical insights, on the use of voice and speech as nonverbal behaviors to convey vocal attractiveness in political speech. This chapter reports investigations on the perceptual features that characterize vocal attractiveness in charismatic political discourse. This work highlights the features of charisma conveyed by the speakers and its social attractiveness on listeners speaking several languages. In the following sections, I first present a tool developed to measure the differences between vocal qualities of speaking individual political leaders. I later introduce studies that aimed to distinguish various kinds of charisma while singling out the features of voice that are responsible for their perception.

3.2 Charismatic Voices

3.2.1 Cultural- and Language-Based Descriptors of Charisma

In contemporary literature about the perception of charisma from voice, scholars ask participants to rate voices in terms of adjectives that in previous studies had been connected to charisma (e.g., Rosenberg & Hirschberg, 2009). In our research, studies testing how people describe the charisma of group leaders in different languages and cultures were carried out in order to make a scale for the rating of charisma (Signorello et al., 2012a, 2012b). Through an empirical and non-biased approach, positive and negative traits of charisma in several languages (American English, French, Italian, and Brazilian Portuguese) were collected to develop the "Multi-dimensional Adjective-based Scale of others' Charisma Perception" (MASCharP) (Signorello, 2014), a psychometric tool to be used in research on the perception of charismatic traits from individuals' perceivable behaviors, such as voice. This approach entailed three experimental phases.

The first phase involved the collection of lexical and semantic descriptions of charismatic traits communicated through an individual's perceivable behaviors from subjects of the languages being studied. This part entailed the gathering of adjectives that describe charismatic, as well as noncharismatic prototypes of leadership. It is fundamental to understand that the language in question is inseparable from its culture. These two factors act as filters in the attribution of an individual's traits.

The second phase involved dimensions of theoretical classification of the adjectives gathered. As in Di Blas and Forzi (1998), the adjectives were selected by their frequency of usage. Only the most frequently used terms that are representative and descriptive of charismatic traits in the participants' language were retained. In the first stage of data sorting, adjectives with a frequency higher than 1 were retained, indicating a cognitive commonality between at least two individuals who agree on a semantic-representational connection that designates the adjective as a trait of charisma. The adjectives used most frequently to describe charisma were then categorized in dimensions that were deduced from aspects of the persuasive process illustrated in the Sect. 3.2 of this chapter. The data were then organized according to semantic closeness, as in the cases of Saucier (2009) and Di Blas and Forzi (1998), corresponding to the dimensions of Poggi's theory of persuasion (Poggi, 2005). An example of the definitive selection of adjectives and dimensional classification constitutes the MASCharP as represented in Table 3.1 (American English).

The third phase involved the creation of a psychometric tool to perform the perceptual tests and measure the perception of charisma from voice. Each adjective from MASCharP could be evaluated through a Likert scale (Likert, 1932). An interface based on the server-side software Limesurvey® (The LimeSurvey project team, (2011)) was developed to collect the data. This software is written in PHP and uses a MySQL database to store data. The interface features the combination of the MASCharP with the 7-point Likert scale. The use of this tool has already been validated in several studies to measure the traits and types of charismatic leadership conveyed by voice (Signorello et al. 2012a, 2012b, 2014b; D'Errico et al., 2012, 2013).

Table 3.1 Positive and negative interpersonal traits of perceived other's charisma in American English. Classification according to Signorello (2014)

Positive Charisma Traits	Negative Charisma Traits
Caring, Passionate, Kind, Enthusiastic, Understanding	Rude, Mean, Cold, Unkind, Egotistical
Extroverted, Optimistic, Trustworthy, Outspoken, Friendly, Genuine, Sociable	Introverted, Pessimistic, Dishonest, Selfish, Hostile, Aloof
Intelligent, Witty, Humble, Brave, Determined, Bold, Respectful, Assertive, Well-spoken	Ignorant, Stubborn, Closed-minded, Arrogant, Reserved
Dynamic, Confident, Energetic, Strong, Leader, Engaging, Persuasive	Aggressive, Angry, Apathetic, Shy, Weak, Overbearing, Dull, Obnoxious, Intimidating
Charming, Funny, Attractive, Humorous, Interesting, Relatable, Personable	Boring, Annoying, Uninteresting, Depressing

3.2.2 Charisma Perception in Cross-Language Settings

The following study was conducted to understand what in the voice perceptual domain could be considered as universal versus language and culture-based. The perception of charismatic speaker identity from voice might be influenced unpredictably by one vocal characteristic or by a whole complex pattern resulting from source and filter characteristics, mode of vocal fold vibration, temporal characteristics, articulatory settings and characteristics, degree of nasality, prosodic line, and syllable structure (Kreiman & Sidtis, 2011).

To do so, this study first assessed how listeners use the vocal pitch as a biological cue to detect speakers' charismatic traits from voice and how they use this cue to assess leadership fitness and choose their leader. In several studies vocal pitch has emerged as a feature that serves as an important biological cue that signals social and physical dominance (e.g., Ohala, 1982, 1983, 1984, 1994, 1996; Puts et al. 2007), conveys leadership (Klofstad et al., 2012; Anderson & Klofstad, 2012), and that influences the choice of a leader (Darwin, 1871; Tigue et al., 2012). In an experiment, 40 French listeners evaluated the dominance conveyed by different voice quality patterns in the voice of an Italian speaker and political leader (Umberto Bossi, former leader of the Lega Nord party from 1980 to 2012). The results showed significant negative correlations between the perceived dominant type of charismatic leadership and average F0 ($r = -0.19$, $p < 0.05$, linear regression), wide F0 range ($r = -0.18$, $p < 0.05$), and maximum F0 ($r = -0.18$, $p < 0.05$). Meanwhile, higher F0 mean ($r = 0.52$, $p < 0.01$), minimum F0 ($r = 0.49$, $p < 0.01$), maximum F0 ($r = 0.55$, $p < 0.01$), and the F0 range ($r = 0.53$, $p < 0.01$) are significantly and positively correlated with a nondominant type of charismatic leadership.

To confirm and extend these results, the investigations were repeated with the manipulation of F0 for vocal stimuli from two different leaders (Luigi de Magistris, an Italian leader; François Hollande, a French leader). Forty-eight Italians were then asked to rate vocal stimuli from the French leader and 48 French listeners were asked to rate vocal stimuli from the Italian leader. Results show that French and Italian listeners perceive leaders as having a less dominant charisma when they use a high F0 (average of 200 Hz for the French speaker; 212 Hz for the Italian speaker) and a wide F0 range (16 semitones for French listeners; 12 semitones for Italian listeners). This experiment studied the way in which listeners assess leadership fitness from voice. A voice sounding more dominant (low frequencies of F0 and a narrow F0 range) would be perceived as more effective by Italian listeners ($r = 0.61$, $p < 0.0001$; simple linear regression), whereas French participants perceive effective leadership from higher pitched voices ($r = 0.41$, $p = 0.004$). Results from the two experiments imply that low frequencies of F0 and a narrow F0 range convey a dominant charismatic leadership and that higher F0 average and wider F0 range, cause the perception of a nondominant charismatic leader. These different types of leadership would be perceived as more or less effective in different cultures.

Finally, the perception of specific charismatic traits from overall vocal characteristics was studied taking into account the role of the language and the culture of listeners. The study first assessed the way in which different patterns of voice quality convey the different charismatic traits of leaders. Forty French participants assessed the charisma of the Italian leader Umberto Bossi from natural voice samples. Detailed profiles based on the correlation between voice acoustics, perception of charismatic traits, emotional states aroused, and choice of leader were created. A profile with a voice pattern characterized by a medium pitch range (13 semitones), moderate falling pitch contour movements, modal phonation, phrase-final harsh-high (middle-range) vowels and long inter-word pauses (~1 s) communicate an Authoritarian-Threatening type of charisma where the leader is perceived as individualistic, untrustworthy, influential, confident, organized, resolute, egocentric, determined, authoritarian, menacing, scary, and cold (see Table 3.2), and moreover arouses negative emotional states in the listeners like anxiety. A second profile shows that a voice pattern characterized by a wide pitch range (16 semitones) from very low to very high frequencies, abrupt pitch contour movements, harsh or modal phonation, and sentence-final vowels in creaky phonation communicate a Proactive-Attractive type of charisma. Listeners who perceived the Proactive-Attractive type of charismatic leadership described the leadership of the speakers as vigorous, active, dynamic, charming, and attractive (see Table 3.2), arousing positive emotions like amusement, admiration, enthusiasm, reassertion, and calmness. French listeners would be most likely to choose a leader perceived as Proactive-Attractive. The third profile shows a voice pattern characterized by a narrow pitch range from low to medium-high frequencies (9–13 semitones), but not as high as the two vocal patterns above, smooth pitch contour movements, harsh-low, harsh-mid, or modal phonation types, and an increasing duration of the vocalization (from ~1 s to 6.5 s). This pattern communicates the Competent-Benevolent type of charismatic leadership, characterized by participant-selected adjectives such as wise, prudent, calm, trustworthy, fair, intelligent, easygoing, honest, sagacious, and sincere (see Table 3.2), arousing amusement but not calmness emotions. This type of leadership communicates the image of a

Table 3.2 Charisma types and interpersonal traits. Speaker: Umberto Bossi. Assessed perceptually through the MASCharP tool. Exploratory Factor Analysis: Varimax Rotation that extracted three factors which explained 45% of the variance; significant Bartlett's test of sphericity ($p = 0.000$); Kaiser–Mayer Olkin (KMO) measure of Sampling Adequacy (0.83); high level of reliability (Proactive-Attractive: $\alpha = 0.92$, i.i. $= 0.52$; Calm-Benevolent: $\alpha = 0.87$, i.i. $= 0.44$; Authoritarian-Threatening: $\alpha = 0.90$, i.i. $= 0.41$)

Authoritarian-Threatening		**Proactive-Attractive**		**Calm-Benevolent**	
Determined	0.508	Vigorous	0.837	Wise	0.825
Menacing	0.775	Active	0.767	Prudent	0.737
Who scares	0.767	Dynamic	0.766	Calm	0.731
Dishonest	0.762	Charming	0.738	Trustworthy	0.689
Cold	0.679	Attractive	0.709	Fair	0.645
Individualistic	0.642	Courageous	0.701	Intelligent	0.605
Authoritarian	0.585	Convincing	0.687	Easygoing	0.585
Leader	0.578	Captivating	0.676	Honest	0.576
Untrustworthy	0.563	Seductive	0.642	Sagacious	0.527
Influent	0.552	Bewitching	0.604	Sincere	0.514
Confident	0.523	Sexy	0.592		
Organized	0.509	Eloquent	0.553		
Resolute	0.506	Determined	0.54		
Egocentric	0.485	Who propose	0.54		
		Visionary	0.472		
Variance	22.52%		12.6%		10.83%

leader competent enough to access vital resources and benevolent enough to share those resources with other individuals. French listeners in the sample studied would not choose this type of leadership.

3.3 Conclusions

3.3.1 Leaders' Social Attractiveness

Since Weber (1920), first launched the notion of charisma, the definition has gone through various changes. The notion itself may have seemed too difficult to operationalize, while the literature has fluctuated from serious investigation to skeptical consideration. This may be partly due to the very nature of charisma, which lives at the crossroad of various psychosocial dimensions and takes very different forms (Shamir, 2000). This work has defined charisma as a set of internal and physical qualities of a person that make him or her capable of influencing other people by wakening their most positive emotions, and hence inducing them to do what she/he wants very willingly and exploiting their internal motivation. These qualities are related to various perceived aspects of the group leaders persona (moral, intellectual, affective), of power management, as well as esthetic and even erotic aspects. Charisma is a multidimensional psychosocial notion: the studies presented in this chapter tried to discover and disentangle its dimensions from participants' description of charismatic and noncharismatic persons using a scale of charisma perception. The present research found out that dimensions may combine to give rise to different types of charisma. The type of perceived charisma depends on whether the esthetic and dynamic dimensions prevail, resulting in a Proactive-Attractive charisma, or whether they are moderated by the intellectual and ethical side, thus enhancing a calm-benevolent charisma; or finally whether the dimensions of dominance and deliberate influence cluster in an Authoritarian-Threatening charisma.

Besides discovering these internal features and their combinations, this investigation focused on a peculiar property of charismatic political leaders, their vocal communication, showing that charisma resides in particular types of speech acts, but also in particular parameters of the leader's voice that, depending on given variations, may become less charismatic, or take up a different type of charisma. Two issues we specifically investigated in this connection were the change in charisma caused by a switch from modal to dysphonic voice, and the different perception of charisma caused, in the French and the Italian culture, by a change in pitch and pause duration.

Results on the former issue—that the modal voice conveys a proactive-attractive, or even an authoritarian-threatening charisma, whereas the disordered one bears a calm-benevolent one—may be accounted for by an evolutionary perspective that views a dynamic leader as more functional to the effectiveness of the group.

As to the issue of whether charisma perception is universal or cultural, our results may be interpreted as follows: The single traits attributed to a charismatic leader tend to be different between cultures and may arise at two levels: first, the single properties may cluster in different ways for two cultures, in that a type of charisma may be more salient in one culture and dispersed in single properties in another; second, as seen in the third phase of study, each specific type of charisma may be evoked by some vocal parameters in one language or culture and by other parameters in another.

These results may help answer some questions concerning charisma. For instance, one possible objection to the very existence of such a notion is that a person may appear as charismatic to some people but not to others. In other words, is it true that -beauty is in the eyes of the beholder-? In our view, this is not so. Different perceptions of charisma may well be accounted for by its multidimensionality. In this sense, interactive accounts that view charisma as determined by the intertangling between a leader and their followers may be sound. -Charismatic leadership- may hold per se, but also, followers can contribute their perceptual preferences to its emergence (Shamir, 2000).

In the same vein, the multidimensionality account might answer the question whether and why the perception of charisma varies across cultures. Since cultures definitely attribute different importance to different dimensions of life, cognitive functioning and social interaction, two cultures may well see the same leader as charismatic or not, depending on the dimensions they value the most. Yet, this leads to another question: aren't there any aspects of charisma that are universal, that is, any characteristics of a leader (or of a person) that are perceived as charismatic by people of all cultures?

An answer in line with the "emotional culture" approach above (Ekman & Friesen, 1971; Turner, 1976; Gordon, 1989; Matsumoto, 1990; Bagozzi, Verbeke, & Gavino, 2003) would be that leaders are perceived as charismatic to the extent to which they adapt to the communicative norms of their culture. Yet, we might contend that, on the contrary, the charismatic leader, does not "adapt to", but rather, "leads" his followers, imposing new norms and values, and thus also changing the relative preference of the charismatic dimensions. Therefore, a primary and possibly universal dimension of charisma might be just the visionary skill that makes a leader point at something new.

A final issue, among others, that is raised by our investigation is how the notion of charisma proposed here can be applied not only to political leaders but to a broader domain: not only social leaders can be charismatic, but actors, singers, managers, and teachers. Our theoretical explanation of charisma could be applied generally to all charismatic individuals.

3.3.2 The Charismatic Voice

The present research demonstrates how a specific vocal pattern used by leaders can convey different traits and types of their charisma, and also how several patterns can influence the perception of the same type of characteristic leadership when perceived by different individuals or social groups. The acoustics of voice in political speech is a cue to the perception of charisma in leaders. We used a cross-cultural approach to assess and distinguish the physiological/anatomical and cultural influence in the production and perception of voice in charismatic leadership.

In the perceptual domain, the research described above, first found evidence that vocal pitch is a cross-cultural signal to distinguish dominant versus less dominant charisma. This result is consistent with previous studies on the perception of dominance versus submission related to vocal pitch (e.g., Collins, 2000; Feinberg et al., 2006). Higher fundamental frequency and wider range are used by the speaker while addressing a more diverse audience (in terms of sex, age and social status). Lower fundamental frequency and narrower range are used by the leader-speaker when addressing an audience of similar social status (other leaders). Healthy vocal range is used by leaders in informal contexts of communication (during which no political topics are addressed and the leadership is not questioned).

This work then found that certain vocal quality patterns used by the speaker-leader fit the listener's expectations about the vocal style that best conveys charisma in a given language and culture. The same vocal pattern can convey both an Authoritarian-Threatening and a Proactive-Attractive charisma that are perceptually distinguished in different languages and cultures. Competent-Benevolent charismatic leadership can be conveyed by several vocal quality patterns.

These results may help to better distinguish between the biological components on the one hand, and language and cultural components on the other, present in voice behavior that fit listeners' expectations and influences the choice of the social group's leader. Listeners seem capable of accurately distinguishing these vocal features of the charismatic leader and these results might explain why some leaders have been found to be endowed with a cross-language and cultural charisma (e.g., Barack Obama was found to be the most charismatic leader in the general sense in several cultures), and some other leaders not endowed with effective speaking (Bligh & Robinson, 2010), are mostly endowed with a circumscribed charisma restricted within social groups and languages (Gandhi is only charismatic if we understand English or if it is translated).

References

Anderson, R. C., & Klofstad, C. A. (2012). Preference for leaders with masculine voices holds in the case of feminine leadership roles. *PLoS ONE*, *7*(12), e51216. https://doi.org/10.1371/journal.pone.0051216.

Aristotle (1991). *Rhetoric*, translated by George A. Kennedy. Location: Acheron Press, Kindle ed.

Atkinson, M. (1984). *Our Masters' voices. The language and body language of politics*. London: Routledge.

Bagozzi, R. P., Verbeke, W., & Gavino, J. C, Jr. (2003). Culture moderates the self-regulation of shame and its effects on performance: The case of salespersons in The Netherlands and the Philippines. *Journal of Applied Psychology*, *88*(2), 219.

Baken, R., & Orlikoff, R. (2000). *Clinical measurement of speech and voice*, 2nd (rev ed.). San Diego: Singular Publishing Group.

Bass, B. M. (1985). *Leadership and performance beyond expectations*. New York: Free Press.

Bass, B. M. (1990). *Bass and Stogdill's handbook of leadership: Theory, research, and managerial applications* (3rd ed.). New York: Free Press.

Biadsy, F., Hirschberg, J., Rosenberg, A., & Dakka, W. (2007). Comparing American and Palestinian perceptions of charisma using acoustic-prosodic and lexical analysis. In *Proceedings of Interspeech*

Biadsy, F., Rosenberg, A., Carlson, R., Hirschberg, J., & Strangert, E. (2008). A cross-cultural comparison of American, Palestinian, and Swedish perception of charismatic speech. In *Proceedings of the 4th Conference on Speech Prosody*, (pp. 579–582). Campinas, Brazil.

Bligh, M. C. and Kohles, J. C. (2009). The enduring allure of charisma: How barack obama won the historic 2008 presidential election. *The Leadership Quarterly*, 20(3), 483–492.

Bligh, M. C., & Robinson, J. L. (2010). Was Gandhi 'Charismatic'? Exploring the rhetorical leadership of Mahatma Gandhi. *The Leadership Quarterly*, *21*(5), 844–55.

Bono, J. E. and Judge, T. A. (2004). Personality and transformational and trans- actional leadership: A meta-analysis. *Journal of Applied Psychology*, 89(5): 901–910.

Boss, P. (1976). Essential attributes of charisma. *Southern Speech Communication Journal*, *41*(3), 300–13.

Bull, P. (1986). The use of hand gesture in political speeches: Some case studies. *Journal of Language and Social Psychology*, *5*(2), 103–118.

Burns, J. (1978). *Leadership*. New York, NY, USA: Harper & Row.

Castelfranchi, C. & Falcone, R. (2000). Trust is much more than subjective probability: Mental components and sources of Trust. In *32nd Hawaii International Conference on System Sciences—Mini Track on Software Agents*. Maui: IEEE Press.
Cavalli, L. (1995). *Carisma: la qualità straordinaria del leader*. Rome, Italy: Laterza.
Cicero (1967). *De Oratore*, translated by E. W. Sutton. Cambridge, MA: Harvard University Press.
Collins, S. A. (2000). Men's voices and Women's choices. *Animal Behavior*, *60*(6), 773–80.
Darwin, C. (1871). *The descent of man, and selection in relation to sex*. Murray, J., London, UK.
Dastur, Y. (2016). *Charisma perception in the Japanese language*. Department of Linguistics: University of Southern California.
Den Hartog, D. N., House, R. J., Hanges, P. J., Ruiz-Quintanilla, S. A., & Dorfman, P. W. (1999). Culture specific and cross-culturally generalizable implicit leadership theories: Are attributes of charismatic/transformational leadership universally endorsed? *The Leadership Quarterly*, *10*(2), 219–56.
D'Errico, F., Signorello, R., Demolin, D., & Poggi, I. (2013). The perception of charisma from voice: A cross-cultural study. In *Proceedings of the 2013 Humaine Association Conference on Affective Computing and Intelligent Interaction* (pp. 552–557). IEEE Computer Society.
D'Errico, F., Signorello, R., & Poggi, I. (2012). Le Dimensioni del Carisma. In M. Cruciani & F. Cecconi (Eds.), *IX Convegno Annuale dell'Associazione Italiana di Scienze Cognitive–AISC* (pp. 245–52). Rome: Università di Trento.
Di Blas, L., & Forzi, M. (1998). The circumplex model for interpersonal trait adjectives in Italian. *Personality and Individual Differences*, *24*(1), 47–57.
Ekman, P., & Friesen, W. V. (1971). Constants across cultures in the face and emotion. *Journal of Personality and Social Psychology*, *17*(2), 124.
Emerich, K. A., Titze, I. R., Švec, J. G., Popolo, P. S., & Logan, G. (2005). Vocal range and intensity in actors: A studio versus stage comparison. *Journal of Voice*, *19*(1), 78–83.
Esling, J. (2006). Voice quality. In K. Brown (Ed.), *The Encyclopedia of Language and Linguistics* (2nd ed., pp. 470–4). Oxford: Elsevier.
Feinberg, D. R., Jones, B. C., Law Smith, M. J., Moore, F. R., DeBruine, L. M., Cornwell, R. E., et al. (2006). Menstrual cycle, trait estrogen level, and masculinity preferences in the human voice. *Hormones and Behavior*, *49*(2), 215–22.
Foti, R. J., & Luch, C. H. (1992). The influence of individual differences on the perception and categorization of leaders. *Leadership Quarterly*, *3*, 55–66.
Garvin, P., & Ladefoged, P. (1963). Speaker identification and message identification in speech recognition. *Phonetica*, *9*, 193–199.
Gordon, S. L. (1989). *Institutional and impulsive orientations in selectively appropriating emotions to self The sociology of emotions: Original essays and research papers* (pp. 115–35).
Haslam, S. A., Reicher, S. D., & Platow, M. J. (2011). *The New psychology of leadership*. Hove, UK: Psychology Press.
Hofstede, G. (1993). Cultural constraints in management theories. *Academy of Management Executive*, *7*(1), 81–94.
Hollander, E. P., & Julian, J. W. (1970). Studies in leader legitimacy, influence, and innovation. *Advances in Experimental Social Psychology*, *5*, 33–69.
House, R. J. and Howell, J. M. (1992). Personality and charismatic leadership. *The Leadership Quarterly*, 3(2), 81–108.
Kaiser, H. F. (1974). An index of factorial simplicity. *Psychometrika*, *39*, 31–6.
Klofstad, C. A., Anderson, R. C., & Peters, S. (2012). Sounds like a winner: Voice pitch influences perception of leadership capacity in both men and women. *Proceedings of the Royal Society B: Biological Sciences*, *279*(1738), 2698–2704.
Kreiman, J., & Sidtis, D. (2011). *Foundations of voice studies: An interdisciplinary approach to voice production and perception*. Oxford, UK: Wiley-Blackwell.
Kruskal, W. H., & Wallis, W. A. (1952). Use of ranks in one-criterion variance analysis. *Journal of the American Statistical Association*, *47*(260), 583–621.

Lamarche, A., Ternström, S., & Hertegård, S. (2009). Not just sound: Supplementing the voice range profile with the singer's own perceptions of vocal challenges. *Logopedics, Phoniatrics, Vocology, 34*(1), 3–10.

Lewin, K. (1952). *Field theory in social science: Selected theoretical papers*. London: Tavistock.

Likert, R. (1932). A technique for the measurement of attitudes. *Archives of Psychology, 140*, 1–55.

Lowe, K. B., Kroeck, K. G., & Sivasubramaniam, N. (1996). Effectiveness correlates of transformational and transactional leadership: A meta-analytic review of the Mlq Literature. *The Leadership Quarterly, 7*(3), 385–425.

Matsumoto, D. (1990). Cultural similarities and differences in display rules. *Motivation and Emotion, 14*(3), 195–214.

Offermann, L. R., Kennedy, J. K., & Wirtz, P. W. (1994). Implicit leadership theories: Content structure, and generalizability. *Leadership Quarterly, 5*(1), 43–55.

Ohala, J. (1996). Ethological theory and the expression of emotion in the voice. In *Proceedings of the 4th International Conference on Spoken Language Processing (ICSLP 96)* (vol. 3, pp. 1812–1815), Philadelphia, PA, USA.

Ohala, J. J. (1982). The voice of dominance. *Journal of the Acoustical Society of America, 72*(S1), S66–S66.

Ohala, J. J. (1983). Cross-language use of pitch: An ethological view. *Phonetica, 40*(1), 1–18.

Ohala, J. J. (1984). An ethological perspective on common cross-language utilization of F0 of voice. *Phonetica, 41*(1), 1–16.

Ohala, J. J. (1994). The frequency codes underlies the sound symbolic use of voice pitch. *Sound Symbolism* (pp. 325–347). Cambridge, MA, USA: Cambridge University Press.

Patel, S., Scherer, K. R., Björkner, E., & Sundberg, J. (2011). Mapping emotions into acoustic space: The role of voice production. *Biological Psychology, 87*(1), 93–98.

Pisanski, K., Fraccaro, P. J., Tigue, C. C., O'Connor, J. J. M., Röder, S., Andrews, P. W., et al. (2014). Vocal indicators of body size in men and women: A meta-analysis. *Animal Behaviour, 95*, 89–99.

Poggi, I., D'Errico, F., & Vincze, L. (2011). Discrediting moves in political debate. In P. Ricci-Bitti (Ed.), *Proceedings of UMMS* (pp. 84–99). Heidelberg: Springer.

Poggi, I. (2005). The goals of persuasion. *Pragmatics & Cognition, 13*(2), 297–336.

Puts, D. A., Hodges, C. R., Cárdenas, R. A., & Gaulin, S. J. C. (2007). Men's voices as dominance signals: Vocal fundamental and formant frequencies influence dominance attributions among men. *Evolution and Human Behavior, 28*(5), 340–344.

Raymond, M. (2008). *Cro-magnon toi-même! Petit Guide Darwinien de la vie Quotidienne*. Paris: Seuil.

Reboul, O. (1998). *Introduction à la Rhétorique* (3rd ed.). Paris: Presses Universitaires de France.

Reicher, S., Haslam, S. A., & Hopkins, N. (2005). Social identity and the dynamics of leadership: Leaders and followers as collaborative agents in the transformation of social reality. *The Leadership Quarterly, 16*(4), 547–68.

Rhee, N., & Signorello, R. (2016). The acoustics of charismatic voices in Korean political speech: A cross-gender study. *Journal of the Acoustical Society of America, 139*(4), 2123–3.

Riggio, R. E., Chaleff, I., & Lipman-Blumen, J. (2008). *The art of followership: How great followers create great leaders and organizations*. San Francisco: Jossey-Bass.

Rosenberg, A., & Hirschberg, J. (2009). Charisma perception from text and speech. *Speech Communication, 51*(7), 640–55.

Rudolph, S. H., & Rudolph, L. I. (1983). *Gandhi: The traditional roots of charisma*. Chicago: University of Chicago Press.

Saucier, G. (2009). Semantic and linguistic aspects of personality. In P. J. Corr & G. Matthews (Eds.), *The Cambridge handbook of personality psychology* (pp. 379–99). Cambridge: Cambridge University Press.

Scherer, K. R. (2010). Voice appeal and its role in political persuasion. In *International Workshop on Political Speech, Rome*.

Scherer, K. R. (1972). Judging personality from voice: A cross-cultural approach to an old issue in interpersonal perception. *Journal of Personality*, *40*, 191–210.
Shamir, B. (2000). Taming charisma for better understanding and greater usefulness: A response to Beyer. *The Leadership Quarterly*, *10*(4), 555–562.
Shamir, B., House, R. J., & Arthur, M. B. (1993). The motivational effects of charismatic leadership: A self-concept based theory. *Organization Science, A Journal of the Institute of Management Sciences*, *4*(4), 577.
Shamir, B., Zakay, E., Breinin, E., & Popper, M. (1998). Correlates of charismatic leader behavior in military units: Subordinates' attitudes, unit characteristics, and Superiors' appraisals of leader performance. *Academy of Management Journal*, *41*(4), 387–409.
Signorello, R. (2014). *La Voix Charismatique: Aspects Psychologiques et Caractéristiques Acoustiques*. Ph.D diss. Université de Grenoble, France and Università degli Studi Roma Tre, Italy.
Signorello, R. (2014b). The biological function of fundamental frequency in leaders' charis-matic voices. *The Journal of the Acoustical Society of America*, 136(4), 2295–2295.
Signorello, R., & Demolin, D. (2013). The physiological use of the charismatic voice in political speech. In *Proceedings of the 14th Annual Conference of the International Speech Communication Association (Interspeech)* (pp. 987–991).
Signorello, R., D'Errico, F., Poggi, I., & Demolin, D. (2012b). How charisma is perceived from speech: a multidimensional approach. *ASE/IEEE International Conference on Social Computing* (pp. 435–440).
Signorello, R.; D'Errico, F.; Poggi, I.; Demolin, D. & Mairano, P. (2012a). "Charisma Perception in Political Speech: A Case Study." In *Proceedings of the VIIth GSCP International Conference: Speech and Corpora*, edited by H. Mello, M. Pettorino, and T. Raso, 343–8. Firenze: Firenze University Press.
Signorello, R., & Rhee, N. (2016). The voice acoustics of the 2016 United States presidential election candidates: A cross-gender study. *Journal of the Acoustical Society of America*, *139*(4), 2123–3.
Signorello, R., Demolin, I., Poggi, D., & D'Errico, F. (2011). *Il Carisma del Corpo: Caratteristiche Acustiche della Voce Carismatica*. In X Giornate della Ricerca: Universita degli Studi Roma Tre.
Strangert, E. & Gustafson, J. (2008). "What Makes a Good Speaker? Subject Ratings, Acoustic Measurements and Perceptual Evaluations." In *Proceedings of the 9th Annual Conference of the International Speech Communication Association (Interspeech 2008)*: 1688–91.
Surawski, M. K., & Ossoff, E. P. (2006). The effects of physical and vocal attractiveness on impression formation of politicians. *Current Psychology*, *25*(1), 15–27.
The LimeSurvey project team (2011). LimeSurvey. Web-based com-puter program. Retrieved June 26, 2011, from http://www.LimeSurvey.org/.
Tigue, C. C., Borak, D. J., O'Connor, J. J. M., Schandl, C., & Feinberg, D. R. (2012). Voice Pitch Influences Voting Behavior. *Evolution and Human Behavior*, *33*(3), 210–216.
Touati, P. (1993). Prosodic Aspects of Political Rhetoric. *ESCA Workshop on Prosody*, 168–71.
Tuppen, C. (1974). Dimensions of Communicator Credibility: an Oblique Solution. *Speech Monographs*, *41*(3), 253–60.
Turner, R. H. (1976). The Real Self: From institution to impulse. *American Journal of Sociology*, 989–1016.
Weber, M. (1920). The theory of social and economic organization. Oxford University Press, New York, USA.
Zuckerman, M., & Driver, R. (1989). What sounds beautiful is good: The vocal attractiveness stereotype. *Journal of Nonverbal Behavior*, *13*(2), 67–82.

Chapter 4
Vocal Preferences in Humans: A Systematic Review

Melissa Barkat-Defradas, Michel Raymond, and Alexandre Suire

Abstract Surprisingly, the study of human voice evolution has long been conducted without any reference to its biological function. Yet, following Darwin's original concept, John Ohala was the first linguist to assume the functional role of sexual selection to explain vocal dimorphism in humans. Nevertheless, it is only at the very beginning of the millennial that the study of voice attractiveness developed, revealing that beyond its linguistic role, voice also conveys important psycho-socio-biological information that have a significant effect on the speaker's mating and reproductive success. In this review article, our aim is to synthesize 20 years of research dedicated to the study of vocal preferences and to present the evolutionary benefits associated with such preferences.

Keywords Vocal preferences · Perception · Language evolution · Sexual selection · Evolutionary biology · Acoustics · Voice · Fundamental frequency · Formant dispersion · Voice attractiveness

4.1 Introduction

Darwin thought of mate choice as a purely aesthetic experience, a selection of beauty for its own sake (Darwin, 1871). However, his view has not been embraced by modern evolutionary biology, for which mate choice results from human adaptive preferences, a mechanism that has evolved because of dimorphic physical features or sexual ornaments (such as the female waist-to-hip ratio, the male shoulder-to-

M. Barkat-Defradas (✉) · M. Raymond · A. Suire
Institut des Sciences de l'Evolution de Montpellier, University of Montpellier, Centre National de la Recherche Scientifique, Institut pour la Recherche et le Développement, Ecole Pratique des Hautes Etudes – Place Eugène Bataillon, 34095 Montpellier, France
e-mail: melissa.barkat-defradas@umontpellier.fr

M. Raymond
e-mail: michel.raymond@umontpellier.fr

A. Suire
e-mail: alexandre.suire@umontpellier.fr

B. Weiss et al. (eds.), *Voice Attractiveness*, Prosody, Phonology and Phonetics,
https://doi.org/10.1007/978-981-15-6627-1_4

hip ratio, facial traits, breast size, voice, and so on) that are assumed to be reliable indicators of mate quality (Arak & Enquist, 1993). Indeed, the mere sound of a person's voice contains important, embedded biological information. Consequently, a large amount of research has been dedicated to identifying men's preferences for women's secondary sexual characteristics and vice versa, as well as the evolutionary benefits associated with such preferences.

Preferences partly proceed from an unconscious mechanism: an individual may be aware of the factors that have led him to choose one sexual partner instead of another, but it does not necessarily mean s/he is conscious of the link existing between his or her preference and the property conveyed by the cue itself. A good example to illustrate this statement rests on women's preference for masculine low-pitched voices. Though female subjects are often conscious of their attraction for this type of vocal attribute in males, they are hardly aware that it indicates men's phenotypic quality as well as part of their heritable genotypic value as potential mates (Apicella, Feinberg, & Marlowe, 2007). In human species, mate's selective value includes several phenotypic qualities among which: state of health, fertility, age, intelligence, social status, and so on … (Buss, 1989; Geary, Vigil, & Byrd-Craven, 2004; Sugiyama, 2015). All these qualities are displayed through the face, the body, and the voice. For example, health is indicated by skin complexion, the body shape is a proxy of nutritional status, and the vocal height is determined by testosterone level. Therefore, it is reasonable to assume that female typical preference for men exhibiting deep voices has been shaped by evolution as an honest signal of masculinity related to an increased level of androgens, a high physical strength, a good immune system, etc., all of these features favoring men's—and thus women's—fitness. However, masculine versus feminine preferences for the ornaments exhibited by the other sex are not the same since some of the traits that are associated to desirable qualities in men may differ from those linked to desirable phenotypic qualities in women. Consequently, men and women do not grant the same importance to the different socio-biological cues driving mate choice. Generally speaking, and at least in Western industrialized societies, men tend to attach a great importance to women's beauty, and as early as Ancient Greece, the concept of beauty has been closely associated with physical attractiveness, especially feminine physical attractiveness (for a detailed review of the evolution of feminine beauty see Bovet, 2018). But when choosing a mate, men and women also use non-physical features, such as smell, movements, behaviors, and voice. Although these traits are not all equally weighted in mating decisions, they all likely contribute to the general evaluation of a potential partner.

Our aim here is not to explore the diverse effects of physical attractiveness but rather to examine the role of voice in the mating context by showing which vocal features are considered attractive by men and/or women and why. Previous research on vocal attractiveness (i.e., the perceived attractiveness of voices when isolated from other cues, such as visual or olfactory cues) has suggested that vocal attractiveness plays a role in mate choice in humans (e.g., Apicella et al., 2007; Hill et al., 2013; Leongomez et al., 2014). For example, individuals possessing vocal characteristics that are correlated with attractiveness report greater reproductive potential (as indexed by reported number of sexual partners, Kordsmeyer, Hunt, Puts, Ostner, & Penke,

2018; Hill et al., 2013) and, at least in hunter-gatherers, have greater reproductive fitness (Apicella et al., 2007). People also alter their vocal attractiveness in mating contexts, such as when interacting with an attractive potential mate (Leongomez et al., 2014; Pisanski, Bhardwaj, & Reby, 2018; Suire, Raymond, & Barkat-Defradas, 2018). In accordance to the runaway selection mechanism,[1] we assume preferences may contribute to the shaping of attractiveness in human voices. Our goal therefore is to show that preferences for some vocal attributes are likely the result of sexual selection. Although the acoustic features associated with vocal attractiveness are not exhaustively studied here (i.e., the prosodic dimension, in particular, could be further developed), we propose an exhaustive review of the different studies (n = 37, over a period of 40 years covering the years 1979–2020) that tackled the issue of vocal preferences for men and women (see Table 4.1). Subsequently, we will focus on the evolutionary mechanisms driving our preferences. Before fully entering our topic, it should be noted that only the studies that have clearly identified the acoustic correlates behind vocal preferences were considered.

Overall, a first remarkable point appears to be the importance ascribed to the study of F0 and the formant position. Secondly, one will immediately notice that English speakers are overrepresented in comparison with speakers of other languages. From a methodological point of view, it appears that the number and the nature of vocal stimuli used in the perceptual experiments are quite variable (i.e., spontaneous speech, isolated words or vowels, reading versus oral speech ...). Likewise the number of auditory judges is extremely heterogeneous from one study to another. As for the acoustic analyses themselves, we distinguish between two types of approaches: on the one hand, there are correlational studies, which basically aim at relating acoustic characteristics and vocal attractivity from auditory judge's scores on Likert's scales and on the other hand, there are experimental studies that try to establish causal relations between acoustic features. All these studies help us pinpoint some general trends about human vocal preferences.

A brief overview in Table 4.1 reveals that among the different measures that were investigated for qualifying vocal attractiveness across studies, it is undoubtedly vocal height (i.e., F0) that has most often aroused the authors' interest. Nevertheless some other articulatory and acoustic features have lead to interesting results suggesting vocal attractiveness is not confined to the realm of fundamental frequency but also extend to other aspects, which effects on perceived vocal attractiveness are also reviewed in the next sections.

[1]Runaway selection is a mechanism whereby a secondary sexual trait expressed in one sex is correlated with a preference for the trait in the other sex. The genetic coupling of the trait and the preference leads to self-reinforcing loops of coevolution between the trait and preference for the trait (Travers, 2017).

Table 4.1 Studies are characterized by the language under study, the number and nature of tested the stimuli, the number and gender of auditory judges, the methodology (Likert's scale versus forced choice), results by gender and the direction of observed correlations. For Likert's scales, the lowest score (i.e., 1) corresponds to the less attractive stimuli, the highest to the most attractive voice; when forced choice is used the judge has to choose between two stimuli the one he perceives the most attractive. By "manipulated speech" we mean that the subjects were recorded after they were asked to modify their voices following the experimentaters' instructions. Note that for studies based on modified stimuli (whether naturally or not) forced choice is often used since it allows judges to select the most attractive stimuli between two versions of the same voice (i.e., natural versus modified). NB: CA stands for Canadian, AU for Australian, U.S. for American, and U.K. for British variants of English

References	Language(s) under study	No. of stimuli	No. of auditory judges	Type of stimuli	Method and no. of evaluated stimuli	Acoustic features under study and direction of the observed correlations
Tuomi and Fisher (1979)	English (CA)	10♀ 5♂	10♀ 10♂	Spontaneous sentences	Likert's scale n = 15 (i.e., all)	– Low F0 +attractive ♀ and ♂ (n.s.)
Zuckerman and Miyake (1993)	English (U.S.)	62♀ 48♂	8♀ 9♂	Speech reading	Likert's scale n = 110 (i.e., all)	– Low F0 +attractive ♂ – Low F0min +attractive ♂ – Low energy +attractive ♂ – Less pausing time +attractive ♂ – n.s. ♀
Oguchi and Kikuchi (1997)	(i) Japanese – Experience 1	4♂	25♀	Read sentences	Likert's scale n = 4 (i.e., all)	– Low F0 +attractive ♀ and ♂ – Low F0-SD +attractive ♀ and ♂
	(ii) Japanese –Experience 1	8♀ 8♂	42♀ 20♂	Read sentences	Likert's scale n = 16 (i.e., all)	– Low F0 +attractive ♀ and ♂ – Low F0-SD +attractive ♀ and ♂
Collins (2000)	Dutch	34♂	54♀	Isolated vowels (natural speech)	Likert's scale n = between 10 & 14	– Spectral distribution in low frequencies +attractive♂ – Low formant spacing +attractive ♂
Collins and Missing (2003)	English (U.K.)	30♀	30♂	Isolated vowels (natural speech)	Likert's scale n = 10	– Spectral distribution in high frequencies +attractive♂ – High formants +attractive ♂ – High formant spacing +attractive♂
Feinberg et al. (2005)	English (CA)	10♂	68♀	Isolated vowels (manipulated speech)	Likert's scale n = 10 (i.e., all)	– Low F0 +attractive ♂ – Lower formant spacing + attractive ♂

(continued)

Table 4.1 (continued)

References	Language(s) under study	No. of stimuli	No. of auditory judges	Type of stimuli	Method and no. of evaluated stimuli	Acoustic features under study and direction of the observed correlations
Bruckert et al. (2006)	French	26♂	102♀	Isolated vowels (natural speech)	Likert's scale n = 6	– Low F0 +attractive ♂ – High F0-SD +attractive ♂
Riding et al. (2006)	English (U.S.)	9♂	54♀	Spontaneous then manipulated speech	Likert's scale n = 11	– High F0 -attractive ♂ – F0-SD n.s. ♂
Saxton et al. (2006)	English (U.K.)	12♂	40♀ 7–10 y.o 40♀ 12–15 y.o 40♀ 20–34 y.o	Number counting (natural speech)	Forced choice n = 6 or 12	– Low F0 +attractive for 12–15 and 20–34 y.o.
Feinberg, DeBruine, Jones, and Little (2008a, 2008b)	–Experience 1 English (CA)	123♀	10♂	Isolated vowels (manipulated speech)	Likert's scale n = 61 or 62	– High F0 +attractive ♂
	–Experience 2 English (CA)	15♂	263♀ 342♂	Number counting (from 1 to 10)	Forced choice n = 15 pairs	– High F0 +attractive ♂
Hughes et al. (2008)	English (U.S.)	31♀ 40♂	50♀ 51♂	Numbers recitation (from 1 to 10)	Likert's scale n = 71 (each voice being evaluated by 13 or 15 judges)	– Low F0min +attractive ♂ – F0, F0max, F0 range, median F0, Intensity, Duration, Jitter, Shimmer, HNR n.s. ♂ – n.s. ♀
Leaderbrand et al. (2008)	English (U.S.)	1♀ 1♂	39♀ 9♂	Sentences (manipulated speech)	Likert's scale n = 4	– Low F0min +attractive ♂ and ♀
Vukovic et al. (2008)	English (U.K.)	36♂	58♀ (+contraceptive) 65♀ (-contraceptive)	Sentences (manipulated speech)	– Forced choice n = 16 pairs – Likert's scale for each preferred voice	– Low F0 +attractive ♂ – No effect of contraception on vocal preferences

(continued)

Table 4.1 (continued)

References	Language(s) under study	No. of stimuli	No. of auditory judges	Type of stimuli	Method and no. of evaluated stimuli	Acoustic features under study and direction of the observed correlations
Saxton et al. (2009)	English (U.K.)	6♀ 6♂ 11–13 y.o 6♀ 6♂ 13–15y.o 6♂ 13–15 y.o	148♀ 177♂ (same category of age)	Isolated vowels (manipulated speech)	– Forced choice n = 6 pairs – Likert's scale for each preferred voice	– High F0 +attractive ♀ (for 11–13 y.o. ♂ only) – Low F0 +attractive ♂ (for 13–15 y.o. ♀ only)
Jones et al. (2010)	English (U.K.)	4♀	30♀ 30♂	Spontaneous sentences (natural speech)	Forced choice n = 16 pairs	– High F0 +attractive ♀
Fraccaro et al. (2011)	English (CA)	6♀	178♂	Isolated vowels (manipulated speech)	Forced choice n = 6 pairs	– High F0 +attractive ♀ (long-term relationship condition only)
Hodges-Simeon et al. (2010)	English (U.S.)	111♂	142♀	Spontaneous speech	Likert's scale n = 30 or 31	– Low F0 +attractive ♂ (no effect of short vs. long relationship condition, no effect of ♀ menstrual cycle phase) – Low F0-SD +attractive ♂ (in long term + fertile context and in short term + unfertile context) – Spectral distribution in the low frequencies +attractive ♂ (in short/long term + fertile conditions)
Hughes, Farley and Rhodes (2010)	English (U.S.)	25♀ 20♂	27♀ 12♂	Truncated phone calls + speech manipulation	Forced choice n = 45 (i.e., all)	– Low F0 +attractive ♂ and ♀
Jones et al. (2018)	English (U.K.)	6♀ 6♂	100♀ 100♂	Isolated vowels (manipulated speech)	Forced choice n = 6 pairs	– High F0 +attractive ♀ – Low F0 +attractive ♂

(continued)

Table 4.1 (continued)

References	Language(s) under study	No. of stimuli	No. of auditory judges	Type of stimuli	Method and no. of evaluated stimuli	Acoustic features under study and direction of the observed correlations
Borkowska and Pawlowski (2011)	Polish	58♀	144♂	Isolated vowels (manipulated speech)	Likert's scale n = 13 voices	– High F0 +attractive ♀ (non linear relation)
Pisanski and Rendall (2011)	English (CA)	2♀ 6♂	30♀ 31♂	Words list (natural then manipulated speech)	Likert's scale n = 40 voices	– Low F0 and formants +attractive ♂ (same trend observed for natural and manipulated speech) – Low F0 and formants-attractive ♀ (same trend observed for natural and manipulated speech)
Puts et al. (2011)	English (U.S.)	72♀	63♂	Text reading (manipulated speech)	Likert's scale n = 18 pairs	– High F0 +attractive ♀ – Spectral distribution in the high frequencies +attractive ♀
Liu and Xu (2011)	English (U.K.)	1♀	10♂	3 repetitions of 1 single emotion-free sentence (natural then manipulated speech)	Likert's scale n = 81 (i.e., all)	– High F0 +attractive ♀ – Small vocal length tract +attractive ♀
Simmons et al. (2011)	English (AU)	54♂	15♀	Isolated vowels (natural speech)	Likert's scale n = 54 (i.e., all)	– Low F0 +attractive ♂
Barkat-Defradas et al. (2012)	French	62♂	92♀	Text reading + Isolated vowel (natural speech)	Likert's scale n = 34	- F0 n.s. – Mild roughness degree +attractive ♂ – Low breathiness +attractive ♂
Re et al. (2012)	English (CA)	1♀ 1♂	9♀ 10♂	Isolated vowels (manipulated speech)	Forced choice n = 50 pairs + supplementary pairs 6 ♂ and 42 ♀	– High F0 +attractive ♀ – Low F0 +attractive ♂

(continued)

Table 4.1 (continued)

References	Language(s) under study	No. of stimuli	No. of auditory judges	Type of stimuli	Method and no. of evaluated stimuli	Acoustic features under study and direction of the observed correlations
Fraccaro et al. (2013)	English (CA)	4♀ 4♂	104♀ 110♂	Isolated vowels (manipulated speech)	Forced choice n = 16 pairs	– High F0 +attractive ♀ – Low F0 +attractive ♂
O'Connor et al. (2013)	English (CA)	4♀ 4♂	128♂	Words (manipulated speech)	Likert's scale n = 40	– High F0 +attractive ♀ – Low F0 +attractive ♂
Xu et al. (2013)	English (U.K.)	– Experience 1 1♀	10♂	Sentences (manipulated speech)	Likert's scale n = 81 (i.e., all)	– Low F0 +attractive ♂ – Formants: n.s.
		– Experiences 2–5 1♀ 1♂	16♀ 16♂	Sentences (synthetized speech)	Likert's scale n = 81 (i.e., all)	– High F0 +attractive ♀ – Low F0 +attractive ♂ – High breathiness +attractive ♀ ♂ – Low formants +attractive ♂ – Formants n.s. ♀
Babel et al. (2014)	English (U.S.)	30♀ 30♂	15♀ 15 ♂	Words (natural speech)	Likert's scale n = 15 (one single voice for each trial)	– Spectral distribution in high frequencies +attractive ♀ – Low F0 +attractive ♀ (n.s.) – Breathy voices +attractive ♀ – Spectral distribution in low frequencies +attractive ♂ – Shorter duration +attractive ♂
Hughes et al. (2014)	English (U.S.)	20♀ 20♂	20♀ 20♂	Number recitation (from 1 to 10) (manipulated speech)	Likert's scale n = 40 voices	– High hoarseness +attractive ♀ n.s. ♂ – Longer duration +attractive ♂ ♀ – Low F0 +attractive ♂ n.s. ♂ – Loudness n.s. ♂ ♀
Skrinda et al. (2014)	Latvian	60♂	29♂	Isolated vowels (natural speech)	Likert's scale n = unspecified	– Low F0 +attractive ♂ – Low F2 values +attractive ♂ – Other formants n.s. ♂

(continued)

Table 4.1 (continued)

References	Language(s) under study	No. of stimuli	No. of auditory judges	Type of stimuli	Method and no. of evaluated stimuli	Acoustic features under study and direction of the observed correlations
Tsantani et al. (2016)	English (U.K.)	10 ♀ 9♂	183♀ 57♂	"Hello" (manipulated speech)	Forced choice n = 40 pairs	– Low F0 +attractive ♂ n.s. ♀
Sebesta et al. (2017)	Cross-linguistic	45♂ (Cameroonians) 48♂ (Namibians)	62 ♀ Czechs	Sentence (natural speech)	Likert's scale n = 45 – 48 pairs	– Low F0 +attractive Cameroonian ♂ – Low formant position + attractive Namibian ♂ – High breathiness +attractive Namibian ♂
Shirazi et al. (2018)	Cross-linguistic	6♂ (EN U.S)	20 breastfeeding + 20 nulliparous Filipinos ♀	Sentences (manipulated speech)	Likert's scale n = 12 stimuli	– High F0 +attractive ♂ – n.s. between the 2 groups of ♀
Suire et al. (2018)	French	58♂	137♀	Sentence	Forced choice n = 11 pairs	– Low F0 +attractive ♂ – High F0-SD +attractive ♂ – Other acoustic measures n.s.
Suire et al. (2019)	French	13♀	135♂ 2 conditions: short- versus long-term relationship	Read sentences (natural speech)	Forced choice n = 13 pairs	– High speaking rate +attractive ♀ – Low F0 +attractive ♀ – Spectral distribution in high frequencies +attractive ♀ – High roughness (high Jitter values) +attractive ♀ – Low breathiness (high HNR values) +attractive♀
Zheng et al. (2020)	Chinese	80♀ 35♂	88♀ 79♂	Isolated vowels (natural speech)	3 conditions: – habitual voice – raised in pitch (+20 Hz) – lowered in pitch (−20 Hz)	– High F0 +attractive ♂ both for male and female raters – High F0 +attractive ♂ for male raters – Low F0 +attractive ♂ for female raters (but very low pitches are perceived less attractive > cue for laryngeal damage or below the intelligibility threshold)

4.2 Preferences for Vocal Height

Most of the previous studies, whether they are correlational or experimental, have revealed a negative correlation between vocal height and attractivity of men. Such a regular trend shows that women, whatever their linguistic environments and/or cultural backgrounds, are predominantly attracted to men exhibiting deep low voices (Bruckert, Lienard, Lacroix, Kreutzer, & Leboucher, 2006; Feinberg et al., 2005; Hodges-Simeon, Gaulin, & Puts, 2010; Hughes, Farley, & Rhodes, 2010; Jones, Feinberg, DeBruine, Little, & Vukovic, 2010; Pisanski & Rendall, 2011; Vukovic et al., 2008; Xu, Lee, Wu, Liu, & Birkholz, 2013; Suire, Raymond, & Barkat-Defradas, 2019). Still, a few exceptions are to be considered. As a matter of fact, Babel, McGuire, and King (2014) and Hughes, Mogilski, and Harrison (2014) reported no significant correlation between vocal height and attractivity in American men. Likewise, Barkat-Defradas et al. (2012) demonstrated F0 does not seem to be the most salient perceptual feature to assess masculine voice attractiveness as compared to roughness at least in clinical context when patients range into a comparable vocal height category (i.e., ±125 Hz) irrespective of their global dysphonic grade. Lastly, Shirazi, Puts, and Escasa-Dorne (2018) obtained an unexpected opposite result with Filipino women judging male vocal samples produced in English by American speakers. As for women vocal attractiveness, the vast majority of studies reach the same results with men being consistently attracted by high-pitched feminine voices (Borkowska & Pawlowski, 2011; Collins & Missing, 2003; Feinberg et al., 2008a, 2008b; Jones et al., 2010; Puts, Barndt, Welling, Dawood, & Burriss, 2011; Re et al., 2012). But here again, the results obtained by Leaderbrand, Dekam, Morey, and Tuma (2008), Oguchi and Kikuchi (1997) go in the opposite direction when those by Hughes et al. (2010, 2014) reveal interesting trends. In Hughes et al. (2010), the authors show that women tend to lower their voices when interacting with men they consider as particularly attractive while they significantly raise their pitch when facing men they are not attracted to. The same kind of unexpected result is observed for men who judge those low-pitched women as sexier. More recently, Pisanski et al. (2018) replicated the same results. In a second study, in which female subjects were asked to modify their voice so as they might be perceived as more attractive by male auditory judges, it has been shown that in such an evoked seductive context, women are also inclined to deepen their voices, and interestingly the subsequent perceptual study revealed that the female voices attesting the lower pitch values are also those that were perceived as the most attractive by the group of male auditory judges (Hughes et al., 2014). The results launched by Zheng, Compton, Heyman, and Jiang (2020) in what must be to our knowledge the most recent available study tackling the subject aimed at determining more precisely the effect of raised versus lowered pitch on voice perceived attractiveness. In order to answer this question, the authors used a method based on voluntarily pitch-shifted voices. Their findings suggest that indeed pitch shifts do affect voice attractiveness in the sense that female voices are perceived—both for male and female raters—as more attractive when vocal pitch is raised (+20 Hz from a digitally computed average pitch at 237 Hz). As for male voices, they typically show that lowered pitch lead to better evaluations by female raters (up to certain limits beneath which low voices are perceived either as pathological or unintelligible). But surprisingly, they also come to the result that their male raters consider high-pitched masculine voices as more attractive. According to the authors this may be explained by the fact that in real-life conditions, men are more often placed into the position of evaluating sex-opposite attributes using morphological signals, like waist-to-hip ratio,[2] but also vocal cues so as to find information of phenotypical compatibility, which makes their perceptual evaluation biased either by a lack of experience or by the unconscious usage of a perceptual grid of evaluation that is structured around feminine vocal references and which is consequently quiet unsuitable for the evaluation of male voices.

4.3 Preferences for Vocal Modulation

If studies dealing with the effect of mean F0 on vocal attractiveness are relatively numerous, those based on the measure of F0-SD (i.e., the increased versus reduced mean fundamental frequency variations, which the listener perceives, respectively, as rather flat versus highly modulated speech) are rather scarce. Yet, Hodges-Simeon et al. (2010) have shown that male speakers producing speech with very little variations in F0 are perceived as more masculine and attractive by female raters. Given that the extent of F0-excursions is affected by attitudinal and emotional factors (Traunmöller & Eriksson, 1995), such a trend appears to be kind of difficult to explain at first glance. Indeed, as it is well admitted the non-verbal characteristics of voices can play a significant role in signaling emotional as well as health state, like for the latter, major depression that is regularly reflected through reduced vocal modulation, female preferences for small melodic variations in male voices may be explained both by vocal dimorphism (since it has been regularly shown lively speech is related with feminine talking style (Polce-Lynch, Myers, Kilmartin, Forssmann-Falck, & Kliewer 1998; Hall, 1978) and social factors (as the extensive vocal expression of emotions is more often associated with female behavior (Fischer & Manstead, 2000). Therefore, assuming pitch variations are perceived along a continuum (from monotonic to highly expressive speech), the receivers may have assigned monotonous voices to masculinity and, reversely, dynamic speech to feminity. Besides, Suire et al. (2020) have shown males' sexual orientation can be inferred more accurately from F0-SD than mean F0, suggesting vocal modulation is a more reliable acoustic cue for gays' vocal feminization than vocal height. Moreover, though previous studies assessed

[2]The WHR has been used as an indicator of health and the risk of developing serious health conditions. WHR correlates with fertility (with different optimal values in males and females). The concept and significance of WHR as an indicator of attractiveness has been theorized by Singh (1993) who argued the WHR is a consistent estrogen marker, and thus a reliable proxy of fertility. Women with a 0.7 WHR are usually rated as more attractive by men from Indo-European cultures (Singh & Young 2001), but preferences may vary according to the culture under study (Fisher & Voracek, 2006).

that reduced fundamental frequency variations are rather linked to vocal masculinity, two other studies lead to unexpected opposite results. According to Bruckert et al. (2006), monotonous voices are judged as significantly less attractive for men while Leongómez et al. (2014) found modulated voices are rated as more attractive for both sexes. Further researches are thus needed to disentangle these inconsistent results. But yet for now, it is interesting to notice that the same criterion may lead to different auditory impressions, which valences are somehow contradictory. For example, although perceived as more attractive, those masculine speakers exhibiting monotonous, low-pitched voices are also perceived as being less cooperative (Tognetti et al., 2019), more threatening, and their likelihood to have extramarital affairs is considered as higher. This claim does not result from unfounded subjective impressions since there is also evidence that suggest men with masculine voices report a higher number of extra-pair sex partners and are more often chosen by women as extra-pair partners (Hughes et al., 2004).

The above suggests that men with relatively more masculine voices—that are negatively correlated with testosterone levels (Evans, Neave, Wakelin, & Hamilton, 2008)—may present a greater infidelity risk to their partners, though it is still unclear whether observers assess infidelity risk via vocal cues to underlying testosterone levels. Likewise, women with relatively high-pitched, modulated voices—that are linked both with youth, higher fertility, and increased perceived attractivity—are also seen as more conspicuous and more likely to commit adultery (O'Connor, Re, & Feinberg, 2011). But, while there is substantial evidence for a positive relationship between testosterone, deep voice, and "unbridled" sexuality among men, the relationship between women's sexuality and feminine vocal features is more complex (for a review, see Bancroft, 2005). We should therefore be cautious and presume that women with attractive voices may be more likely to be unfaithful due to a greater opportunity for extra-pair sex given their desirability as a mate as their attractive voices are more often chosen by paired men as extra-pair partners (Hughes, Dispenza, & Gallup, 2004).

4.4 Preferences for Timbre

Sounding vocalizations are the product of multiple acoustic parameters, including formant position and formant dispersion. Formant dispersion is a measure of the average spacing between the formants (Fitch, 1997). It is a function of the length and shape of the vocal tract and corresponds to the space through which sound waves must travel from the vocal folds to the oral cavity. Until sexual maturity, vocal tract length grows without any sexual dimorphism between boys and girls (Vorperian et al., 2005), but at puberty, under the influence of androgens, males' larynges descend farther than females' (Fitch & Giedd, 1999). Indeed, working through hormone receptors in the epithelial cells of the laryngeal tissue, testosterone enlarges the larynx on the one hand and lengthens and thickens the vocal folds on the other. The consequence of these remarkable anatomic modifications is a longer vocal tract and the acoustic result is a lower vocal height and a deeper and more resonating voice in adult males. On average, the vocal tract is about 15% longer in men than women (Fant, 1960) and this results in perceptible sex differences in formant dispersion, with males exhibiting formants of lower frequency (measured through formant position) as well as lower formant dispersion (Hanson, 1997).

Studies trying to correlate vocal resonances and perceived attractiveness have lead to controversial results. For instance, Hodges-Simeon et al. (2010), Pisanski and Rendall (2011) showed that the lower the formant dispersion, the more attractive the masculine voices. The same tendency was observed by Sebesta et al. (2017) for whom the formant position was the acoustic variable of interest. Conversely, Skrinda et al. (2014) and Xu et al. (2013) found no correlation between low resonances and male voice attractiveness. Interestingly, two other studies led to original results. Using formant dispersion, Babel et al. (2014) showed that only tall women tend to prefer low resonances in males' voices. Likewise, Feinberg et al. (2005) observed the same preferences but only for the two high vowels /i/ and /u/, which are perceived more attractive when the spacing between F1 and F2 is reduced. Such a result may be explained by basic acoustic principles. Indeed, Holmberg et al. (1995) showed that the relative amplitude of the harmonics is closely related with the adduction of the vocal folds, with the higher the adduction, the lower the harmonics at the glottal exit. Moreover, using fiberscopy to characterize vocal closure as function of speakers' gender, Södersten, Lindestad, and Hammarberg (1991) showed female speakers' higher degree of incomplete closure is correlated with increased harmonics. Therefore, the results of Feinberg et al. are in line with theoretical analysis and observations in experimental acoustics, since sounds with greater low-frequency and weaker high-frequency components are recognized to result from more adducted glottal considerations that are, themselves, more typical of male speakers (Hanson, 1996).

Collins and Missing (2003) investigated the relationship between male human vocal characteristics and female judgments about the speaker and showed that, in general, women found men's voices with harmonics that are closer together and lower in frequency more attractive. This corroborates the findings of earlier studies where less masculine sounding speakers were described as having higher formant frequencies (Avery & Liss, 1996). In their study aiming in testing listeners' weighting of F0 and/or formant frequency for the rating of vocal attractiveness, Pisanski and Rendall (2011) reached the same conclusion, that is, voices with relatively low F0 and/or low formant frequencies rated as more attractive if male and less attractive if female. Interestingly, the authors also showed that, in assessing attractivity, listeners appeared to weigh formant frequency cues more heavily than F0, an unpredicted result which suggests female listeners might interpret lower frequency cues as indicating greater masculinity and thus greater attractiveness in male voices. Finally, the results obtained by Xu et al. (2013) also showed male voices sounded more attractive when they are low pitched and with densely distributed formants associating such characteristics with the large body size projected.

4.5 Preferences for Voice Quality

Among the various complex acoustic features that give a voice its quality, the variations of the glottal source waveform hold a special place. The values of the parameters that describe the glottal waveform can vary depending on the glottal configuration and/or the quality of the vocal fold vibrations, and it is expected that these variations may lead to different voice qualities. Some voice qualities are usually associated with disordered voice, such as harshness (also referred to as vocal roughness or hoarseness), but since our main concern here is vocal attractiveness, we will focus on those that may occur for voices that are not perceived to be pathological. Voice qualities that occur frequently in normal speech are described to be "modal," that is, smooth and acoustically brilliant voices (Laver, 1980; Titze, 1994), but there are also some voice qualities that are commonly related to dysphonia but may also occur in normal (i.e., non-pathological) conversational speech and still be perceived attractive (Barkat-Defradas et al. 2012). It is typically the case for both moderately breathy and rough voices. According to Fairbanks (1960: 179), "breathy quality" (also called murmured voice or whispery voice) is described as an inefficient laryngeal vibration: *"(...) In the coordination of normal voice quality the vibrating vocal folds approximate in the midline once per cycle, closing the glottis and interrupting the airflow. In breathy quality the vocal folds vibrate, but the intermittent closure fails and the airflow is continuous."* Interestingly, the author also underlines breathy voice lowers voice pitch and is almost invariably accompanied by limited vocal intensity. As for vocal roughness, or "harsh quality," it is defined as an *"irregular, aperiodic noise in the vocal fold spectrum caused by an excessive laryngeal tension"* (Fairbanks, 1960: 179; Laver, 1980: 133, 1994: 477). Though the indication of psychological attributes conveyed through voice quality has aroused researchers' attention since ancient times (Laver, 2009: 38), this belief has long found rather eccentric and impressionistic assertions. For example, a breathy quality was supposed to show that men were "aesthetic" and women "pretty and callow"; flat that men are "distant" and women "hard and lethargic"; nasal that men are "unattractive and self-effacing" and women the same; tense that men are "cantankerous" and women "high-strung"; throaty that men are "stable" and women "oafish"; orotund (or loud) that men are "suave" and women "aggressive"; and so on. The idea that personality characteristics are correlated with voice quality has recently been tested more scientifically, and although some controversy remains, it must be admitted some correlations do exist. Among the few studies that have tackled the topic of vocal breath and/or vocal roughness and their effects on perceived voice attractiveness, it has been shown that harsh voices are regularly correlated with more aggressive, dominant, and authoritative personalities while breathy ones are more frequently associated with self-effacing, submissive, and weak temperaments. A way to quantify breathiness—which is caused by glottal air leakage—is to measure harmonics-to-noise ratio (henceforth HNR), a measure that quantifies the relative amount of additive noise.[3] As for vocal roughness, it results from irregular vocal fold vibrations. These vibratory perturbations have come to be more commonly referred to as vocal jitter. As a matter of fact, a number of investigators have demonstrated a significant correlation between increased levels of jitter and perceived roughness (Lieberman, 1963; Moore & Thomson, 1965). For example, Babel et al. (2014) and van Borsel et al. (2009) found female voices were perceived more attractive when breathy. Unexpectedly, Sebesta et al. (2017) and Xu et al. (2013) showed significant relations between vocal breath and attractivity for both sexes. A plausible explanation for male vocal attractiveness unexpectedly enhanced by breathiness in this particular study lies in the fact that this predominantly feminine vocal feature may presumably soften the aggressiveness regularly associated with low deep voices.

Though some other phonetic characteristics could be addressed so as to characterize vocal attractiveness (e.g., preferences for speech tempo), the above overview offers an exhaustive assessment of the state of the art regarding the topic and underlines the necessity to question both understudied acoustic parameters that may be relevant for vocal pleasantness and the effect of language/culture on perceived attractiveness.

4.6 Sources of Variations in Vocal Preferences

Though some general tendencies emerge from studies dealing with vocal preferences, some sources of variations should be mentioned. These are mainly of two different natures. Some sources of variation seem to be due to physiological matters (i.e., variations in hormonal levels) while some others are more concerned with cultural arguments (i.e., social representations).

4.6.1 The Effect of Menstrual Cycle on Females' Vocal Preferences

It has been suggested that women's preferences maybe affected both by menstrual cycle (i.e., whether they are in their ovulatory versus follicular and/or luteal phase) and the context of mating they are looking for (i.e., short- versus long-term relationships). Feinberg et al. (2006), Pisanski et al. (2014), and Puts (2005) have put forward the hypothesis of *"good genes ovulatory shift"* which suggests that women

[3] At the physiological level, low HNR values are believed to be related to insufficient vocal fold adduction during the so-called "closed" interval of the phonatory cycle. Insufficient closure would allow excessive airflow through the glottis, giving rise to a turbulence noise component in the quasi-periodic source signal. This friction noise would result in a higher noise level in the spectrum, especially in the higher frequencies.

in ovulatory phase tend to prefer more masculine men (higher masculinity being associated with a better genotypic quality according to the theory of immunocompetence handicap[4]) more particularly in the context of short-term relationships (Jünger et al., 2018). Conversely, in the context of long-term relationships, women in their follicular and/or luteal phases tend to prefer men exhibiting less masculine traits, indicating they are more likely to invest themselves in parental care. Such variability in females' preferences would account for an adaptive strategy allowing women to optimize their fitness (i.e., reproductive success) in function of their menstrual cycle.

As for vocal preferences specifically, Puts (2005) noted that for the same vocal stimulus, women in their ovulatory phase judge low-pitched masculine voices (i.e., low F0) more attractive when looking for a short-lived relationship. Likewise, Feinberg et al. (2006) and Pisanski et al. (2014) observed this choice is even more marked for women in their fertility window. Hodges-Simeon et al. (2010) also investigated the effect of vocal resonance (i.e., formant dispersion) on females' vocal preferences and, though they could not find any effect of the type of relationship (i.e., short or long) specifically linked to this feature, they showed women are more likely to judge attractive masculine voices that exhibit a low dispersion of formants (i.e., deep voices). They also notice a shift in women's preferences as function of both menstrual cycle and duration commitment: monotonous masculine voices (low F0-SD) being judged as more attractive by unfertile women in the context of short-term liaisons while the same vocal stimuli are perceived as more attractive for fertile women who are engaged in a long-term relationship. Those somehow inconsistent results lead some authors to question the validity of menstrual cycle as a reliable explanatory factor for women's variations in their attractiveness preferences. For example, Jones et al. (2018) and Marcinkowska, Galbarczyk, and Jasienska (2018) found no effect of female's menstrual cycle on body and face attractiveness evaluations of men. Likewise, Jünger et al. (2018)—using a robust methodology—could not confirm any effect neither of cycle phases nor of steroids to explain females' variations in their choices. As for feminine voices, since laryngeal epithelial cells are known to be highly sensitive to hormonal variations (Haselton, Mortezaie, Pillsworth, Bleske-Rechek, & Frederick, 2007; Miller et al., 2007; Higgins & Saxman, 1989; Abitbol et al., 1999; Amir & Biron-Shental, 2003; Bryant & Haselton, 2009; Fischer et al., 2011), women's voices undergo perceivable variation in their quality. As a matter of fact, Pipitone and Gallup (2008) have shown that feminine voices—which are higher pitched when women approach their fertile period—are perceived as more attractive by men whereas they sound lower pitched outside the ovulatory phase and are, consequently, judged less appealing (Bryant & Haselton, 2009; Fischer et al.,

---

[4]The theory of immunocompetence handicap (Zahavi, 1975) suggests that androgen-mediated traits accurately signal condition due to the immunosuppressive effects of androgens. This immunosuppression may be either because testosterone alters the allocation of limited resources between the development of ornamental traits and the immune system or because heightened immune system activity has a propensity to launch autoimmune attacks against gametes, such that suppression of the immune system enhances fertility. Therefore, only healthy individuals can afford to suppress their immune system by raising their testosterone levels, which also augments secondary sexual traits and displays (among which low deep voices for men).

2011). These variations in females' vocal quality are essentially due to changes in estrogens and progesterone levels across the menstrual cycle, which lead to physiological modifications in the mass, the tension, and the viscosity of the vocal folds, which in turn modify their oscillatory properties. It has been suggested these cyclic vocal quality variations could have been adaptive since they could contribute to the enhancement of women's attractiveness and facilitate mating when the risk of conception is higher and, therefore, the chance to conceive higher (Fischer et al., 2011; Pipitone & Gallup, 2008; Puts et al., 2013).

4.6.2 The Effect of Sociocultural Environment on Vocal Quality

Though they are remarkably scarce, the few existing studies that have investigated the effect of sociocultural environment on vocal preferences have shown they are not universal but language/culture dependent. For example, van Bezooijen (1995) demonstrated that Japanese women exhibited the highest vocal pitch among a large sample of natural languages (i.e., 232 Hz) while the mean fundamental frequency of American women is around 214 Hz and that of Dutchwomen close to 196 Hz. Vaissière (2015) found French women's voice are even lower pitched with a mean F0 close to 190 Hz. It has been suggested that these significant differences in female vocal height could be constrained by specific cultural requirements that are themselves shaped by social values and expectations that are linked to the roles allocated to women versus men and, more generally, to the stereotypes of feminity versus masculinity defined by the culture in question. Stereotypes of gender therefore vary among different cultures as well as among different ethnic groups (Landrine, 1985; Harris, 1994). In this way, the figure of feminity in Japanese culture is traditionally related to modesty, innocence, gentleness, subordination, physical fragility, and psychological submission (Sughira & Katsurada, 1999); these personality traits being vocally signalized to Japanese men who share the same cultural background through that famous *"voix de petite fille"* which has been subtly described by Léon (1981). Conversely, in the Netherlands—a country described as more egalitarian—women exhibit more masculine (i.e., low pitched) voices since their culture favors psychological traits that are associated with female independence. In conclusion, it seems that the acoustic features that are typical of feminine versus masculine voices are not only due to anatomical and/or physiological criteria (i.e., vocal length tract and hormonal level) but also to cultural aspects depending on the social values attributed to sex roles in a given society. Besides, the studies conducted by Sebesta et al. (2017) and Shirazi et al. (2018) have shown that cultural expectations do not only concern vocal height. For example, in a Namibian population, male attractiveness is not predicted by F0 but by the degree of vocal breathiness they exhibit. Likewise, in the Philippines, females tend to prefer men with higher pitched voices. Though the effect of sociocultural representations on voice has been focused on, there is, to our knowledge, no study that aimed at identifying the factors of this variation. Yet, it does not seem to occur randomly in the same way as it has been observed for the evolution of the waist-to-hip ratio (Bovet & Raymond, 2015; Bovet, 2019), the body mass index, or the stature, in which variations have been shown to be partly due to the ecology (see Pisanski & Feinberg, 2013 for a discussion), and that is why cross-cultural surveys are still needed to evaluate the weight of culture on vocal preferences. The scope of research dealing with voice attractiveness should also consider the issue of preferences limitations. As a matter of fact, there are very few studies that tackle the topic of superior and/or inferior limits above/below which a voice is no longer perceived as attractive. Among these, Re et al. (2012) have shown women's preferences do not vary when male vocal pitch is below 96 Hz, but when they have to choose between two stimuli above this value, they regularly prefer the lower voice. As for men, to our knowledge, two studies were interested in determining a vocal height threshold (in the range 160–300 Hz) below/above which female voices would no longer be perceived as attractive (Feinberg et al., 2008a, 2008b; Re et al., 2012). Results show men always consider high-pitched voices as more attractive for women. Moreover, Borkowska and Pawlowski (2011) reported a non-linear relation between vocal height and attractivity, the latter starting to decrease when F0 is close to 260 Hz. According to the authors, this may be due to the fact that high-pitched voices are commonly associated to sexually immature females. Though works dealing with the determination of perceptive thresholds from which vocal attractiveness is affected are still in the pipeline, several studies have shown that straight after a voice is perceived as too distant from the norm, it is often categorized as pathological and associated with negative personality traits (Barkat-Defradas et al., 2015; Revis, 2017).

Conversely, vocal attractiveness has a profound influence on listeners—a bias known as the *"what sounds beautiful is good"* vocal attractiveness stereotype—with tangible impact on a voice owner's success at mating, job applications, and/or elections (Zuckerman & Driver, 1989). This led some authors, like Bruckert et al. (2010), to test the effect of averaging voices via auditory morphing on perceived attractivity. Overall, their results reveal that the larger the number of voices averaged, the more attractive the result. This is partly because composite voices have a smoother, more regular texture and also because they sound more like the average voice and reflect norm-based encoding of vocal stimuli. Preferences for some voices may also be explained by the principle of sparseness. It has been demonstrated that human perceptual systems (visual, auditory, and olfactory) have been selected so as to code the information efficiently that is to say quickly and as parsimoniously as possible to be in line with the principle of least effort (Renoult, Bovet, & Raymond, 2016). Such a cognitive process relies on the elimination of the redundant components of a signal, by which processing is consequently more accurate and less costly while the storage and the retrieval of relevant information is more efficient. Nevertheless, the neuropsychological mechanisms driving the coding of acoustic signals in relation with vocal attractivity has received little scientific attention and, to our knowledge, there is no study investigating these aspects specifically. Yet, since clear evidence for interference between facial and vocal information has been observed (Aben, Pflügera, Koppensteiner, Coquerellee, & Grammer, 2015), it seems reasonable to claim that vocal and facial cues convey redundant information about a speaker's mate value and thus may serve as a backup signal for human mate choice decisions.

4.7 How Evolution Shaped Human Voice via Opposite Sex's Preferences

Though it is easy to understand how morpho-anatomical, physiological as well as behavioral differences between species result from natural selection and environmental adaptations, in some famous cases, those well-known mechanisms fail to explain the existence of certain remarkable features (Darwin, 1871). The iconic example that is traditionally invoked to illustrate this point is the male peacock's tail (*Pavo cristatus*), which is adorned with iridescent feathers. Darwin himself recognized this extravagant ornament contradicted his theory of natural selection. As a matter of fact, no doubt the male peacock's tail represents a critical bulk for his flight, and its outstanding colors has the disadvantage to attract his predators' attention. Besides, noting their absence in females and juveniles, the author concludes such an ornament cannot serve the animal's survival. Indeed, if peacocks' tail feathers were useful against predators then females and juveniles would exhibit the same. Therefore, he suggests the presence of some morphological characteristics cannot be explained solely by the advantages they provide to their bearers in terms of survival (which refers to "natural selection" itself) but also in terms of mating and fitness (which refers to a complementary concept, he defines as "sexual selection"). According to Darwin, sexual selection is restricted to secondary sex characteristics[5]—among which body size—and explains why many species exhibit sexual dimorphism at sexual maturity through the spectacular feathers of the birds-of-paradise, the impressive antlers of the male members of the deer family and, last but not least, vocal dimorphism in humans, among other dimorphic traits. The theory of Ohala's frequency code (1984)—inspired by Morton (1977)[6])—indicates that despite the development of highly complex language capable of conveying fine subtleties in meaning, humans still use an encoding strategy similar to the one widely used by nonhuman animals, namely, (i) by using relatively low-frequency sounds to indicate they are likely to attack versus (ii) more high-frequency sounds to indicate they are submissive,

[5]Secondary sex characteristics are features that appear during puberty in humans, and at sexual maturity in other animals. Secondary sex characteristics include, for example, the manes of male lions, the bright facial and rump coloration of male mandrills, and horns in many goats and/or antelopes. In humans, visible secondary sex characteristics include pubic hair, enlarged breasts and widened hips of females, facial hair, Adam's apples on males, etc.

[6]In a famous article dealing with vocal communication in animals, Morton (1977) introduces his 'motivation-structural rules' theory, which suggests physical proprieties of acoustic signals (sounds of high versus low frequencies) are motivated since they reflect the vocalizer's body size and inform about his/her intentions and/or emotional state. He argues a large number of birds and mammals use low-frequency sounds to express hostility, threat, and aggression whereas high-frequency sounds are rather used to express fear, submission, and "amicability.".

appeasing, or fearful. Here pattern (i) is to project a large body size so as to threaten the receiver, because a larger animal has a better chance at winning a physical confrontation. Pattern (ii) is to project a small body size to attract the receiver, because a smaller animal is less likely to be a threat (Morton, 1977). Following this reasoning, Ohala (1984) argues the longer vocal folds of human males may have evolved under a selection pressure to compete with other males in achieving dominance for the sake of gaining access to female mates (i.e., intra-sexual selection). Likewise, the longer vocal tract of males may have evolved under the same pressure, as it may also reflect a larger body size and attract females (i.e., inter-sexual selection, see Puts et al., 2006 for an exhaustive presentation of the role of intra-selection in males). Extending the mechanism further, the shorter vocal folds and vocal tract of females may have developed under a pressure in the opposite direction, i.e., to project a small body size in order to attract male mates. To sum it up, by making an analogy between, on the one hand, the appearance of antlers in male deers, which develop when they attain sexual maturity and, on the other hand, voice change in pubescent boys, Ohala was a pioneer in assessing the functional role of sexual selection for the emergence of vocal dimorphism in humans.

I think the enlargement of the vocal apparatus also occurs to enhance aggressive displays. Males, by their role in the family unit and the fact that they compete for the favors of the female—i.e, they are subject to what Darwin called sexual selection—would be the ones to develop such deviations from the 'norm'. However, they would only need these aggressive decorations when they are ready to compete and retain the favors of a female, that is, at the time of sexual maturity (Ohala, 1984: 14).

4.8 Conclusion

This contribution aimed at showing the mechanism of sexual selection formalized by Darwin as early as 1871 constitutes a crucial force in the evolution of voice, which directly intervenes in reproductive strategies. Though such an argument has been considered as obvious for many species, it is only at the very beginning of the 2000s that the phenomenon of vocal dimorphism has been tackled in relationship with Darwin's theory. As a matter of fact, it is surprising that the study of language activity has long been conducted without any reference to its biological function. Traditionally, humanities (anthropology, linguistics ...) used to consider language as a pure cultural product, which had been created by humans in the same ways as writing or art (Levi-Strauss, in Charbonnier, 1959: 48; Noble and Davidson, 1996: 214; Tomassello, 1999: 94), and which developed irrespective of any selective pressure (Chomsky, 1975: 75). In this purely cultural conception, the study of ultimate (or distal) causes explaining the existence of vocal dimorphism in terms of evolutionary forces has been left aside for the benefit of extensive analyses of proximal mechanisms, which explain its biological function in terms of immediate physiological or environmental factors. Yet, a transdisciplinary approach—at the crossroad of linguistics and evolutionary biology—is of a great interest to better understand the whys and wherefores of the evolution of articulated language in the human lineage. Indeed, beyond its evidenced social function (Dunbar, Duncan, & Nettle, 1995), vocal behavior should undoubtedly be regarded as a reliable way to display one's phenotypic value (Puts, 2010). Moreover, the existence of a low laryngeal configuration—an indispensable condition for language—in many non-speaking species undermines the hypothesis of a specific adaptation to language in humans (Fitch & Reby, 2001). Reversely, considering such a disposition is present in several animals of different species clearly indicates it has evolved during phylogenesis to respond to other functions.

References

Aben, P., Pflügera, L., Koppensteiner, M., Coquerellee, M., & Grammer, K. (2015). Sound of female shape: A redundant signal of vocal and facial attractiveness. *Evolution and Human Behavior, 36*(3), 174–181.

Abitbol, J., Abitbol, P., & Abitbol, B. (1999). Sex hormones and the female voice. *Journal of Voice, 13*, 424–446.

Amir, O., & Biron-Shental, T. (2003). The impact of hormonal fluctuations on female vocal folds. *Current Opinion in Otolaryngology & Head and Neck Surgery, 12*, 180–184.

Apicella, C. L., Feinberg, D. R., & Marlowe, F. W. (2007). Voice pitch predicts reproductive success in male hunter-gatherers. *Biology Letters, 3*(6), 682–684.

Arak, A., & Enquist, M. (1993). Hidden preferences and the evolution of signals. *Philosophical Transactions of the Royal Society of London B, 340*, 207–213. https://doi.org/10.1098/rstb.1993.0059.

Avery, J. D., & Liss, J. M. (1996). Acoustic characteristics of less masculine-sounding male speech. *The Journal of the Acoustical Society of America, 99*, 3738–3748.

Babel, M., McGuire, G., & King, J. (2014). Towards a more nuanced view of vocal attractiveness. *PLoS One, 9*(2), e88616.

Bancroft, J. (2005). The endocrinology of sexual arousal. *Journal of Endocrinology, 186*, 411–427.

Barkat-Defradas, M., Busseuil, C., Chauvy, O., Hirsch, F., Fauth, C., Revis, J., & Amy de la Bretéque, B. (2012). Dimension esthétique des voix normales et dysphoniques: Approches perceptive et acoustique. *TIPA* 28.

Barkat-Defradas, M., Fauth, C., Didirkova, F., Amy de la Breteque, B., Hirsch, F., Dodane, C., & Sauvage, J. (2015). Dysphonia is beautiful: A perceptual and acoustic analysis of vocal roughness. *International Congress of Phonetic Sciences*, 18th ICPhS, Glasgow 10–14 August 2015, Scotland, UK.

Borkowska, B., & Pawlowski, B. (2011). Female voice frequency in the context of dominance and attractiveness perception. *Animal Behaviour, 82*(1), 55–59.

Bovet, J. (2018). The evolution of feminine beauty. In Z. Kapoula et al. (Eds.), *Exploring Trans disciplinarity in Art and Sciences* (pp. 327–348). Springer International Publishing AG, Part of Springer Nature.

Bovet, J. (2019). Evolutionary theories and men's preferences for women's waist-to-hip ratio: Which hypotheses remain? *A Systematic Review. Frontiers in Psychology, 10*, https://doi.org/10.3389/fpsyg.2019.01221.

Bovet, J., & Raymond, M. (2015). Preferred women's waist-to-hip ratio variation over the last 2,500 years. *PLoS One, 10*, e0123284.

Bruckert, L., Bestelmeyer, P., Latinus, M., Rouger, J., Charest, I., Rousselet, G. A., et al. (2010). Vocal attractiveness increases by averaging. *Current Biology, 20*(2), 116–120.

Bruckert, L., Lienard, J.-S., Lacroix, A., Kreutzer, M., & Leboucher, G. (2006). Women use voice parameters to assess men's characteristics. *Proceedings of the Royal Society B: Biological Sciences, 273*(1582), 83–89.
Bryant, Gregory A., & Haselton, Martie G. (2009). Vocal cues of ovulation in human females. *Biology Letters, 5*, 12–15.
Buss, D. M. (1989). Sex differences in human mate preferences: Evolutionary hypotheses tested in 37 cultures. *Behavioral and Brain Sciences, 12*(1), 1–14.
Charbonnier, G. (1959). Entretiens avec Claude Levi-Strauss. https://www.jpbu.com/philo/.../Levi-Strauss_Charbonnier_Culture-langage.rtf.
Chomsky, A. N. (1975). *Reflections on Language*. New York: Pantheon Books.
Collins, S. A. (2000). Men's voices and women's choices. *Animal Behaviour, 60*(6), 773–780.
Collins, S. A., & Missing, C. (2003). Vocal and visual attractiveness are related in women. *Animal Behaviour, 65*(5), 997–1004.
Darwin, C. (1871). *The descent of man, and selection in relation to sex*. London, UK: John Murray.
Dunbar, R., Duncan, N., & Nettle, D. (1995). Size and structure of freely forming conversational groups. *Human nature, 6*(1), 67–78.
Evans, S., Neave, N., Wakelin, D., & Hamilton, C. (2008). The relationship between testosterone and vocal frequencies in human males. *Physiology and Behavior, 93*, 783–788.
Fairbanks, G. (1960). *Voice and articulation*. J. Cotler Books, 2nd revised edition.
Fant, G. (1960). *Acoustic Theory of Speech Production*. Berlin: De Gruyter.
Feinberg, D. R., DeBruine, L. M., Jones, B. C., & Little, A. C. (2008a). Correlated preferences for men's facial and vocal masculinity. *Evolution and Human Behavior, 29*(4), 233–241.
Feinberg, D. R., DeBruine, L. M., Jones, B. C., & Perrett, D. I. (2008b). The role of femininity and averageness of voice pitch in aesthetic judgments of women's voices. *Perception, 37*(4), 615–623.
Feinberg, D. R., Jones, B. C., DeBruine, L. M., Moore, F. R., Law Smith, M. J., Cornwell, R. E., et al. (2005). The voice and face of woman: One ornament that signals quality? *Evolution and Human Behavior, 26*(5), 398–408.
Feinberg, D. R., Jones, B. C., Law Smith, M. J., Moore, F. R., DeBruine, L. M., Cornwell, R. E., et al. (2006). Menstrual cycle, trait estrogen level, and masculinity preferences in the human voice. *Hormones and Behavior, 49*(2), 215–222.
Fischer, J., Semple, S., Fickenscher, G., Jürgens, R., Kruse, E., Heistermann, M., et al. (2011). Do women's voices provide cues of the likelihood of ovulation? The importance of sampling regime. *PLoS One, 6*(9).
Fischer, A. H., & Manstead, A. S. (2000). The relation between gender and emotions in different cultures. *Gender and emotion: Social psychological perspectives, 1*, 71–94.
Fisher, M. L., & Voracek, M. (2006). The shape of beauty: determinants of female physical attractiveness. *Journal of Cosmetic Dermatology, 5*(2), 190–4.
Fitch, W. T., & Reby, D. (2001). The descended larynx is not uniquely human. *Proceedings of the Royal Society of London. Series B: Biological Sciences, 268*(1477), 1669–1675.
Fitch, W. T. (1997). Vocal tract length and format frequency dispersion correlated with body size in rhesus macaques. *The Journal of the American Society of America, 102*, 1213–1222.
Fitch, W. T., & Giedd, J. (1999). Morphology and development of the human vocal tract: A study using magnetic resonance imaging. *The Journal of the American Society of America, 106*, 1511–1522.
Fraccaro, P. J., Jones, B. C., Vukovic, J., Smith, F. G., Watkins, C. D., Feinberg, D. R., et al. (2011). Experimental evidence that women speak in a higher voice pitch to men they find attractive. *Journal of Evolutionary Psychology, 9*(1), 57–67.
Fraccaro, P. J., O'Connor, J. J. M., Re, D. E., Jones, B. C., DeBruine, L. M., & Feinberg, D. R. (2013). Faking it: Deliberately altered voice pitch and vocal attractiveness. *Animal Behaviour, 85*(1), 127–136.
Geary, D. C., Vigil, J., & Byrd-Craven, J. (2004). Evolution of human mate choice. *Journal of Sex Research, 41*(1), 27–42.
Hall, J. A. (1978). Gender effects in decoding nonverbal cues. *Psychological Bulletin, 85*, 845–857.

Handson, H. (1997). Glottal characteristics of female speakers: Acoustic correlates. *The Journal of the American Society of America, 101*(1), 466–81.

Hanson, H. M. (1997). Glottal characteristics of female speakers: Acoustic correlates. *The Journal of the American Society of America, 101*(1), 466–481.

Harris, A. C. (1994). Ethnicity as a determinant of sex role identity: A replication study of item selection for the Bem Sex Role Inventory. *Sex Roles, 31*, 241–273.

Haselton, M. G., Mortezaie, M., Pillsworth, E. G., Bleske-Rechek, A., & Frederick, D. A. (2007). Ovulatory shifts in human female ornamentation: Near ovulation, women dress to impress. *Hormones and Behavior, 51*, 40–45.

Higgins, M. B., & Saxman, J. H. (1989). Variations in vocal frequency perturbation across the menstrual cycle. *Journal of Voice, 3*, 233–243.

Hill, A. K., Hunt, J., Welling, L. L. M., Cárdenas, R. A., Rotella, M. A., Wheatley, J. R., et al. (2013). Quantifying the strength and form of sexual selection on men's traits. *Evolution and Human Behavior, 34*(5), 334–341.

Hodges-Simeon, C. R., Gaulin, S. J. C., & Puts, D. A. (2010). Different vocal parameters predict perceptions of dominance and attractiveness. *Human Nature, 21*(4), 406–427.

Holmberg, E. B., Hillman, R. E., Perkell, J. S., Guiod, P. C., & Goldman, S. L. (1995). Comparisons among aerodynamic, electroglottographic, andacoustic spectral measures of female voice. *Journal of Speech, Language, and Hearing Research, 38*, 1212–1223.

Hughes, S. M., Dispenza, F., & Gallup, G. G. (2004). Ratings of voice attractiveness predict sexual behavior and body configuration. *Evolution and Human Behavior, 25*, 295–304.

Hughes, S. M., Farley, S. D., & Rhodes, B. C. (2010). Vocal and physiological changes in response to the physical attractiveness of conversational partners. *Journal of Nonverbal Behavior, 34*(3), 155–167.

Hughes, S. M., Mogilski, J. K., & Harrison, M. A. (2014). The perception and parameters of intentional voice manipulation. *Journal of Nonverbal Behavior, 38*(1), 107–127.

Hughes, S. M., Pastizzo, M. J., & Gallup, G. G. (2008). The sound of symmetry revisited: Subjective and objective analyses of voice. *Journal of Nonverbal Behavior, 32*(2), 95–108.

Jones, B. C., Feinberg, D. R., DeBruine, L. M., Little, A. C., & Vukovic, J. (2010). A domain-specific opposite-sex bias in human preferences for manipulated voice pitch. *Animal Behaviour, 79*(1), 57–62.

Jones, B. C., Hahn, A. C., Fisher, C. I., Wang, H., Kandrik, M., Han, C., et al. (2018). No compelling evidence that preferences for facial masculinity track changes in women's hormonal status. *Psychological Science, 29*(6), 10.

Jünger, J., Motta-Mena, N. V., Cardenas, R., Bailey, D., Rosenfield, K. A., Schild, C., et al. (2018). Do women's preferences for masculine voices shift across the ovulatory cycle? *Hormones and Behavior, 106*, 122–134.

Kordsmeyer, T. L., Hunt, J., Puts, D. A., Ostner, J., & Penke, L. (2018). The relative importance of intra- and intersexual selection on human male sexually dimorphic traits. *Evolution and Human Behavior, 39*(4), 424–436.

Landrine, H. (1985). Race x class stereotypes of women. *Sex Roles, 13*, 65–75.

Laver, J. (1980). *The phonetic description of voice quality*, (vol. 2009, 1st edn). Cambridge University Press.

Laver, J. (1980). *The phonetic description of voice quality*. Cambridge University Press.

Laver, J. (1994). *Principles of phonetics*. Cambridge University Press.

Leaderbrand, K., Dekam, J., Morey, A., & Tuma, L. (2008). The effects of voice pitch on perceptions of attractiveness: Do you sound hot or not? *Winona State University Psychology Student Journal, 6*.

Léon, P. (1981). BB ou la voix charmeuse, petite fille et coquette. *Studia Phonetica, 18*, 159–171.

Leongomez, J. D., Binter, J., Kubicova, L., Stolařova, P., Klapilova, K., Havlíček, J., et al. (2014). Vocal modulation during courtship increases proceptivity even in naive listeners. *Evolution and Human Behavior, 35*(6), 489–496.

Lieberman, P. (1963). Some acoustic measures of the fundamental periodicity of normal and pathologic larynges. *The Journal of the American Society of America*, *35*(3), 344–353.

Liu, X., & Xu, Y. (2011). What makes a female voice attractive? In *Proceedings of ICPhS. Hong-Kong* (pp. 1274–1277).

Marcinkowska, U. M., Galbarczyk, A., & Jasienska, G. (2018). La donna è mobile? Lack of cyclical shifts in facial symmetry, and facial and body masculinity preferences—A hormone based study. *Psychoneuroendocrinology*, *88*, 47–53.

Miller, G. F., Tybur, J., & Jordan, B. (2007). Ovulatory cycle effects on tip earnings by lap-dancers: Economic evidence for human estrus? *Evolution and Human Behavior*, *6*, 375–381.

Moore, P., & Thomson, C. L. (1965). Comments on physiology of hoarseness. *Archives of Otolaryngology Head and Neck Surgery*, *81*(1), 97–102. https://doi.org/10.1001/archotol.1965.00750050102022.

Morton, E. S. (1977). On the occurrence and significance of motivation-structural rules in some bird and mammal sounds. *The America Naturalist*, *111*(981), 855–869.

Noble, W., & Davidson, I. (1996). *Human evolution, language and mind*. Cambridge: Cambridge University Press.

O'Connor, J. J. M., Fraccaro, P. J., Pisanski, K., Tigue, C. C., & Feinberg, D. R. (2013). Men's preferences for women's femininity in dynamic cross-modal stimuli. *PLoS One*, *8*(7), e69531.

O'Connor, J. J., Re, D. E., & Feinberg, D. R. (2011). Voice pitch influences perceptions of sexual infidelity. *Evolutionary Psychology*, *9*(1), 147470491100900109.

Oguchi, T., & Kikuchi, H. (1997). Voice and interpersonal attraction. *Japanese Psychological Research*, *39*(1), 56–61.

Ohala, J. J. (1984). An ethological perspective on common cross- language utilisation of F0 of voice. *Phonetica*, *41*(1), 1–16.

Pipitone, R. N., & Gallup, G. G. (2008). Women's voice attractiveness varies across the menstrual cycle. *Evolution and Human Behavior*, *29*(4), 268–274.

Pisanski, K., Bhardwaj, K., & Reby, D. (2018). Women's voice pitch lowers after pregnancy. *Evolution and Human Behavior*, *39*(4), 457–463.

Pisanski, K., Fraccaro, P. J., Tigue, C. C., O'Connor, J. J. M., Röder, S., Andrews, P. W., et al. (2014). Vocal indicators of body size in men and women: A meta-analysis. *Animal Behaviour*, *95*, 89–99.

Pisanski, K., & Rendall, D. (2011). The prioritization of voice fundamental frequency or formants in listeners' assessments of speaker size, masculinity and attractiveness. *The Journal of the American Society of America*, *129*(4), 2201–2212.

Polce-Lynch, M., Myers, B. J., Kilmartin, C. T., Forssmann-Falck, R., & Kliewer, W. (1998). Gender and age patterns in emotional expression, body image, and self-esteem: A qualitative analysis. *Sex Roles*, *38*(11–12), 1025–1048.

Puts, D. A. (2005). Mating context and menstrual phase affect women's preferences for male voice pitch. *Evolution and Human Behavior*, *26*(5), 388–397.

Puts, D. A. (2010). Beauty and the beast: Mechanisms of sexual selection in humans. *Evolution and Human Behavior*, *31*(3), 157–175.

Puts, D. A., Bailey, D. H., Cárdenas, R. A., Burriss, R. P., Welling, L. L. M., Wheatley, J. R., et al. (2013). Women's attractiveness changes with estradiol and progesterone across the ovulatory cycle. *Hormones and Behavior*, *63*(1), 13–19.

Puts, D. A., Barndt, J. L., Welling, L. L. M., Dawood, K., & Burriss, R. P. (2011). Intrasexual competition among women: Vocal femininity affects perceptions of attractiveness and flirtatiousness. *Personality and Individual Differences*, *50*(1), 111–115.

Puts, D. A., Gaulin, S. J. C., & Verdolini, K. (2006). Dominance and the evolution of sexual dimorphism in human voice pitch. *Evolution and Human Behavior*, *27*(4), 283–296.

Renoult, J. P., Bovet, J., & Raymond, M. (2016). Beauty is in the efficient coding of the beholder. *Royal Society Open Science*, *3*(3), 160027.

Re, D. E., O'Connor, J. J. M., Bennett, P. J., & Feinberg, D. R. (2012). Preferences for very low and very high voice pitch in humans. *PLoS One*, *7*(3), e32719.

Revis, J. (2017). *La voix et soi: Ce que notre voix dit de nous*. France: Solal.

Riding, D., Lonsdale, D., & Brown, B. (2006). The effects of average fundamental frequency and variance of fundamental frequency on male vocal attractiveness to women. *Journal of Nonverbal Behavior*, *30*(2), 55–61.

Saxton, T. K., Caryl, P. G., & Craig R. S. (2006). Vocal and facial attractiveness judgments of children, adolescents and adults: The ontogeny of mate choice. *Ethology*, *112*(12), 1179–1185.

Saxton, T. K., Debruine, L. M., Jones, B. C., Little, A. C., & Roberts, S. C. (2009). Face and voice attractiveness judgments change during adolescence. *Evolution and Human Behavior*, *30*(6), 398–408.

Sebesta, P., Kleisner, K., Turecek, P., Kočnar, T., Akoko, R. M., Trebicky, V., et al. (2017). Voices of Africa: Acoustic predictors of human male vocal attractiveness. *Animal Behaviour*, *127*, 205–211.

Shirazi, T. N., Puts, D. A., & Escasa-Dorne, M. J. (2018). Filipino women's preferences for male voice pitch: Intra-individual, life history, and hormonal predictors. *Adaptive Human Behavior and Physiology*, *4*(2), 188–206.

Simmons, L. W., Peters, M., & Rhodes, G. (2011). Low pitched voices are perceived as masculine and attractive but do they predict semen quality in men? *PLoS One*, *6*(12).

Singh, D. (1993). Adaptive significance of female physical attractiveness: Role of waist-to-hip ratio. *Journal of Personality and Social Psychology*, *65*(2), 293–307.

Singh, D., & Young, R. K. (2001). Body weight, waist-to-hip ratio, breasts, and hips: Role in judgments of female attractiveness and desirability for relationships. *Ethology and Sociobiology.*, *16*(6), 483–507.

Skrinda, I., Krama, T., Kecko, S., Moore, F. R., Kaasik, A., Meija, L., et al. (2014). Body height, immunity, facial and vocal attractiveness in young men. *Naturwissenschaften*, *101*(12), 1017–1025.

Södersten, M., Lindestad, P. A., & Hammarberg, B. (1991). Vocal fold closure, perceived breathiness, and acoustic characteristics in normal adult speakers. In J. Gauffin & B. Hammarberg (Eds.), *Vocal fold physiology: Acoustic, perceptual, and physiological aspects of voice mechanisms* (pp. 217–224).

Sughira, Y., & Katsurada, E. (1999). Masculinity and feminity in Japanese culture. *Sex Roles*, *40*(718), 635–646.

Sugiyama, L. S. (2015). Physical attractiveness in adaptationist perspective. In D. M. Buss (Ed.), *The handbook of evolutionary psychology* (pp. 292–343). Wiley.

Suire, A., Raymond, M., & Barkat-Defradas, M. (2019). Male vocal quality and its relation to females' preferences. *Evolutionary Psychology*, 1–12, https://doi.org/10.1177/1474704919874675.

Suire, A., Tognetti, A., Durand, V., Raymond, M., & Barkat-Defradas, M. (2020). The influence of sexual orientation and circulating testosterone levels on speech acoustic features. *Archives of Sexual Behavior*, *10*, 1–9. https://doi.org/10.1007/s10508-020-01665-3.

Suire, A., Raymond, M., & Barkat-Defradas, M. (2018). Vocal behavior within competitive and courtship contexts and its relation to mating success in humans. *Evolution and Human Behavior*, *39*, 684–691.

Titze, I. R. (1994). *Principles of voice production*. Englewood Cliffs, N.J.: Prentice Hall.

Tognetti, A., Durand, V., Barkat-Defradas, M., & Hopfensitz, A. (2019). Does he sound cooperative? Acoustic correlates of cooperativeness. *British Journal of Psychology*, 1–17. https://doi.org/10.1111/bjop.12437.

Tomasello, M. (1999). *The cultural origins of human cognition*. Harvard University Press.

Traunmüller, H., & Eriksson, A. (1995). *The frequency range of the voice fundamental in the speech of male and female adults*. Unpublished manuscript.

Travers, L. M. (2017). Runaway selection. In J. Vonk & T. Shackelford (Eds.), *Encyclopedia of animal cognition and behavior*. Springer.

Tsantani, M. S., Belin, P., Paterson, H. M., & McAleer, P. (2016). Low vocal pitch preference drives first impressions irrespective of context in male voices but not in female voices. *Perception*, *45*(8), 946–963.

Tuomi, S. K., & Fisher, J. E. (1979). Characteristics of simulated sexy voice. *Folia Phoniatrica and Logopaedica*, *31*(4), 242–249.
Vaissière, J. (2015). *La phonétique*. Paris: Presses Universitaires de France.
van Bezooijen, R. (1995). Sociocultural aspects of pitch differences between Japanese and Dutch women. *Language and Speech*, *38*(3), 253–265.
van Borsel, J., Janssens, J., & De Bodt, M. (2009). Breathiness as a feminine voice characteristic: A perceptual approach. *Journal of Voice*, *23*(3), 291–294.
Vorperian, H. K., Kent, R. D., Lindstrom, M. J., Kalina, C. M., Gentry, L. R., & Yandell, B. S. (2005). Development of vocal tract length during early childhood: A magnetic resonance imaging study. *The Journal of the American Society of America*, *117*, 338–350.
Vukovic, J., Feinberg, D. R., Jones, B. C., DeBruine, L. M., Welling, L. L. M., Little, A. C., et al. (2008). Self-rated attractiveness predicts individual differences in women's preferences for masculine men's voices. *Personality and Individual Differences*, *45*(6), 451–456.
Xu, Y., Lee, A., Wu, W.-L., Liu, X., & Birkholz, P. (2013). Human vocal attractiveness as signaled by body size projection. *PLoS One*, *8*(4), e62397.
Zahavi, A. (1975). Mate selection—A selection for handicap. *Journal of Theoretical Biology*, *53*, 205–214.
Zheng, Y., Compton, B. J., Heyman, G. D., & Jiang, Z. (2020). Vocal attractiveness and voluntarily pitch-shifted voices. *Evolution and Human Behavior*.
Zuckerman, M., Driver, R. E. (1989). What sounds beautiful is good: The vocal attractiveness stereotype. *Journal of nonverbal behavior*, *13*, 67–82, https://doi.org/10.1007/BF00990791
Zuckerman, M., & Miyake, K. (1993). The attractive voice: What makes it so? *Journal of Nonverbal Behavior*, *17*(2), 119–135.

Part II
Voice

Chapter 5
What Does It Mean for a Voice to Sound "Normal"?

Jody Kreiman, Anita Auszmann, and Bruce R. Gerratt

Abstract It is rather unclear what is meant by "normal" voice quality, just as it is often unclear what is meant by "voice quality" in general. To shed light on this matter, listeners heard 1-sec sustained vowels produced by 100 female speakers, half of whom were recorded as part of a clinical voice evaluation and half of whom were undergraduate students who reported no vocal disorder. Listeners compared 20 voices at a time in a series of sort-and-rate trials, ordering the samples on a line according to the severity of perceived pathology. Any voices perceived as normal were placed in a box at one end of the line. Judgments of "normal" versus "not-normal" status were at chance. Listeners were relatively self-consistent, but disagreed with one another, especially about what counts as normal. Agreement was better, but still limited, about what counts as "not normal." Strategies for separating "normal" from "not normal" differed widely across individual listeners, as did strategies for determining how much a given voice deviated from normal. However, acoustic modeling of listeners' responses showed that several acoustic measures—F0, F1 and F2, and F0 coefficient of variation—appeared more often than others as significant predictors of both categorical judgments and of scalar normalness ratings. These variables did not account for most of the variance in these analyses, and did not appear together in the perceptual models for even half of the listeners, but they did appear individually in most analyses, suggesting that in practice the concept of "normal" may have some small core of meaning based on F0 and vowel quality. Thus, the answer to our initial question of what it means for a voice to sound normal is a complex one that depends on the listener, the context, the purpose of the judgment, and other factors as well as on the voice.

J. Kreiman (✉) · A. Auszmann
Department of Head and Neck Surgery and Linguistics, University of California, Los Angeles, CA, USA
e-mail: jkreiman@ucla.edu

A. Auszmann
e-mail: auszmannanita@gmail.com

B. R. Gerratt
Department of Head and Neck Surgery, University of California, Los Angeles, CA, USA
e-mail: bgerratt@ucla.edu

B. Weiss et al. (eds.), *Voice Attractiveness*, Prosody, Phonology and Phonetics,
https://doi.org/10.1007/978-981-15-6627-1_5

Keywords Voice quality · Normal voice · Dysphonia · Voice perception · Voice disorders · Listener · Agreement

5.1 Introduction

The voice literature provides surprisingly little insight into what it means for a voice to be "normal," despite the fact that much depends on the concept of a normal voice. Many studies have shown that a listener's perception of vocal abnormality may lead to negative assessments of the personality, health, intelligence, or social desirability/social attractiveness of the speaker. For example, Amir and Levine-Yundof (2013) found significant differences between speakers with voice disorders and non-dysphonic speakers with respect to listeners' judgments of attractiveness, agreeableness, reliability, potency, aggressiveness, and tenseness. Similarly, Maryn and Debo (2015) found a correlation of $r = 0.85$ between clinicians' ratings of severity of dysphonia and naïve listeners' ratings of healthiness. Similar results have been reported for adult or child listeners, and for expert and naïve judges (Table 5.1). Results also appear to apply to both child and adult speakers, and are robust cross-culturally (e.g., Altenberg & Ferrand, 2006; Irani et al., 2014). These kinds of effects can cause embarrassment and interfere with job performance; in the worst case, they can lead to reduced career opportunities and social isolation.

In clinical settings, a clear understanding of "normal" voice would seem to underlie the entire diagnosis-and-treatment enterprise. A sense that a voice does not sound normal leads patients to initiate treatment, and "normal" serves as a target for determining when therapy is complete. Studies of treatment efficacy logically depend on defining a normal voice as a target, and the practice of establishing normative values for instrumental measures of voice assumes that "normal" has at least a relatively constant meaning.

Despite the importance of "normal" in understanding voice and voice disorders, authors discussing the nature of normal voice have typically emphasized the difficulty of pinning down exactly what it is, echoing Sundberg's (1988) lament that everyone knows what voice is until they try to be specific. Discussions of normal quality have focused on two main themes. The first and more common one describes a normal voice as one that properly presents the person speaking—their age, sex, emotional state—and that adequately meets the speaker's occupational and social communication needs (e.g., Behlau & Murry, 2012; Dehqan et al., 2010; Greene & Mathieson, 1992; Johnson et al., 1965; Aronson & Bless, 2009). Such definitions emphasize the functionality of a voice. For example, Greene and Mathieson (1992) wrote:

> The simplest definition of normal voice is it is 'ordinary': it is inconspicuous with nothing out of the ordinary in its sound. To achieve this standard of acceptability, the voice must be loud enough to be heard, and appropriate for the age and sex of the speaker. It must be reasonably pleasing to the ear of the listener, modulated and clear, not droning and flat or hoarse and breathy. It must be appropriate to the context and not too loud or assertive. (p. 43)

Table 5.1 Representative studies showing perceptual and social sequelae of perceived disordered voice or speech

Speakers	Listeners	Attribute judged	Result	References
Normal and hypernasal children	Children	Social acceptance	Negative responses increased with increasing hypernasality	Blood and Hymen (1977)
Normal and hypernasal children	Children	Social acceptance	Even mild-to-moderate hypernasality decreased social acceptance	Watterson et al., (2013)
Normal and dysphonic female adolescents	Teachers	Personality	Voice disorders increased negative perceptions	Zacharias et al., (2013)
Normal and dysphonic adult females	Adults; monolingual and bilingual, younger and older	Personality	Even mild voice disorders led to negative impressions, for all listener groups	Altenberg and Ferrand (2006)
Normal and dysphonic adult females	Adults	Personality, attractiveness	Nasality and breathy/harsh quality both associated with worse perceptions	Blood et al., (1979)
Normal, dysphonic, and hypernasal females	Students with and without information about voice disorders	Social desirability	Ratings were more negative for speakers with voice disorders	Lallh and Rochet (2000)
Normal and dysphonic adults and children	Adult SLPs; naïve listeners	Healthiness	Even slight dysphonia produced the perception of unhealthiness	Maryn and Debo (2015)
Normal and dysphonic speakers of Hebrew	Young and older adults	Personality	Dysphonia associated with negative perceptions, for women more than for men	Amir and Levine-Yundof (2013)

It follows from this definition that standards and judgments will vary across listeners and contexts. For example, Moore (1971) wrote:

> It is apparent that the voice is abnormal for a particular individual when he or she judges it to be so regardless of the circumstances. Judgment implies a set of standards that are learned through experience and that are related to the judge's own aesthetic and cultural criteria. Judgment also implies that standards are not fixed, that there is opportunity for more than one conclusion. This flexibility in determining the defectiveness of voices does not alter the validity of the basic definition of voice disorders, but it does underscore the observation that vocal standards are culturally based and environmentally determined. (p. 535)

However, to our knowledge the nature and extent of this variability have not been studied, nor have the factors conditioning variability in perceived vocal abnormality.

A second definitional approach emphasizes physical normalness, without particular concern for vocal quality or for use of the voice in communication. For example, normal voice can be characterized as the acoustic product of a normal vocal tract that is functioning normally (Mathieson, 2000) or as a voice produced by a speaker with no current or previous voice complaint and that passes a perceptual evaluation by a speech-language pathologist (Bonilha & Deliyski, 2008).

To our knowledge, no empirical data exist in support of either of these views. In the face of the importance a perceived voice disorder can have for a speaker, clinicians and scientists have proceeded as if "normal" unambiguously exists. For example, numerous studies propose algorithms devised to automatically separate normal from pathological phonation, arguing that such algorithms bring needed objectivity to clinical voice evaluation (e.g., Arias-Londoño et al., 2011; Orozco-Arroyave et al., 2015; Wang et al., 2011; Moro-Velázquez et al., 2016). "Normal" in these studies remains an unexamined concept, and algorithms typically show good classification accuracy (usually >90% correct), suggesting this approach is not unreasonable. Similarly, many more studies have reported normative values for acoustic (e.g., Goy, Fernandes et al., 2013; Wuyts et al., 2002), physiological (e.g., Xue & Hao, 2006), and/or aerodynamic measures of voice (e.g., Lewandowski et al., 2017), again implying that it is possible to define "normal" as a quality with clear boundaries. The voice literature thus presents a paradox. Clinical concerns combined with the demonstrated social and personal importance of sounding normal lead researchers to design studies that assume a clear boundary between normal and not-normal phonation, while at the same time arguing that no such boundaries exist in theory, all of this in the absence of empirical evidence about what sounds normal or not normal to listeners.

This study is intended to address this situation. Our goals are to gather listeners' assessments of the extent to which voices sound normal, and to seek insight into the factors that determine whether a voice sounds better or worse to a particular listener.

5.2 Methods

5.2.1 Speakers and Voice Samples

The voices of 100 female speakers were used in this experiment. Females (as opposed to males) were selected for this preliminary study because of recent research interest in the perception of normal versus abnormal female voice quality, particularly with respect to vocal fry and "creaky voice" (Yuasa, 2010; Anderson et al., 2014; Oliveira et al., 2016). Fifty voice samples were drawn from an existing database of recordings of speakers who had a diagnosis from an otolaryngologist ("not normal"). Voices were unselected with respect to diagnoses, which included functional and neurogenic disorders, mass lesions, reflux, and age-related dysphonia. Samples ranged from extremely mild to very severe vocal pathology. An additional 50 voices were drawn from the UCLA Speaker Variability Database (Keating et al. 2018), which includes multiple voice samples from over 200 male and female UCLA undergraduate students, all of whom reported no history of voice or speech complaints ("normal"). Note that although voices were categorized as ± normal based on diagnostic status, no assumptions were made about the normal or abnormal quality of the voices, and no attempt was made to select "normal" or "not-normal" voices that sounded more or less normal, beyond informally ensuring that the "not-normal" samples represented a broad range of severity of perceived pathology.

All speakers sustained the vowel /a/ as part of their recording sessions, and all were recorded with a Brüel and Kjær 1/2" microphone. Steady-state vowels were studied rather than continuous speech, to allow listeners to focus on voice quality and not on articulation or native/nonnative status of the speakers. Previous studies (e.g., Gerratt et al., 2016) have shown negligible effects of stimulus type on quality assessment. Samples were directly digitized at a 20 kHz (clinical samples) or 22 kHz (normal samples) sampling rate, edited to 1 s duration, and then downsampled to 10 kHz prior to acoustic analyses and testing.

5.2.2 Listeners and Listening Task

Stimuli were assembled into blocks of 20 voices each, which in turn were assembled into five sets of nine trials (each trial comprising one 20-voice block), such that across the five sets of trials, every voice was compared at least once to every other voice and every voice received a total of 90 ratings. Each listener heard 9, 20-voice trials, for a total of 180 judgements/listener: each stimulus voice was judged at least once/listener, with 80 voices repeated in 2 different trials so that test–retest reliability could be assessed. (No voices were repeated within a single trial.)

All experimental procedures were approved by the UCLA Institutional Review Board. Ten UCLA students and staff (aged 18–68; mean age = 21.5; sd = 9.67) heard each set of trials, for a total of 50 listeners. All listeners reported normal

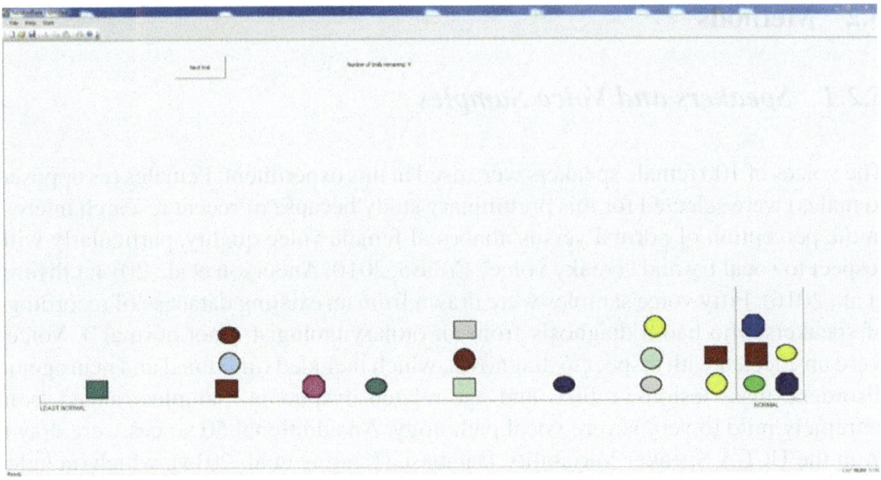

Fig. 5.1 The testing interface for the sort-and-rate task. Listeners played each voice by clicking its icon, and then dragged the icon to indicate (1) whether the voice sounded normal, in which case the icon was placed in the box on the right and (2) if it did not sound normal, how close to normal it sounded. The most abnormal-sounding voices were placed toward the left end of the line; those that approached normal were placed near the box

hearing and received course credit in return for their participation. Clinicians were not targeted separately during subject selection because evidence indicates they do not differ significantly from naïve listeners when judging the severity of dysphonia (Eadie et al. 2010).

Subjects heard the stimuli over Etymotic insert earphones (model ER-1) at a comfortable constant listening level. The testing interface is shown in Fig. 5.1. Each icon in the figure represents a single voice token, randomly assigned to that icon. Listeners played each voice by clicking its icon, then dragged the icon to a line to indicate (1) whether the voice sounded normal, in which case the icon was placed in the box on the right end of the line and (2) if it did not sound normal, how close to normal it sounded (a visual sort-and-rate task; Granqvist, 2003). The most abnormal-sounding voices were placed toward the left end of the line; those that approached normal were placed near the box. Voices judged as equally dysphonic were to be stacked on the line. Because the box for "normal" voices appeared rather small on the screen, listeners were explicitly instructed that box size did not mean that there were only a few normal voices in the set, and that they could place as many or as few icons as desired in the box. Listeners were encouraged to play the voices as often as required, in any order, until they were satisfied with their sort, after which testing advanced to the next trial. The experiment was self-paced and listeners were allowed to take breaks as needed. They were not told how many total speakers were included in the experiment. Total testing time was less than 1 h.

Table 5.2 Acoustic variables. Means and coefficients of variation were calculated for all measures using VoiceSauce software

Variable	Definition and reference
H1*-H2*	Relative amplitudes of the first and second harmonics, corrected for the effects of formants on amplitude (Hanson, 1997; Iseli & Alwan, 2004)
H2*-H4*	Relative amplitudes of the second and fourth harmonics, corrected for the effects of formants on amplitude
H4*-H2kHz*	Relative amplitudes of the fourth harmonic and the harmonic nearest 2 kHz, corrected for the effects of formants on amplitude
H2kHz*-H5kHz	Relative amplitudes of the harmonic nearest 2 kHz and that nearest 5 kHz; H2kHz is corrected for the effects of formants on amplitude
Cepstral peak prominence (CPP)	The relative amplitude of the cepstral peak in relation to the expected amplitude as derived via linear regression; a measure of aperiodicity (Hillenbrand et al., 1994)
Energy Root Mean Square (RMS)	Energy, calculated over five pitch pulses.
Subharmonic-to-harmonic ratio (SHR)	The amplitude ratio between subharmonics and harmonics; characterizes speech with alternating pulse cycles (period-doubling; Sun, 2002)
Fundamental frequency (F0)	The frequency of the first harmonic
F1, F2, F3, F4	Center frequencies of the first four formants

5.2.3 Acoustic Analyses

Acoustic measurements (Table 5.2) were made on all stimuli to facilitate interpretation of listeners' perceptual strategies. As a set, these measures constitute a psychoacoustic model of voice quality (Kreiman et al., 2014) and were chosen because as a set they are sufficient to model the perceived quality of virtually any sustained phonation. Variables were measured every 5 ms using VoiceSauce software (Shue et al., 2011), and then averaged across the entire utterance. Coefficients of variation were also calculated as estimates of signal variability.

5.3 Results

Analyses fall into two groups, corresponding to the two approaches to defining "normal" discussed in the Introduction. The first analyses treated "normal" (i.e., placed in the box by a listener; Fig. 5.1) and "not-normal" (placed on the line outside the box) responses as straightforwardly categorical, consistent with definitions of nor-

Fig. 5.2 Accuracy of classification judgments as a function of the mean rating (where a larger rating = a more normal voice). Open circles represent "not-normal" voices; filled circles represent "normal" voices.

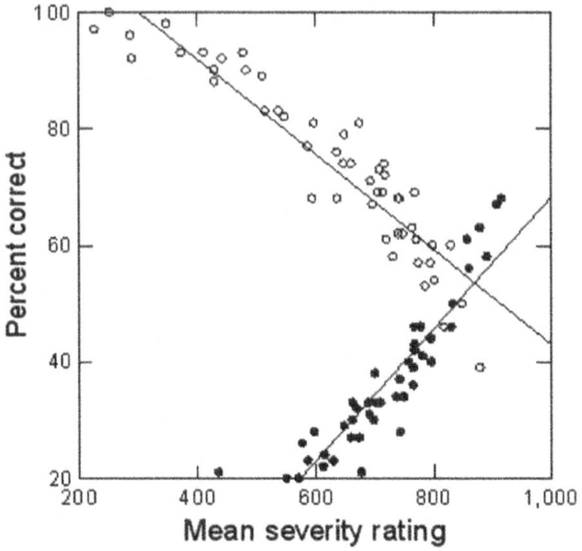

mal as "lacking a diagnosis." The second set of analyses treated ratings as forming a continuum from most severe (=0), to normal (=1000), consistent with the idea that perceived normalness varies continuously as a function of listening context (Gerratt et al., 1993), social and/or communicative context, and other such factors. Both sets examined (1) listener agreement about (the degree of) perceived deviation and (2) the acoustic cues that explained listeners' judgments.

Figure 5.2 shows the relationship between these two measurement approaches in a plot of categorization accuracy as a function of mean normalness ratings. In this figure, a priori "normal" voices are plotted as filled circles and a priori "not-normal" voices are plotted as open circles. Note that accuracy is greater for "not-normal" voices than for "normal" voices: It is apparent from this figure that many voices with diagnoses sound quite normal, and many nominally normal voices sound rather abnormal on average. The majority of "normal" voices were judged normal less than 50% of the time, while only a few "not-normal" voices were incorrectly categorized more than 50% of the time. Also note that the range of severity ratings for "normal" voices completely overlaps that for "not-normal" voices, but not vice versa. This pattern occurs because the normal end of the scale has an absolute ending point—a voice cannot be more normal than normal—but one can always imagine a worse voice, so that the left end of the scale can extend infinitely.

5.3.1 *Categorical Judgments of "Normal" Versus "Not-Normal" Voice Quality*

5.3.1.1 Can Listeners Accurately Separate Nominally Normal from Nominally Not-Normal Voices?

If the boundary between normal and not-normal voice quality is ill-defined, as suggested by the papers reviewed in the Introduction, then it should be difficult for listeners to make categorical decisions regarding the status of a voice sample. This proved to be the case. For voices deemed normal a priori, listener performance ranged from 1.1 to 67.8% correct classification, with a mean of 34.1% correct (sd = 14.64%), where chance is 50%. Performance was somewhat better for a priori not-normal voices, which were correctly classified an average of 73.6% of the time (sd = 14.99%), with a range of 45.6–100%. Chi square analyses indicated that listeners heard only 2/50 a priori normal voices as normal at above chance levels, but agreed at above chance levels that 30/50 normal voices were not normal. For a priori not-normal voices, 35/50 were significantly often classified as not normal, and none was incorrectly classified as normal.

Finally, d′ analysis (e.g., Green & Swets, 1966) was used to assess overall categorization accuracy across the entire group of listeners. In this context, d′ measures listeners' ability to correctly identify voices as normal or not normal, independent of response biases in favor of "normal" or "not-normal" responses. Ratings on the normal/not-normal scale were quantized to range from 1 to 10, where 1 represented the worst voice quality and 10 meant the voice had been classified as normal. These rescaled values were then used to calculate d′ for each listener and for the group as a whole (Macmillan & Creelman, 2005). Results for both the pooled listeners and for individuals indicated that performance was at chance levels. For the pooled listeners, d′ equaled 0.21, while across individual listeners, values averaged 0.24, with a range of −0.27–0.81 (sd = 0.28). We conclude from these data that listeners were unable to distinguish nominally normal from nominally not-normal voices at above chance rates, due to misclassifications both of normal voices as dysphonic and of not-normal voices as normal.

5.3.1.2 Do Listeners Agree with One Another in Their Categorical Judgments?

Although listeners were inaccurate in their categorical responses, it is possible that this occurred because some of the clinical voice samples were very mildly deviant, and some of the nominally normal voices were characterized by high or low F0, vibrato, vocal fry, and/or breathiness, which could be interpreted as abnormal. This is especially possible when not normal is defined entirely in terms of physiology, because abnormal-appearing vocal folds can sometimes occur without any perceptual consequences. If this is the case, listeners might agree in their normal/not-normal judgments, even though these do not correspond to the clinically defined state of affairs.

To assess this possibility, we examined listener agreement about vocal status, independent of the existence of a diagnosis. Listeners did not agree unanimously that any voice was normal; they were unanimous regarding only a single not-normal voice. Significant agreement was almost as uncommon as unanimous agreement. Chi square analyses showed that listeners agreed at above chance levels that only 2/100 voices were normal (both of which were in fact normal; $p < 0.05$); they agreed at above chance levels that 65/100 voices were not normal (30 of which were nominally normal, as noted above; $p < 0.05$). We conclude that listeners are no more in agreement than they are accurate when asked to judge whether or not a voice is normal.

5.3.1.3 Are Listeners Self-consistent in Their Judgments?

Two possibilities emerge from the findings that listeners are highly inaccurate and disagree widely when asked to judge whether a voice is or is not normal. First, it is possible that "normal" is truly meaningless in practice. However, it is also possible that every listener has his/her own consistent idea of what "normal" is, but that these ideas differ from listener to listener. To examine these possibilities, we calculated intrarater agreement in normal/not-normal judgments for the 80 repeated ratings each listener provided. Average intrarater agreement equaled 75.8%, with a range from 57.5 to 94.4% (sd = 9.22%; chance = 50%). Three of 50 listeners were self-consistent at rates below 60%; 30/50 were self-consistent at rates of 75% or above. These results indicate that most listeners are reasonably reliable when they report that a voice is or is not normal, but suggest that the basis for these judgments may vary across listeners, leading to self-consistency but low interrater agreement. We pursue this possibility in the next section.

5.3.2 Can We Predict Listeners' Categorical Responses from Voice Acoustics?

Linear discriminant (LD) analysis was used to determine how well listeners' categorical "normal" versus "not-normal" judgments could be predicted from voice acoustics (regardless of the existence/non-existence of a diagnosis). All variables from the psychoacoustic model were entered simultaneously into the analysis. One eigenfunction accounted for 100% of the variance in the data (canonical correlation = 0.263; Wilks' lambda = 0.931; chi square = 642.72, df = 14, $p < 0.001$). 70% of stimuli were correctly classified as perceptually normal or not normal. Predictors with weights >= 0.3 (~10% variance accounted for) included F2 (weight = −0.52), F0 (weight = 0.33), and F0 cv (weight = −0.30). These results suggest that, even

Table 5.3 Patterns of weights on eigenfunctions resulting from LD analyses relating individual listeners' categorical normal/not-normal judgments to acoustic variables

Primary predictor variable	Additional significant predictors	Number of listeners
Variability		14
Vowel quality		7
Vowel quality	Variability	2
Vowel quality	Noise	5
Vowel quality	F0	5
F0		1
F0	Noise	3
F0	Spectral shape	5
Noise		6
Spectral shape		2

when considered as a group, listeners are not responding randomly, but also show that only a few rather simple variables (vowel quality, pitch, and pitch variability) are apparently shared across listeners.

To examine differences among listeners, we repeated the LD analyses for each of the 50 individual listeners. Results showed significant classification based on acoustic measures for all but 1 listener; across individuals, voices were correctly categorized as "perceived to be normal" or "perceived to be not normal" 81.35% of the time (sd = 6.64; range = 67.8–96.7%). However, listeners differed widely in the measures that emerged from these analyses. For brevity of presentation, the acoustic parameters were grouped into five categories: variability (coefficients of variability for all measures), vowel quality (F1, F2, F3, F4); spectral noise (CPP, energy, SHR), F0, and source spectral shape (H1*-H2*, H2*-H4*, H4*-H2kHz*, H2kHz*-H5kHz). Variables that weighted at 0.3 or higher on the eigenvector for each listener are tallied in Table 5.3. As in the group analyses just described, F0 and vowel quality were important for explaining individual listeners' normal/not-normal decisions, but overall acoustic variability and noise also emerged as important predictors.

Finally, context effects are well known to affect ratings of vocal severity. For example, a given voice will sound rougher in the context of normal voices, and less rough in the context of voices with severe vocal pathology (Gerratt et al., 1993). To examine the influence such effects might have had on perceptual strategies in the present task, we repeated the LD analyses separately for each of the five groups of listeners. Recall that all listeners heard all the voices, but voices were grouped into different sets of 20, so the context in which each voice was judged varied from group to group. Results appear in Table 5.4. Groups did differ somewhat in the acoustic variables that predict overall categorical response patterns. Notably, spectral shape parameters appear in the solutions for two groups, and CPP appears in two other solutions. However, the increased complexity of the sets of predictor variables did

Table 5.4 Discriminant analysis results for the five groups of listeners. All analyses $p < 0.001$; only weights exceeding 0.3 are listed

Listener group	Variables (weights)	% Correct classification
1	CPP (0.46), CPP cv (−0.41), F2 (−0.35), F1 (0.34)	70.3
2	F2 (−0.50), H4*-2kHz* (−37), H2*-H4* (0.31), F0 cv (−0.30)	66.7
3	F2 (−0.49)	70.8
4	F0 (0.52), CPP cv (−0.48), F0 cv (−0.44)	77.8
5	F2 (−0.66), H4*-2kHz* (−0.30)	64.0

not result in improved correct classification rates, which generally remained well below those observed for individual listeners. This suggests that, although context effects exist, individuals in even small groups ($n = 10$) vary enough in perceptual strategies that controlling context effects does not improve correct classification to any measurable extent.

To summarize, across all listeners, parameters associated with F0, F0 variability, and vowel quality appear to be important for separating normal from not-normal voices for many, but not most, listeners, and thus provide at best moderate prediction of how a voice will be judged. Listeners' strategies vary with listening context, but modeling this aspect of variation does not improve overall prediction. However, LD analyses indicated that individual listeners' strategies can be well predicted from acoustics, but that listeners differ widely from one another. We conclude that listeners disagree because they are using rather different perceptual strategies, which are more idiosyncratic than they are context dependent. We examine this possibility further in the next section.

5.3.3 Do Listeners at Least Sort Voices in Similar Fashions?

A final possible explanation for our findings is that listeners rank the voices similarly on a scale from normal to maximally not normal, but differ in where they place the dividing line between categories. This could also have occurred if listeners differed in their interpretation of the size of the "normal" box in the experimental interface. To investigate these possibilities, we calculated Spearman correlations between scalar ratings for all pairs of listeners within a group. Rank-order correlations averaged only 0.267 (sd = 0.107; range = −0.093–0.583), indicating that listeners do not agree even about the relative normalness/not-normalness of the voices.

5.3.4 Can We Predict the Extent to Which a Voice Sounds Not Normal? What Parameters Are Associated with Increasing Perceived Vocal Deviance for Individual Listeners?

Analyses in previous sections have demonstrated that listeners are individually self-consistent but inaccurate and in disagreement when separating normal from not-normal voices. To investigate this further, we modeled each listener's perceptual strategy with a series of correlation and multiple regression analyses using only the voices categorized as not normal. First, for each listener, we calculated a multiple regression between the scalar not-normal ratings and the complete set of acoustic measures, entered into the equation in five blocks (F1, F2, F3, and F4; the coefficients of variation; F0; CPP, energy, and SHR; and the four spectral shape parameters). Order of entry was determined by the overall importance of the sets of variables in the LD analyses (Table 5.3). Next, for each listener, we calculated Pearson's correlation between each acoustic measure and the scalar rating on the normal/not-normal scale for that listener, again including only the voices that the listener categorized as not normal. Finally, we calculated additional multiple regressions again relating ratings to acoustic measures for each listener, but this time using only the variables that were significant predictors in the first regression for that listener plus any additional variables that were significantly correlated with that listener's not-normal ratings.

Results are shown in Table 5.5. All the regressions were statistically significant ($p < 0.01$), but all accounted for rather small amounts of variance in listeners' judgments (mean $r = 0.477$; $sd = 0.126$; range $= 0.227$–0.699). As Table 5.5 shows, every variable contributed significantly to predicting ratings for at least one listener, but F0, F1, F2, and F0 cv stand out as more important across listeners than the rest. Recall that these same variables were associated with categorical normal/not-normal judgments for many listeners, as described above. This suggests that, for at least some listeners, deciding whether or not a voice sounds normal and establishing exactly how not normal it sounds depend on the same cues and thus are essentially the same process.

5.4 Discussion and Conclusions

To summarize our findings, judgments of diagnostically "normal" versus "not-normal" status were at chance. Listeners were relatively self-consistent in their judgments, but disagreed with one another, especially about what counts as normal. Agreement was better, but still limited, about what counts as "not normal." This may have occurred because of differences in the possible ranges of the two labels. As noted above, the range of perceived not-normal quality can extend essentially limitlessly. As a result, there will always be voices that are so far from the boundary between normal and not normal that little or no ambiguity exists with respect to

Table 5.5 The frequency with which each acoustic variable emerged as a significant predictor in multiple regressions relating acoustic variables to the degree of perceived not-normalness. The most important predictors are listed in **bold type**. The maximum possible value is 50 (=the number of listeners)

Variable	# listeners for whom that variable was a significant predictor of perceived not-normalness
H1*-H2*	4
H2*-H4*	7
H4*-H2kHz*	3
H2kHz*-H5kHz	5
CPP	8
Energy	3
SHR	5
F0	**19**
F1	**14**
F2	**24**
F3	3
F4	3
H1*-H2* cv	1
H2*-H4* cv	4
H4*-H2kHz* cv	8
H2kHz*-H5kHz cv	3
CPP cv	9
Energy cv	7
SHR cv	3
F0 cv	**26**
F1 cv	2
F2 cv	2
F3 cv	10
F4 cv	4

their status. In contrast, logically a voice cannot be more normal than "normal," and any deviation in quality, however slight, creates ambiguity (and hence disagreement) about the voice's status. The surprising aspect of our results was how completely the category "normal" was compromised by this process.

The overall picture that emerges from the present data is one of differences between listeners, but less so within listeners, in the attributes they pay attention to when deciding that a voice is or is not normal. Strategies for separating "normal" from "not normal" differed widely across individual listeners, as did strategies for determining how much a given voice deviated from normal, and all variables in the psychoacoustic model played a role in decisions for at least one listener. However, several variables—F0, F1 and F2, and F0 cv—appeared more often than the others as significant predictors of both categorical judgments and of scalar normalness ratings. These variables did not account for most of the variance in these analyses, and did not consistently appear as a set in the perceptual models for even half of the listeners, but they did appear individually in most analyses, suggesting that in practice the concept of "normal" has some small core of meaning based on F0 and vowel quality.

We note that the "core" variables are also important determinants of individual voice quality (see Kreiman & Sidtis, 2011, for review), which is judged in terms of a central category member and idiosyncratic deviations from that "average" voice. Thus, it is possible that (at least some of the time), listeners assess normalness much as they assess individual voice quality in general, with respect to a central pattern and the deviations from that pattern that appear in the particular voice sample at hand. Thus, the answer to our initial question—What does it mean for a voice to sound normal?—is a complex one that depends on the listener, the context, the purpose of the judgment, and other factors as well as on the voice.

A few limitations to this research should be noted. First, stimuli were steady-state vowels rather than connected speech. This means that many details that can characterize disordered speech were not available for consideration, including prosody, articulation, pausing, and other vocal attributes. However, it seems unlikely that inclusion of more complex stimuli would improve overall listener agreement, particularly with respect to which voices sound normal. This study was also restricted to female speakers. While it is likely that different parameters will emerge from studies of normal versus not-normal male voices, the fact that listeners' behavior is consistent with broader models of voice perception makes it unlikely that the overall pattern of results would differ substantially. Studies of male voices are currently underway in our laboratory. Finally, the relatively small size of the response box for "normal" voices in the testing interface (Fig. 5.1) may have discouraged some listeners from categorizing too many voices as normal, despite instructions that any number of voices could be placed in the box. However, we note that correlation analyses showed very poor agreement among listeners, suggesting that the effect of this design issue on the overall pattern of results is minimal.

In conclusion, these results have implications for ongoing efforts to identify acoustic measures to screen for vocal pathology or the provision of normative values for single acoustic measure. The finding that listeners are self-consistent but highly individual in their perceptual strategies for determining what is and is not normal suggest that automatic protocols or screening based on normative values may be of limited clinical or theoretical use. Clear communication between clinicians and patients in a context of cultural awareness would seem to be the straightest path to satisfactory treatment outcomes.

Acknowledgments This work was supported by NIH grant DC01797, and by NSF grants IIS 1704167 and IIS 1450992. A preliminary version was presented at the 175th Meeting of the Acoustical Society of America, Minneapolis, MN, May 2018. We thank Norma Antoñanzas for programming support, Meng Yang for help with acoustic analyses, Jordan Shavalian for assistance with subject testing and data analysis, and Pat Keating and Marc Garellek for helpful comments.

References

Altenberg, E. P., & Ferrand, C. T. (2006). Perception of individuals with voice disorders by monolingual English, bilingual Cantonese-English, and bilingual Russian-English women. *Journal of Speech, Language, and Hearing Research*, *49*, 879–887.

Amir, O., & Levine-Yundof, R. (2013). Listeners' attitude toward people with dysphonia. *Journal of Voice*, *27*, 524.e1–524.e10.

Anderson, R., Klofstad, C., Mayew, W., & Venkatachalam, M. (2014). Vocal fry may undermine the success of young women in the labor market. *PLOS ONE*, *9*, e97506.

Arias-Londoño, J. D., Godino-Llorente, J. I., Markaki, M., & Stylianou, Y. (2011). On combining information from modulation spectra and mel-frequency cepstral coefficients for automatic detection of pathological voices. *Logopedics Phoniatrics Vocology*, *36*, 60–69.

Aronson, A. E., & Bless, D. M. (2009). *Clinical voice disorders*. New York, N.Y.: Thieme.

Behlau, M., & Murry, T. (2012). International and intercultural aspects of voice and voice disorders. In D. E. Battle (Ed.), *Communication disorders in multicultural and international populations* (4th ed., pp. 174–207). St. Louis, MO: Mosby.

Blood, G. W., & Hyman, M. (1977). Children's perception of nasal resonance. *Journal of Speech and Hearing Disorders*, *42*, 446–448.

Blood, G. W., Mahan, B. W., & Hyman, M. (1979). Judging personality and appearance from voice disorders. *Journal of Communication Disorders*, *12*, 63–67.

Bonilha, H. S., & Deliyski, D. D. (2008). Period and glottal width irregularities in vocally normal speakers. *Journal of Voice*, *22*, 699–708.

Bradley, R. A., & Terry, M. E. (1952). Rank analysis of incomplete block desings: I. The method of paired comparisons. *Biometrika*, *39*(3/4), 324–345.

Dehqan, A., Ansari, H., & Bakhtiar, M. (2010). Objective voice analysis of Iranian speakers with normal voices. *Journal of Voice*, *24*, 161–167.

Eadie, T. L., Kapsner, M., Rosenzweig, J., Waugh, P., Hillel, A., & Merati, A. (2010). The role of experience on judgments of dysphonia. *Journal of Voice*, *24*, 564–573.

Gerratt, B. R., Kreiman, J., & Antoñanzas-Barroso. (1993). Comparing internal and external standards in voice quality judgments. *Journal of Speech and Hearing Research*, *36*, 14–20.

Gerratt, B. R., Kreiman, J., & Garellek, M. (2016). Comparing measures of voice quality from sustained phonation and continuous speech. *Journal of Speech Hearing Research*, *59*, 994–1001.

Goy, H., Fernandes, D. N., Pichora-Fuller, M. K., & van Lieshout, P. (2013). Normative voice data for younger and older adults. *Journal of Voice*, *27*, 545–555.

Granqvist, S. (2003). The visual sort and rate method for perceptual evaluation in listening tests. *Logopedics Phoniatrics Vocology*, *28*, 109–116.

Greene, M.C., & Mathieson, L. (1992). *The voice and its disorders*. San Diego, CA: Singular.

Green, D. M., & Swets, J. A. (1966). *Signal detection theory and psychophysics*. New York: Krieger.

Hanson, H. M. (1997). Glottal characteristics of female speakers: Acoustic correlates. *Journal of the Acoustical Society of America*, *101*, 466–481.

Hillenbrand, J., Cleveland, R. A., & Erickson, R. L. (1994). Acoustic correlates of breathy vocal quality. *Journal of Speech, Language, and Hearing Research*, *37*, 769–778.

Irani, F., Abdalla, F., & Hughes, S. (2014). Perceptions of voice disorders: A survey of Arab adults. *Logopedics Phoniatrics Vocology*, *39*, 87–97.

Iseli, M., & Alwan, A. (2004). An improved correction formula for the estimation of harmonic magnitudes and its application to open quotient estimation. In *Proceedings of ICASSP'04* (pp. 669–672), Montreal, Canada.

Johnson, W., Brown, S.F., Curtis, J.F., Edney, C.W., & Keaster, J. (1965). *Speech handicapped school children.* New York: Harper & Brothers. Cited in Aronson & Bless (2009).

Keating, P. A., Kreiman, J., & Alwan, A. (2018). The UCLA speaker variability database. Retrieved July 26, 2018, from https://ucla.box.com/s/81ho14uypkmv2nn1s3tvy1gvdtq70gcf.

Kreiman, J., & Sidtis, D. (2011). *Foundations of voice studies*. Walden, MA: Wiley-Blackwell.

Kreiman, J., Gerratt, B. R., Garellek, M., Samlan, R., & Zhang, Z. (2014). Toward a unified theory of voice production and perception. *Loquens, 1*(1), 1–9. https://doi.org/10.3989/loquens.2014.009.

Lallh, A. K., & Rochet, A. P. (2000). The effect of information on listeners' attitudes toward speakers with voice or resonance disorders. *Journal of Speech, Language, and Hearing Research, 43*, 782–795.

Lewandowski, A., Gillespie, A. I., Kridgen, S., Jeong, K., Yu, L., & Gartner-Schmidt, J. (2018). Adult normative data for phonatory aerodynamics in connected speech. *The Laryngoscope, 128*, 909–914.

Macmillan, N. A., & Creelman, C. D. (2005). *Detection theory: A user's guide* (2nd ed.). Mahwah, NJ: Erlbaum.

Maryn, Y., & Debo, K. (2015). Is perceived dysphonia related to perceived healthiness? *Logopedics Phoniatrics Vocology, 40*, 122–128.

Mathieson, L. (2000). Normal-disordered continuum. In R. D. Kent & M. J. Ball (Eds.), *Voice quality measurement* (pp. 3–12). San Diego, CA: Singular.

Moore, G. P. (1971). Voice disorders organically based. In L. E. Travis (Ed.), *Handbook of speech pathology and audiology* (pp. 535–569). Englewood Cliffs, NJ: Prentice-Hall.

Moro-Velázquez, L., Gómez-García, J., & Godino Llorente, J. (2016). Voice pathology detection using modulation spectrum-optimized metrics. *Frontiers in Bioengineering and Biotechnology, 4*. https://doi.org/10.3389/fbioe.2016.00001.

Oliveira, G., Davidson, A., Holczer, R., Kaplan, S., & Paretzky, A. (2016). A comparison of the use of glottal fry in the spontaneous speech of young and middle-aged American women. *Journal of Voice, 30*, 684–687.

Orozco-Arroyave, J. R., Belalcazar-Bolanos, E. A., Arias-Londoño, J. D., Vargas-Bonilla, J. F., Skodda, S., & Rusz, J.,& Nöth, E.,. (2015). Characterization methods for the detection of multiple voice disorders: Neurological, functional, and laryngeal diseases. *IEEE Journal of Biomedical and Health Informatics, 19*, 1820–1828.

Shue, Y.-L., Keating, P., Vicenik, C., & Yu, K. (2011). VoiceSauce: A program for voice analysis. In *2011 Proceedings of International Congress of Phonetic Sciences (ICPhS) XVII* (pp. 1846–1849), Hong Kong.

Sun, X. (2002). Pitch determination and voice quality analysis using subharmonic-to-harmonic ratio. In *2002 IEEE International Conference on Acoustics, Speech, and Signal Processing (ICASSP)* (Vol. 1, pp. I–333). IEEE. https://doi.org/10.1109/ICASSP.2002.5743722.

Sundberg, J. (1988). *The science of the singing voice*. DeKalb, IL: Northern Illinois University Press.

Wang, X., Zhang, J., & Yan, Y. (2011). Discrimination between pathological and normal voices using GMM-SVM approach. *Journal of Voice, 25*, 38–43.

Watterson, T., Mancini, M., Brancamp, T. U., & Lewis, K. E. (2013). Relationship between the perception of hypernasality and social judgments in school-aged children. *Cleft Palate Craniofacial Journal, 50*, 498–502.

Wuyts, F. L., Heylen, L., Mertens, F., De Bodt, M., & Van de Heyning, P. H. (2002). Normative voice range profiles of untrained boys and girls. *Journal of Voice, 16*, 460–465.

Xue, S. A., & Hao, J. G. (2006). Normative standards for vocal tract dimensions by race as measured by acoustic pharyngometry. *Journal of Voice, 20*, 391–400.

Yuasa, I. P. (2010). Creaky voice: a new feminine voice quality for young urban-oriented upwardly mobile American women? *American Speech, 85*, 315–337.

Zacharias, S. R., Kelchner, L. N., & Creaghead, N. (2013). Teachers' perceptions of adolescent females with voice disorders. *Language, Speech, and Hearing Services in Schools, 44*, 174–182.

Chapter 6
The Role of Voice Evaluation in Voice Recall

Molly Babel, Grant McGuire, and Chloe Willis

Abstract This chapter examines the relationship among a suite of voice evaluation metrics—vocal attractiveness, voice typicality, gender categorization fluency, intelligibility, acoustic similarity, and perceptual similarity—in a set of 60 American English voices with the goal of understanding how these evaluation metrics predict listeners' abilities to accurately recall voices. This question of what makes a voice memorable has been studied from a range of perspectives, as it raises critical theoretical issues about auditory memory and phonetic encoding, in addition to having applied concerns in the context of earwitness testimony. We find that the more subjective voice evaluation measures of stereotypicality and attractiveness predict listeners' ability to recall voices more so than the more objective measures related to voice similarity and processing. These results suggest that listeners' cognitive organization of voices is influenced by social assessments of voices.

Keywords Voice recall · Talker recognition · Voice evaluation · Voice typicality · PCA · Voice organization

---

M. Babel
Department of Linguistics, University of British Columbia, 2613 West Mall, Vancouver, BC V6T 1Z4, Canada
e-mail: molly.babel@ubc.ca

G. McGuire (✉)
Department of Linguistics, University of California Santa Cruz, 1156 High St, Santa Cruz, CA 95060, USA
e-mail: gmcguir1@ucsc.edu

C. Willis
Department of Linguistics, University of California Santa Barbara, Santa Barbara, CA 93106, USA
e-mail: chloemwillis@umail.ucsb.edu

B. Weiss et al. (eds.), *Voice Attractiveness*, Prosody, Phonology and Phonetics,
https://doi.org/10.1007/978-981-15-6627-1_6

6.1 Introduction

This chapter examines the relationship between vocal attractiveness, voice typicality, and other related vocal evaluation metrics along with listeners' ability to recall voices from memory. What makes a voice memorable has been studied from a range of perspectives as it raises critical theoretical issues about auditory memory and phonetic encoding, in addition to having applied concerns in the context of earwitness testimony. In this work, we explore some of the qualities of the voices that improve and detract from voice recall performance.

Talker recognition or listeners' ability to recall voices they have been previously exposed to is highly affected by what is referred to as the *language familiarity effect*. Listeners are more accurate at recalling voices that speak the same language as the listener population (Goggin, Thompson, Strube, & Simental, 1991; Perrachione & Wong, 2007; Thompson, 1987; Winters, Levi, & Pisoni, 2008; Perrachione, Del Tufo, & Gabrieli, 2011; Xie & Myers, 2015; Orchard & Yarmey, 1995; Bregman & Creel, 2014) or speak with a familiar accent (Goggin et al., 1991; Stevenage, Clarke, & McNeill, 2012; Senior et al., 2018; Thompson, 1987; Perrachione, Chiao, & Wong, 2010). The mechanism behind these findings is generally considered to be one of listeners' familiarity with the phonetic distribution of sounds in the language or accent. When listeners are familiar with a language or accent, they are better able to determine which acoustic–phonetic features in the speech stream are language-specific and which are attributes of a particular speaker's voice (Winters et al., 2008; Perrachione, in press).

While this literature has established that voices with familiar languages and accents are generally more accurately recalled, voices within a language variety are not equally memorable. Within a language variety, what makes a voice more or less memorable? Several studies have found that subjective listener ratings of distinctiveness, typicality, memorability, among other evaluative qualities can predict which voices have better recall accuracy (Papcun, Kreiman, & Davis, 1989; Kreiman, & Papcun, 1991; Yarmey, 1991; O'Toole et al., 1998).

For example, Papcun et al. (1989) exposed listeners to 10 voices that had been previously rated on a scale from easy- to hard-to-remember and tested voice recall in an open set task with 1-, 2-, and 4-week delays. Subjects were generally better at rejecting novel voices rather than correctly identifying the voices that they had been exposed to. Specifically, the voices did not differ greatly in accuracy of recall, but did differ in false identifications, such that "hard" voices engendered more false positives. Papcun and colleagues invoke a prototype model to explain these results, hypothesizing that listeners characterize and remember voices in terms of a prototype and deviations therefrom. Thus, more prototypical voices are hard-to-remember as they are more similar to other voices and are more likely to be misidentified as a previously heard voice. Papcun and colleagues propose that easy-to-remember voices are less stable in memory because the voice-specific traits that make a voice easy-to-remember fade as a function of time, as the voice coalesces toward the prototype, resulting in more false alarms in the longer test delays. The authors attribute this to "a psychological analog to statistical regression to the mean" and suggest that hard-to-remember (prototypical) voices are more stable in memory than easy-to-remember (atypical) ones (Papcun et al. 1989, p. 923). In a follow-up study, Kreiman and Papcun (1991) examined the discrimination and recognition accuracy of voices from Papcun et al. (1989). Overall, results were similar to the previous experiment: voices that were rated easier to remember were less likely to be confused with other voices while hard-to-remember voices were easily confused. Of special interest in this study is that the accuracy results were compared with various acoustic and subjective quality predictors (made by a separate group of listeners) that were assessed via a multidimensional-scaling solution. The authors interpret the most predictive dimensions for the discrimination results to be roughly equivalent to "masculinity," "creakiness," "variability," and "mood" while the recognition results were best predicted by what was interpreted as dimensions relating to"masculinity," "breathiness," and "liveliness." These descriptors and their relationship to voice discrimination and recognition are applicable only to the set of 10 voices used in Kreiman and Papcun's studies, but the applicability of these dimensions illustrates the features in which listeners cognitively organize this set of voices.

Voice typicality was the explicit subject evaluation under consideration in Mullennix et al. (2011). Mullennix and colleagues asked listeners to evaluate 40 voices for typicality, using these judgments to prune the larger set for a memory task. The voices with the highest (4 male, 4 female) or lowest (4 male, 4 female) typicality ratings were selected. An independent group of listeners were exposed to the 16 subset voices in a vowel identification task, and were then given a surprise memory task. Overall, listeners were more accurate with the voices they had previously trained on, but showed a bias to make recognition errors when typical voices were used as foils, especially listeners exposed to typical voices. A recurring theme across these studies is that unique or distinctive voices are more easily remembered. What listeners rate when evaluating voices in terms of distinctiveness or typicality is not clear, but it appears to be a measureable quality that listeners exhibit agreement on. Typicality and distinctiveness may be connected to speech clarity and the predictability of phonetic variation. Voices vary in how clearly they produce linguistic contrasts, and this variation in contrast clarity has implications for how listeners process and recognize the speech stream (Bradlow, Torretta, & Pisoni, 1996; Newman, Clouse, & Burnham, 2001). How an individual manifests a phonetic contrast is a talker-specific feature that listeners track and exploit in subsequent processing, spilling over into perceptual events beyond the moment of comprehension (Theodore, Myers, & Lomibao, 2015). Too much phonetic variation can affect listeners' confidence in their categorization of speech sounds (Clayards, Tanenhaus, Aslin, & Jacobs, 2008). Unexpected or unfamiliar phonetic variation associated with accents or dialects that are different from one's own makes comprehension and recognition more challenging (Clopper & Pisoni, 2004; Bradlow & Bent, 2008), and this is often attributed to lack of exposure and experience. While this may be intuitive when thinking about nonnative speakers, the evidence is mixed as to whether nonnative speakers are more variable in their acoustic–phonetic realizations than native speakers (Vaughn et al., 2020; Wade, Jongman, & Sereno, 2007). Talker variability occurs within an accent or speech community as well (Strand, 1999; Bradlow et al., 1996; Babel & McGuire, 2015), resulting in intelligibility and memory benefits for familiar speakers (Nygaard & Pisoni, 1998). Accents that may be less familiar, but are the standard variety, often, however, show similar processing benefits to familiar varieties (Clopper, 2014; Clopper, Tamati, & Pierrehumbert, 2016), suggesting that the cognitive organization of voices is not exclusively tailored to the quantity of experience, but may involve some preferential encoding of socially prestigious exemplars (Babel, 2012; Babel, McGuire, & King, 2014a; Sumner, Kim, King, & McGowan, 2014).

How does the social evaluation of voices affect processing or the cognitive organization of voices? As is clear from the topic of this book, there is extensive evidence that listeners assess voices in terms of their attractiveness. The patterns by which voices are deemed attractive seem to be a combination of culturally acquired (Babel, McGuire, Walters, & Nicholls, 2014b; Bezooijen, 1995) and more strongly evolutionarily encoded (Zuckerman & Miyake, 1993; Puts, Gaulin, & Verdolini, 2006; Riding et al., 2006; Saxton et al., 2006; Feinberg, DeBruine, Jones, & Perrett, 2008; Apicella, Feinberg, & Marlowe, 2007) preferences that tap into acoustic–phonetic dimensions that are related to sexually dimorphic traits. Many of the culturally acquired components appear to stem from what is typical or standard within a speech community. While there may be initial appeal in thinking of typicality or standardness in terms of the pattern that is the most common or at the peak of a community's acoustic–phonetic distribution, linguistic standardness is much more of an imposed concept. Listeners tend to show stronger recognition patterns for pronunciation variants that are standard, despite a different pronunciation variant being far more frequent in the input (Sumner & Samuel, 2005) and listeners exhibit more false memories for a less socially prestigious accent compared to a more prestigious accent, despite equivalence in experience with the two (Sumner & Kataoka, 2013). Media is one means through which standardness and socially conditioned social preferences appear to be formed for speech communities (Kinzler & DeJesus, 2013; Lippi-Green, 2012). Overall, this body of literature makes clear that not all voices are treated equivalently in terms of processing and that both exposure and social preference play a role in voice evaluation.

To better understand the dimensions on which listeners may organize voices and how this organization may affect voice recall, we first report on a set of experiments and analyses intended to quantify the typicality of a set of voices from 60 American English speakers. These experiments provide two response time-based measures—Intelligibility and Categorization Fluency—designed to better reflect exposure by tapping into online frequency effects. Previous research has shown that response latency to voices is a proxy for familiarity; words are more likely to be recognized quickly if heard in a familiar voice rather than an unfamiliar voice (Goldinger, 1996). For the intelligibility task, listeners were asked to shadow voices embedded in noise and in the Categorization Fluency task, listeners identified voices as male or female in a speeded fashion. In both cases, faster responses indicate easier processing for a given voice. Additionally, we provide two subjective assessments, perceived Attractiveness and perceived Stereotypicality. For both of these assessments listeners were asked to subjectively rate the voices on either their attractiveness or typicality. We expect these measures to better tap into social preference. Because previous studies demonstrate that more similar voices are less likely to be remembered and are more likely to be considered a previously heard voice, we also include two measures of similarity, one based on auditory–acoustic measures, Acoustic Similarity, and one based on comparative listener ratings, Perceptual Similarity. After reporting the methods and results of each of these experiments, we examine to what extent these measures tap into similar dimensions in Sect. 6.2.7. Following this, Sect. 6.3 reports on a voice recall experiment, which we analyze with the voice evaluation metrics to assess which voice metrics best predict voice recall performance.

6.2 Voice Evaluation Experiments

6.2.1 Materials for All Experiments

The voice stimuli used in all the experiments reported here were from participants recruited as part of a previous study (Babel, 2012). They consist of 30 female (mean age 24, range 18–57) and 30 male (mean age 24, range 18–47) native speakers of American English reading 50 low-frequency monosyllabic words. For the present study a subset of 15 words which contain /i ɑ u/ as the syllable nucleus were selected for each voice, 5 words per vowel (Table 6.1).

6.2.2 Intelligibility

To quantify the intelligibility of the voices, we used a speeded shadowing task where the response time to the onset of vocalization is taken as a proxy for how easy it was for listeners to understand the utterance.

Table 6.1 Words used in the experiments organized by the vowel category for each item

/i/	/ɑ/	/u/
deed	cot	boot
key	pod	dune
peel	sock	hoop
teal	sod	toot
weave	tot	zoo

6.2.2.1 Participants

Thirty participants (15 male, 15 female) were recruited from the University of California, Santa Cruz, undergraduate population and were compensated with course credit. All were native speakers of American English from the state of California. Ages ranged from 18 to 23, mean 20.4 years.

6.2.2.2 Materials

The same voices and words used in the gender categorization fluency task were used in this task. Each individual sound file was embedded in pink noise at +6 dB signal to noise ratio (SNR). The noise began at the onset of each word and ended at the offset of each word.

6.2.2.3 Procedure

Participants were seated in a sound-attenuated booth at a computer workstation wearing AKG HSC271 model headset with integrated condenser microphone. The stimuli were presented in a randomized order at a comfortable listening volume (approximately 70 dB). Subjects were asked to repeat each word, initiating their repetition as quickly as possible without compromising accuracy. Response times were measured from the onset of the stimulus to the onset of the subject's production as registered by a microphone connected to a PST serial response box. The response time for each trial was displayed on the computer monitor to participants to help motivate fast response times. This feedback screen was displayed for 1000 ms, after which a new trial began. Each word production was recorded as a unique .wav file.

6.2.2.4 Results

Response time was automatically calculated for each production, and the accuracy of each shadowed production was determined by manual coding. A custom-written program brought up each individual sound file and provided an orthographic transcription of the intended word. Each production was categorized as correct or incorrect. Productions with disfluencies, missing phones, or the wrong lexical item were considered incorrect.

Accuracy of the repeated item is a measure of recognition. Female ($M = 81\%$ correct, $SD = 39$) voices achieved higher recognition rates than the male ($M = 76\%$ correct, $SD = 42$) voices [$t\ (51.67) = 2.47$, $p = 0.02$], indicating that female voices were overall more intelligible than the male voices. Correct responses for reaction times within two standard deviations of the group mean were then aggregated across words for each voice. Using response time to correctly identified items as a proxy for intelligibility, we found no significant differences between male and female voices

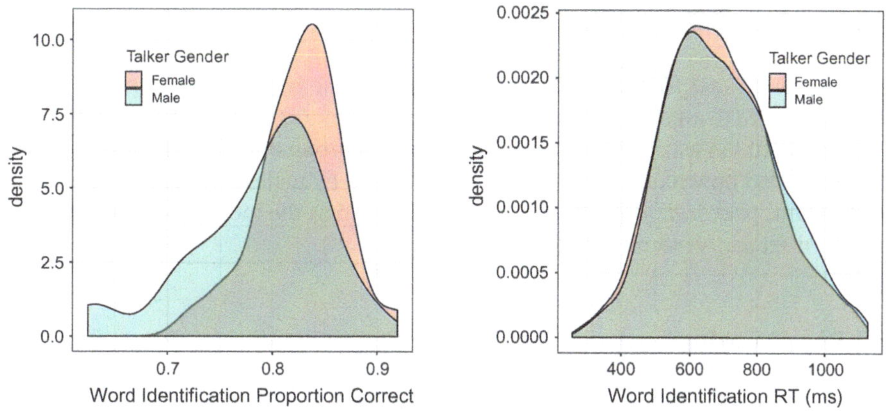

Fig. 6.1 Density plots showing the distribution of accuracy of correctly identifying each item (left panel) in a speeded shadowing task and the distribution of voice intelligibility, as measured by response lag (right panel) in a speeded shadowing task

[t (56.04) = 1.68, p = 0.098]. When items were accurately recognized, there was no difference in the intelligibility of those items for female and male voices. These aggregate measures mask the talker-specific variability of these measures. Figure 6.1 provides density plots to illustrate the range of recognition scores (left panel) and intelligibility (right panel).

6.2.3 Gender Categorization Fluency

In order to have an online estimate of typicality, the voices were assessed using a gender categorization fluency task. This is a speeded classification task where subjects heard a single word and quickly decided the gender of the voice. Previous work has used this for evaluation of typicality for faces (Orena, Theodore, & Polka, 2015) and voices (Strand, 1999).[1]

6.2.3.1 Participants

Thirty participants (15 male, 15 female) were recruited from the University of California, Santa Cruz, undergraduate population and were compensated with course credit. All were native speakers of American English from the state of California. Ages ranged from 18 to 24 years, with a mean of 21.

[1] The data from this experiment were originally reported in Babel and McGuire (2015).

6.2.3.2 Materials

In order for the task to be feasible for the participants to complete in 45 min, the word list was pruned to nine words for each talker (9 words × 60 voices = 540 stimuli). The original word list was presented to an independent group of university students (n = 23) who rated how likely each word was to be used by males or females. The words *teal*, *weave*, *pod*, *sod*, *toot*, and *dune* were identified as the most gender-valenced of the word set and were removed from the list.

6.2.3.3 Procedure

Listeners were presented with the individual words, one per trial. Words and voices were randomized across all voices, and participants were prompted to respond to each word by selecting whether the voice that said the word was male or female. Reaction time feedback was given after each trial and listeners were asked to respond in less than 500 ms. Each trial timed out after 1500 ms if no response was given.

6.2.3.4 Results

Response times for correct responses (98% of the data) made within two standard deviations of the mean were then aggregated across words for each voice. The speed at which listeners identified male ($M = 523$ ms, $SD = 17.5$) and female ($M = 525$ ms, $SD = 14$) voices differed was nonsignificant [$t(55.93) = 0.56$, $p = 0.58$].

6.2.4 *Acoustic Similarity*

To assess the voices in terms of their raw acoustic–auditory similarity, we calculated voice similarity using mel-frequency cepstral coefficients (MFCCs). While MFCCs have no straightforward perceptual interpretations, they provide a global and unbiased acoustic assessment of the speech signal. This type of unbiased acoustic measurement is useful when trying to determine the extent to which listeners' organization of sound patterns are faithful to acoustic–auditory parameters or whether they are influenced by listeners' experiences (Cristiá, Mielke, Daland, & Peperkamp, 2013; Mielke, 2012). The choice to use MFCCs, as opposed to resonant frequencies or other spectral properties more readily connected to listeners' perception of phoneme categories, allows us to side-step any explicit decision about which aspects of the speech spectrum to explicitly measure.

6.2.4.1 Materials

The set of 15 words produced by the 60 talkers was used in this analysis.

6.2.4.2 Procedure

The MFCC acoustic similarity algorithm implemented in Phonological CorpusTools (PCT; Hall, Allen, Fry, Mackie, & McAuliffe, 2018) was used to quantify acoustic vocal distinctiveness within the voice set. In this analysis, twenty-six mel-scaled triangular filters are applied to a windowed signal, and the resulting spectrum is the log of the power of each filter. The mel-frequency cepstrum is calculated using a discrete cosine transform, resulting in twelve coefficients. MFCCs are then compared using a dynamic time warping algorithm, which ultimately returns the summed distances of the best path through the data matrix. This comparison was done between matched words and each voice in the data set. While dynamic time warping may eliminate durational differences among tokens, and thus one cue to gender, it is a reliable way to directly compare the tokens. We chose this method over correlation-based approaches to quantifying spectral similarity because of precedent in the speech literature (Mielke, 2012) and the challenges of correlating signals of different lengths.

6.2.4.3 Results

To compare the acoustic vocal distinctiveness in the voice set, the similarity values for each voice comparison were averaged and used to create a distance matrix. Distance matrices were created separately for male and female voices as a combined analysis resulted in a first dimension that simply separated male and female voices. For both female and male voice sets, a scree plot of stress suggested an elbow at the fourth dimension, therefore a four-dimensional multidimensional-scaling solution was fit to each matrix using isoMDS() from the MASS package in R (Venables & Ripley, Venables and Ripley (2002)). For the female set, the stress of the four-dimensional solution was 8.28, while the stress of the four-dimensional solution was 6.78 for the male set.[2] The visualization of the first two dimensions for both the female and male voices sets are presented in the left panel of Fig. 6.2. We have made no attempt to identify the dimensions.

To use the similarity scores alongside the other voice evaluation metrics, we created a distance score for each voice. Given that talker gender was a robust dimension on which the voices were separated in the MDS space, the voice distance score was calculated separately for female and male voices. Following methods of calculating vowel dispersion (e.g., Ménard et al. 2013), acoustic voice similarity was calculated using the four dimensions of the MDS solution for each gender by taking the

[2]Note that these stress values are not indicative of a particularly strong fit, indicating that more dimensions might ultimately provide a better characterization of the data.

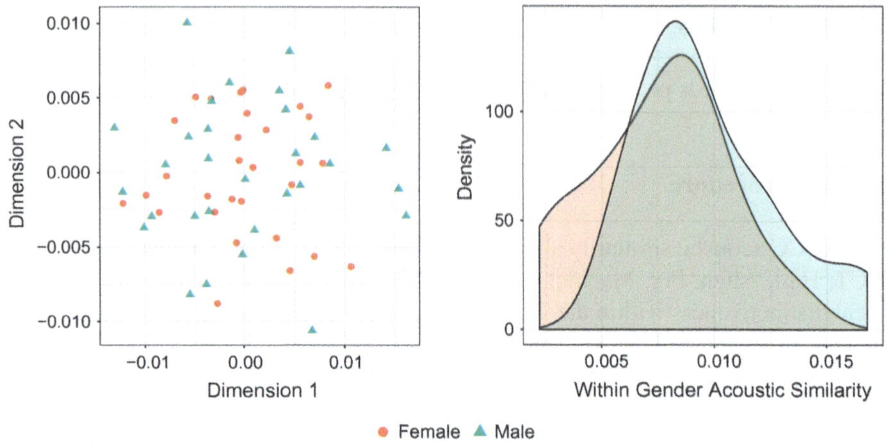

Fig. 6.2 The first two dimensions of the four-dimensional scaling solutions for the MFCC acoustic similarity of the 60 voices (left panel) and a density plot showing the distribution of within-gender acoustic variability for the 60 voices (right panel). Higher values along the x-axis in the density plot indicate more acoustically dissimilar voices. Female data are in red, and male in cyan

Euclidean distance of a voice from the average four-dimensional values for all other voices of that voice's gender. The distribution of these values was relatively normal, and is shown in the right panel in Fig. 6.2.

6.2.5 *Perceptual Similarity*

Even when measures of acoustic similarity use a transformation that models the human auditory system (like the mel-scale used in Sect. 6.2.4), such analyses may not adequately weigh or represent the cues that perceivers rely on when assessing voices. To address this, we conducted a similarity rating experiment using the voice corpus.

6.2.5.1 Participants

A research assistant who was a female native speaker of West Coast English (age = 19) completed this task with all 60 voices.[3]

[3]While having just a single listener does affect the potential generalizability of our conclusions, we ultimately feel this single data point is better than no data point.

6.2.5.2 Materials

The 15 words spoken by the 60 voices were used as stimuli in this task.

6.2.5.3 Procedure

On a given trial, a random selection of nine words (three from each vowel group) from a voice were presented in randomized order with 500 ms interstimuli interval, followed by 1000 ms break, then a second voice comprising the same nine words. After the presentation of the second voice, the participant rated the similarity of the voices on a scale from 1 (very dissimilar) to 9 (very similar) using a computer keyboard. All possible nonidentical pairs were presented in both orders resulting in 3480 trials (60 voices, 60^2 pairs $= 3600$, minus $60 \times 2 = 120$ identical pairs). Given the tedious and repetitive nature of this task, it was conducted at the participant's convenience over the course of several months.

6.2.5.4 Results

The ratings matrix was simplified in a similar way to the acoustic similarity data. Again, a combined analysis demonstrated that the first dimension was based on voice gender, so separate within-gender analyses were fit. A scree plot of stress suggested an elbow at four dimensions for both analyses and thus a four-dimensional nonmetric multidimensional-scaling solution was fit to each matrix using isoMDS() from the MASS package in R (Venables and Ripley, 2002). The stress of the four-dimensional solution was 8.36 for the female set and 7.48 for the male set of voices.[4] The visualization of the first two dimensions for both the female and male voices sets are presented in the left panel of Fig. 6.2.

For comparison with the other measures, perceptual voice similarity was calculated in an identical way to the similarity data. That is, separate distance scores were created for male and female voices by using the four dimensions from the MDS solutions and finding the mean Euclidean distance for each voice by gender. The distribution of these values is shown in right panel of Fig. 6.3.

6.2.6 *Subjective Voice Ratings*

To examine how listeners' subjective impressions of a voice's attractiveness and stereotypicality affect voice memory alongside the more objective measures described above, we collected the metrics described below.[5]

[4] Again, these high stress values suggest that more dimensions could provide a better fit to the data.

[5] These subjective voice ratings were previously reported in Babel and McGuire (2015).

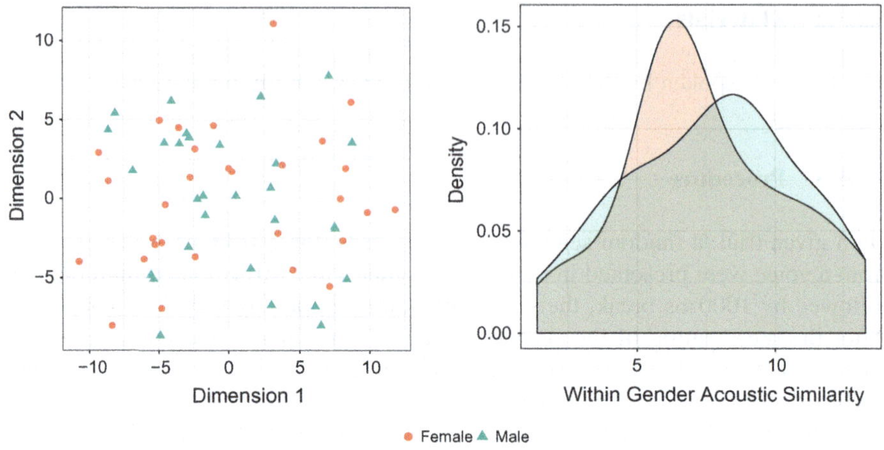

Fig. 6.3 The first two dimensions of the three-dimension multidimensional-scaling solution for the perceptual similarity of the 60 voices (left panel) and a density plot showing the distribution of within-gender perceptual variability for the 60 voices (right panel), where higher values along the x-axis indicate more perceptually dissimilar voices. Female data are in red, and male in cyan

6.2.6.1 Participants

Sixty participants were recruited for explicit rating tasks from the student population of the University of California, Santa Cruz and received course credit or $10 for their participation. Participants were divided into two groups of thirty (15 male, 15 female, each) and assigned to either the Stereotypicality rating group or the Attractiveness rating group.

6.2.6.2 Materials

The full set of 15 words for the 60 talkers were used in the tasks that elicited ratings of stereotypicality and attractiveness.

6.2.6.3 Procedure

For both experiments, subjects heard each voice say each of the 15 words followed by a pause where they were prompted to rate the voice using a 1–9 scale where 1 was “Very Unattractive” or “Very Atypical” and 9 was “Very Attractive” or “Very Typical.” All voices and words were presented in a randomized order. “Attractiveness” was not defined for the participant; they were free to evaluate the voice for sexual attractiveness or pleasantness.

Table 6.2 Mean and standard deviations of the Attractiveness and Stereotypicality Ratings for the male and female voices are shown in the leftmost columns. The Kendall's W values for the ratings are in the rightmost columns

	Female voices	Male voices	Female voices (W)	Male voices (W)
Attractiveness				
Female raters	5.05	4.67	0.274***	0.274***
Male raters	5.07	4.05	0.476***	0.185***
Stereotypicality				
Female raters	6.8	6.62	0.311***	0.255***
Male raters	6.54	6.52	0.325***	0.261***

Values marked with *** indicate p-values <0.001

6.2.6.4 Results

Female voices were overall rated as more Attractive and Stereotypical than male voices. Listeners' ratings were assessed for reliability using Kendall's W, and listeners showed a range of agreement levels. These values are given in Table 6.2.

6.2.7 Global Voice Assessment

While the six voice evaluation metrics are based on unique perception tasks posed to unique groups of listeners or, in the case of the acoustic similarity metric, an independent acoustic–auditory measurement, the metrics may indeed tap into common means of cognitively organizing voices. To assess this, we conducted a principal components analysis (PCA) on a centered and scaled data matrix using the averaged values for each talker's voice using a singular value decomposition strategy.[6] The loadings of the PCA are shown in Table 6.3 and the model summary is presented in Table 6.4. The first principal component accounts for only about 32% of the variance in the data, and the loadings of this component illustrate the positive relationship between perceived attractiveness and stereotypicality along with the negative relationship of these two dimensions with categorization fluency (Babel & McGuire, 2015). The second principal component appears to show a negative relationship between acoustic similarity and intelligibility of the voices. The third component seems to be driven by perceptual similarity.

Somewhat surprisingly, it takes until the fifth principal component for the 95% of the variance to be accounted for. This suggests that not much is achieved through this process of dimensionality reduction and these dimensions, while not completely independent, are not wholly interconnected.

[6]This was done using the prcomp() command in base R.

Table 6.3 Rotation of the six voice evaluation metrics and the principal component loadings

	PC1	PC2	PC3	PC4	PC5	PC6
Attractiveness	0.5774	−0.3100	0.2280	−0.1354	0.2472	0.6625
Stereotypicality	0.6074	−0.03364	0.1849	−0.3875	−0.0630	−0.6645
Categorization fluency	−0.4454	−0.3349	0.4282	−0.2232	0.6449	−0.2021
Intelligibility	−0.2520	−0.6179	−0.1913	−0.4972	−0.5133	0.0863
Perceptual similarity	0.1189	−0.1101	−0.8328	−0.1379	0.5039	−0.0848
Acoustic similarity	0.1466	0.6298	0.0165	−0.7170	0.0456	0.253

Table 6.4 Summary of the PCA on the six voice evaluation metrics

	PC1	PC2	PC3	PC4	PC5	PC6
Standard deviation	1.3931	1.1541	1.0860	0.8931	0.6911	0.5222
Proportion of variance	0.3234	0.2220	0.1966	0.1329	0.0796	0.04546
Cumulative proportion	0.3234	0.5454	0.7420	0.8750	0.9545	1.000

Given this, these metrics will be used below to predict performance in the voice memory task.

6.3 Voice Memory Experiment

The previous sections summarized data evaluating voices using several subjective measures (Stereotypicality, Attractiveness), online processing measures (Categorization Fluency, Intelligibility), and similarity (Acoustic–Auditory, Perceptual). In this section, we turn to the original goal of the paper and use these measures to predict listeners' ability to recall individual voices. Following previous literature, we expect that less typical voices will be easier to recall than more typical voices, and the following experiment will elucidate which of our measures are best at predicting this.

6.3.1 Methods

6.3.1.1 Participants

There were 42 listeners in four counterbalanced groups. All were native speakers of American English and had lived in California since toddlerhood. They were recruited at the University of California, Santa Cruz, and received partial course credit for their participation.

6.3.1.2 Procedure

The voices were divided into two lists of 30 and two word sets for the purposes of balancing. The two voice lists were designed to have an equal number of male and female voices in each and to be roughly equivalent in stereotypicality. The words were randomly assigned to two lists with the constraint that each list had two words for each vowel. In the exposure phase, listeners were presented with one list of 30 voices each saying six words and asked to type each word as accurately as possible. This was similar to Mullennix et al. (2011) in that the exposure phase was a linguistic task rather than a talker-focused one. After a brief self-paced break listeners were given a surprise memory task where they were again presented with voices. This procedure was identical to the exposure phase except that (1) the full set of 60 voices was used and (2) rather than type in the words spoken, subjects were asked to identify each voice as either Old (i.e., previously heard) or New (i.e., not previously heard), logging their response on labeled buttons on a serial response box. Participants were run in groups of up to three at a time in a sound-attenuated booth.

6.3.2 Results

6.3.2.1 Listener-Focused Analysis

To model listeners' decisions regarding the voices, a mixed-effects logistic regression model was used to analyze the probability that listeners could correctly identify the voices as New or Old. Given that the dimensionality reduction of the PCA was not particularly effective (e.g., it took five principal components to account for 95% of the variance when six variables were entered into the model), we also assessed the collinearity of the six voice evaluation metrics via condition number and a variance in inflation (VIF) calculation prior to including these metrics in the model. The condition number analysis, following Baayen (2008), gave a kappa statistic of 22, and the highest VIF value was 2.5. These are both generally considered moderate in terms of collinearity. Given this and the results of the PCA, we opted to include the six metrics in the model. To assist in the interpretability of the model output, however, the six metrics were entered into the model as fixed effects with interactions with New/Old, but not as interactions with each other. New/Old was entered into the model as a fixed effect with dummy coding; New was the reference level. There were random slopes for listeners, along with the random intercepts for New/Old and

Table 6.5 Model output for the listener-focused voice memory analysis

	Estimate	Standard error	z-value	p-value
Intercept	−0.1841	0.1678	−1.097	0.2726
New/Old	0.72821	0.29796	2.444	0.01453*
Attractiveness	−0.1719	0.0999	−1.721	0.08517
Stereotypicality	−0.5680	0.09735	−5.835	< 0.001***
Categorization fluency	−0.0886	0.0781	−1.135	0.2563
Intelligibility	0.02133	0.07426	−0.287	0.7739
Perceptual similarity	0.0379	0.07043	0.539	0.5901
Acoustic similarity	−0.2146	0.07218	−2.97	0.0029**
New/Old:Attractiveness	0.1562	0.12655	1.234	0.2170
New/Old:Stereotypicality	0.7499	0.1279	5.863	< 0.001***
New/Old:Categorization fluency	−0.0071	0.10647	−0.067	0.9468
New/Old:Intelligibility	0.19416	0.0991	1.959	0.0501
New/Old:Perceptual similarity	−0.11858	0.0966	−1.226	0.2201
New/Old:Acoustic similarity	0.24846	0.0980	2.535	0.0603

P-values marked with * indicate values < 0.05, ** indicates values < 0.01, and *** indicates values < 0.001

the voice evaluation metrics. All of the voice evaluation metrics were centered and scaled prior to the regression analysis.[7]

The results of this analysis are summarized in Table 6.5. The lack of a significant intercept indicates that listeners were not very accurate at identifying previously unheard or novel voices as New. The effect of New/Old illustrates that listeners were more accurate at correctly recalling old voices as old than new voices as new. In terms of the voice metrics, Stereotypicality was a significant predictor, and it also surfaced in a significant interaction with New/Old. New voices that had been independently rated as less stereotypical were more accurately identified as new than more stereotypical new voices, and old voices which were more stereotypical were more accurately identified as old than older voices that were less stereotypical. This relationship is shown in the left panel of Fig. 6.4. Acoustic Similarity was also a significant predictor. Listeners were less accurate on new voices that were further from the Euclidean mean of the voice set. That is, listeners were more accurate with voices that were more acoustically typical, somewhat in contradiction with the Stereotypicality results. This relationship is shown in the right panel of Fig. 6.4.

[7]The following code was used: glmer(Accuracy New/Old + Attractiveness + Stereotypicality + Categorization Fluency + Intelligibility + Perceptual Similarity + Acoustic Similarity + New/Old:Attractiveness + New/Old:Stereotypicality + New/Old:Categorization Fluency + New/Old:Intelligibility + New/Old:Perceptual Similarity + New/Old:Acoustic Similarity + (1 + New/Old + Attractiveness + Stereotypicality + Categorization Fluency + Intelligibility + Perceptual Similarity + Acoustic Similarity |Listener)).

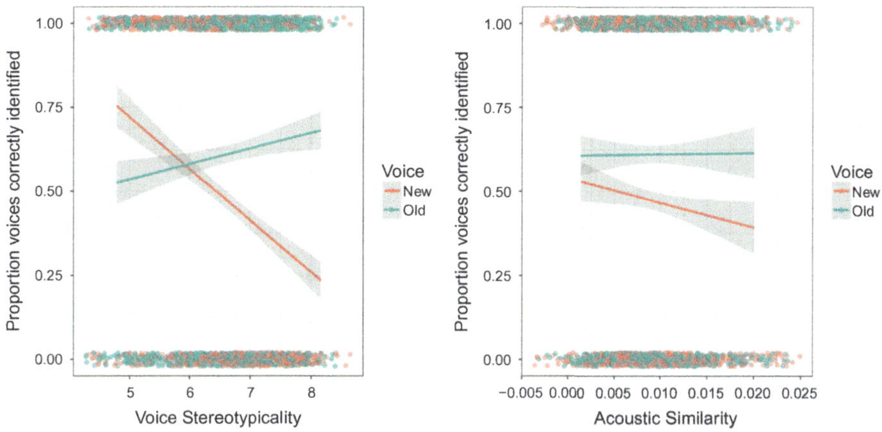

Fig. 6.4 The relationship between voice recall accuracy for Old and New voices and Stereotypicality (left panel) and Acoustic Similarity (right panel). The jittered points represent listener responses

6.3.2.2 Talker-Focused Analysis

To model voice memory with a focus on the talkers' voices, the signal detection theory measures of *d'* (sensitivity) and *c* (bias) were calculated across listeners for each voice (Macmillan & Creelman, 2004). For this analysis the data were averaged across listeners for each voice and correct responses to Old voices were assigned as hits and incorrect Old responses to New voices were assigned as false alarms. This calculation results in positive values of *d'* indicating that listeners correctly identified voices as Old or New, while negative values indicate listeners had more false alarms than hits and, thus, incorrectly classified the voices. The assignment of correct Old responses as hits also means that negative values of *c*, indicate a bias to respond Old and a positive number indicates a bias to respond New. These *d'* and *c* values were used as the dependent measures in simple linear regression models where each voice evaluation measure was entered as an independent variable along with talker gender. Because of the small number of observations one is left with in this style of analysis (n = 60, one data point per talker), we chose to run separate regression models for each voice evaluation metric.

Model results for the *d'* analysis are summarized in Table 6.6. They indicate voices which were lower in attractiveness and stereotypicality had higher *d'* values, indicating listeners were more sensitive to the New/Old decision for voices that were previously rated as less attractive or less stereotypical. The R^2 values indicate that this pattern was more robust along the Stereotypicality than the Attractiveness dimension. Figure 6.5 illustrates these patterns.

The *c* results complement these findings and are summarized in Table 6.7. There was a bias to respond Old to voices that had been rated as Attractive and Stereotypical. Again, there was a larger effect size for the Stereotypicality voice evaluation metric, compared to Attractiveness. These results are visualized in Fig. 6.6.

Table 6.6 Model summaries for the *d'* sensitivity talker-focused voice memory analysis. The Adjusted R^2 for each model's fit is reported in the final column

	Estimate	Standard error	z-value	p-value	Adjusted R^2
Intercept	1.22	0.32	3.83	<0.001***	
Attractiveness	−0.20	0.07	−3.11	0.003**	0.13
Intercept	2.41	0.48	5.06	<0.001 ***	
Stereotypicality	−0.33	0.07	−4.57	<0.001***	0.25
Intercept	−1.37	2.31	−0.59	0.56	
Categorization fluency	0.003	0.004	0.70	0.49	−0.009
Intercept	−1.98	1.30	−1.52	0.13	
Intelligibility	0.003	0.002	1.71	0.09	0.03
Intercept	−0.049	0.08	−1.01	0.08	
Perceptual similarity	2.508	0.10	0.89	0.32	−0.07
Intercept	0.48	0.17	2.87	0.0057**	
Acoustic similarity	−25.68	16.77	−1.531	0.13	0.022

P-values marked with ** indicate values <0.01 and *** indicates values <0.001

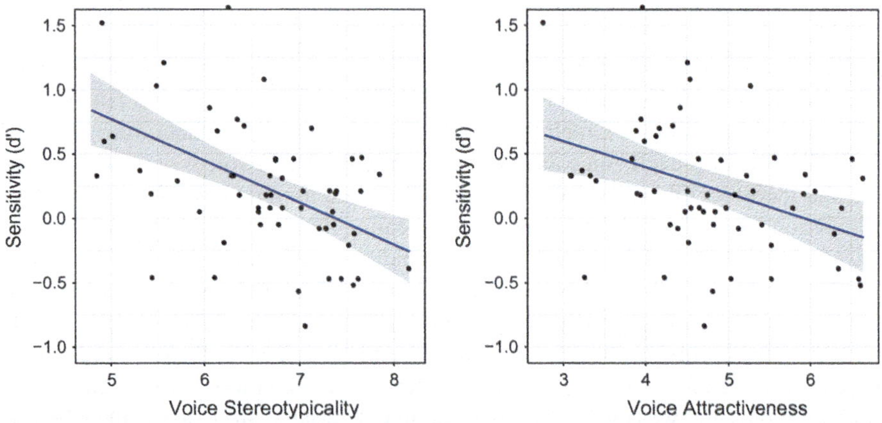

Fig. 6.5 Sensitivity by Stereotypicality (left panel) and Attractiveness (right panel) in the Voice Recall task. Each point represents a talker in the experiment

Together, these results indicate that listeners were more accurate in the voice memory task with voices that were less Attractive and Stereotypical, and there was a strong bias for listeners to respond Old to voices that were more Attractive and Stereotypical.

Table 6.7 Model summaries for the *c* bias talker-focused voice memory analysis. The Adjusted R^2 for the model fit is reported in the final column

	Estimate	Standard error	z-value	p-value	Adjusted R^2
Intercept	0.77	0.24	3.21	0.002 * *	
Attractiveness	−0.18	0.05	−3.68	<0.001***	0.18
Intercept	1.89	0.34	5.54	<0.001***	
Stereotypicality	−0.30	0.05	−5.86	<0.001***	0.36
Intercept	−2.23	1.77	−1.26	0.21	
Categorization fluency	0.004	0.003	1.21	0.23	0.008
Intercept	0.02	1.03	0.02	0.99	
Intelligibility	−0.0002	0.002	−0.11	0.92	−0.02
Intercept	−0.092	0.412	−0.091	0.76	
Perceptual similarity	0.022	0.07	0.71	0.42	−0.01
Intercept	0.0095	0.133	0.071	0.943	
Acoustic similarity	−11.197	13.15	−0.85	0.39	−0.004

P-values marked with ** indicate values <0.01 and *** indicates values <0.001

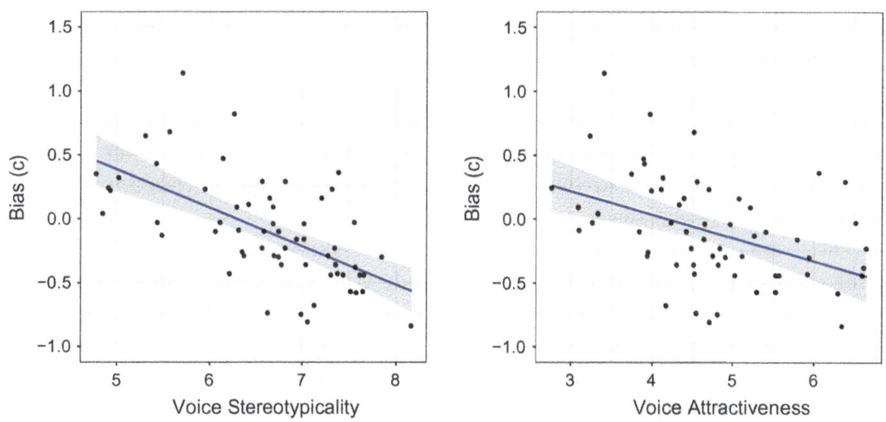

Fig. 6.6 Bias by Stereotypicality (left panel) and Attractiveness (right panel) in the Voice Recall task. Negative values indicate a bias to respond Old, while positive values indicate a bias to respond New

6.4 General Discussion

Listeners process the communicative linguistic signal of a voice while they evaluate it socially (Sumner et al., 2014). In this chapter, we used a combination of online intelligibility and processing measures, measures of acoustic–auditory and perceptual similarity, and subjective voice evaluations to predict voice memory. For decades, it has been established that voice evaluation related to distinctiveness or typicality was a strong predictor of listeners' ability to recall voices (Papcun et al.,

1989; Kreiman & Papcun, 1991; Mullennix et al., 2011). In line with these earlier claims, we find that our subjective measures of voice evaluation—perceived Stereotypicality and Attractiveness, two related dimensions for this set of voices (Babel & McGuire, 2015)—predict performance in a voice recall task, as did our measure of Acoustic–Auditory similarity. Notably, the more online measures of intelligibility and gender categorization fluency do not. Perceptual similarity also did not predict performance, but it is difficult to draw conclusions from one listener.

In this corpus of voices, we can conceive of voices that are more stereotypical and more attractive as being analogous to the voices that Papcun et al. (1989)'s listeners identified as hard-to-remember voices. In our study, these stereotypical and attractive voices are more accurately identified as old voices (i.e., voices previously heard in the experiment) when they are indeed old. Listener accuracy on stereotypical and attractive new voices that listeners were not exposed to was poor. The signal detection theoretic analyses illustrate that listeners had decreased sensitivity to more stereotypical and attractive voices and this was due to listeners having a strong bias to respond "old" to these stereotypical and asesthetically pleasing voices. Papcun, Kreiman, and colleagues (Papcun et al., 1989; Kreiman & Papcun, 1991) argue that their results support a prototype model of voice memory: voices that are typical are well-represented and thus trigger the illusion of experience. Our results complement these findings by providing insight into what voice attributes these prototypes are structured around. In the context of voice memory, it appears that more subjective voice evaluations are at the core of the prototype structure, particularly perceived stereotypicality, as opposed to more objective, online measures like intelligibility or categorization fluency or measures of voice similarity taken from the acoustic–auditory or perceptual space.

The results do raise a contradiction in that listeners were less accurate at identifying acoustically atypical voices as New while voices judged less stereotypical are more accurately identified. These two voice measures, Stereotypicality and Acoustic Similarity are not correlated for our data set [$r = -0.02$, $p = 0.25$]. Moreover, our measure of acoustic similarity is based on MFCCs, which while usefully exploited for automatic talker recognition systems, may not at all adequately capture the phonetic detail around which human listeners organize and distinguish voices. Our attempt to use an online measure of listener-derived voice similarity is stymied by the duration of the task, thus providing us with the perceptual space of a single listener. While the previous research aligns well with our results regarding stereotypicality and attractiveness, more research is necessary to understand the role of voice similarity in the acoustic and perceptual domains.

Sociocultural influences shape listeners' interpretation and social assessment of voices and accents (Hay, Jennifer, Warren, Paul, & Drager, Katie, 2006; Babel & Russell, 2015), in addition to shaping the, for example, gender-specific realization of spoken language (Johnson, 2006; Foulkes, Docherty, & Watt, 2005). Listeners' assessments of what is typical appear not to be based on veridical interpretation of the statistical distributions that listeners are exposed to, but rather are a reflection of a cognitive reorganization that is based on community standards and norms (Sumner et al., 2014; Babel & McGuire, 2015). The results of the voice memory task reported here provide a concrete example of where this has implications: attractive and especially stereotypical voices are recalled less accurately because of a bias to assume they have been previously experienced. Individuals with more typical or attractive voices may thus receive a social benefit in terms of processing advantages that familiar accents experience.

6.5 Conclusion

These results generally support previous research that less typical and more unusual voices are more easily recalled from memory (Papcun et al., 1989; Kreiman & Papcun, 1991; Mullennix et al., 2011). Using several different evaluations of voices we find that stereotypicality and, to a lesser extent, attractiveness and acoustic similarity predict listeners' ability to recall voices, such that less stereotypical voices are recalled more easily, but there is a strong bias to determine that highly stereotypical voices have been previously heard. In contrast, online response time measures do not predict voice recall.

While further research is certainly necessary, a broader conclusion that can be gleaned from this study is that voices are organized and perceived fairly abstractly, with considerable reliance on social factors. This conclusion is a natural extension of the results. If online response time measures, which are typically diagnostic of experiential information and speed of processing, do not predict voice recall, then this negative result suggests that experience plays a more minimal role, or is dwarfed by the social factors that are tapped by asking listeners about attractiveness and stereotypicality. This is perhaps unsurprising as a voice is an aggregate of experiences and words. Many, if not most, exemplar models of speech (Pierrehumbert, 2001; Johnson, 1997) propose words as a basic unit of storage. In this study, participants were asked to recall voices holistically, after hearing six words produced by a voice, not respond "old"/"new" to individual words. Thus, when participants are asked about a voice as a whole, they rely more on abstracted, subjective information.

However, as is clear from diverse work in the speech sciences (Goldinger, 1998, 1996; Nielsen, 2011; Palmeri, Goldinger, & Pisoni, 1993; Theodore & Miller, 2010; Dahan, Drucker, Sarah & Scarborough, 2008) individual instances in memory matter for speech perception. A full theory of voice organization will need to rectify such instances with more abstracted memories. Further research should elucidate this issue.

Acknowledgments We wish to thank Brianne Senior and Stephanie Chung for their contributions to this paper. Funding was, in part, provided by a Hampton Grant from the University of British Columbia and SSHRC to MB, and funding from the UCSC Humanities Institute to GM. Thanks to the audience at LabPhon14 in Tokyo for their feedback on this project.

References

Apicella, C. L., Feinberg, D. R. & Marlowe, F. W. (2007). Voice pitch predicts reproductive success in male hunter-gatherers. *Biology Letters*, *3*(6), 682–684.

Baayen, R. H. (2008). *Analyzing linguistics data: A practical introduction to statistics*. Cambridge: CUP.

Babel, M., McGuire, G., & King, J. (2014b). Towards a more nuanced view of vocal attractiveness. *PloS One*, *9*(2), e88616.

Babel, M., McGuire, G., Walters, S., & Nicholls, A. (2014a). Novelty and social preference in phonetic accommodation. *Laboratory Phonology*, *5*(1), 123–150.

Babel, M. (2012). Evidence for phonetic and social selectivity in spontaneous phonetic imitation. *Journal of Phonetics*, *40*(1), 177–189.

Babel, M., & McGuire, G. (2015). Perceptual fluency and judgments of vocal aesthetics and stereotypicality. *Cognitive Science*, *39*(4), 766–787.

Babel, M., & Russell, J. (2015). Expectations and speech intelligibility. *The Journal of the Acoustical Society of America*, *137*(5), 2823–2833.

Bradlow, A. R., Torretta, G. M., & Pisoni, D. B. (1996). Intelligibility of normal speech I: Global and fine-grained acoustic-phonetic talker characteristics. *Speech Communication*, *20*(3–4), 255–272.

Bradlow, A. R., & Bent, T. (2008). Perceptual adaptation to non-native speech. *Cognition*, *106*(2), 707–729.

Bregman, M. R., & Creel, S. C. (2014). Gradient language dominance affects talker learning. *Cognition*, *130*(1), 85–95.

Clayards, M., Tanenhaus, M. K., Aslin, R. N., &Jacobs, R. A. (2008). Perception of speech reflects optimal use of probabilistic speech cues. *Cognition*, *108*(3), 804–809.

Clopper, C. G., Tamati, T. N., & Pierrehumbert, J. B. (2016). Variation in the strength of lexical encoding across dialects. *Journal of Phonetics*, *58*, 87–103.

Clopper, C. G. (2014). Sound change in the individual: Effects of exposure on cross-dialect speech processing. *Laboratory Phonology*, *5*(1), 69–90.

Clopper, C. G., & Pisoni, D. B. (2004). Effects of talker variability on perceptual learning of dialects. *Language and Speech*, *47*(3), 207–238.

Cristià, A., Mielke, J., Daland, R., & Peperkamp, S. (2013). Similarity in the generalization of implicitly learned sound patterns. *Laboratory Phonology*, *4*(2), 259–285.

Dahan, D., Drucker, S. J., & Scarborough, R. A. (2008). Talker adaptation in speech perception: Adjusting the signal or the representations? *Cognition*, *108*(3), 710–718.

Feinberg, D. R., DeBruine, L. M., Jones, B. C., & Perrett, D. I. (2008). The role of femininity and averageness of voice pitch in aesthetic judgments of women's voices. *Perception*, *37*(4), 615–623.

Foulkes, P., Docherty, G., Watt, D. (2005). Phonological variation in child-directed speech. In *Language*, pp. 177–206.

Goggin, J. P., Thompson, C. P., Strube, G., Simental, L. R. (1991). The role of language familiarity in voice identification. *Memory & Cognition*, *19*(5), 448–458.

Goldinger, S. D. (1996). Words and voices: Episodic traces in spoken word identification and recognition memory. *Journal of Experimental Psychology: Learning, Memory, and Cognition*, *22*(5), 1166.

Goldinger, S. D. (1998). Echoes of echoes? An episodic theory of lexical access. *Psychological Review*, *105*(2), 251.

Hall, K. C., Allen, B., Fry, M., Mackie, S., McAuliffe, M. (2018). *Phonological CorpusTools*, Version 1.2. https://github.com/PhonologicalCorpusTools/CorpusTools/releases.

Hay, J., Warren, P., & Drager, K. (2006). Factors influencing speech perception in the context of a merger-in-progress. *Journal of Phonetics*, *34*(4), 458–484.

Johnson, K. (1997). Speech perception without speaker normalization: An exemplar model. In *Talker Variability in Speech Processing*, pp. 145–165.

Johnson, K. (2006). Resonance in an exemplar-based lexicon: The emergence of social identity and phonology. *Journal of Phonetics*, *34*(4), 485–499.

Kinzler, K. D., & DeJesus, J. M. (2013). Northern = smart and Southern = nice: The development of accent attitudes in the United States. *The Quarterly Journal of Experimental Psychology, 66*(6), 1146–1158.

Kreiman, J., & Papcun, G. (1991). Comparing discrimination and recognition of unfamiliar voices. *Speech Communication, 10*(3), 265–275.

Lippi-Green, R. (2012). *English with an accent: Language, ideology and discrimination in the United States*. Routledge.

Macmillan, N. A. & Creelman, C. D. (2004). *Detection theory: A user's guide*. Psychology Press.

Ménard, L., Toupin, C., Baum, S.R., Drouin, S., Aubin, J., & Tiede, M. (2013). Acoustic and articulatory analysis of French vowels produced by congenitally blind adults and sighted adults. *The Journal of the Acoustical Society of America, 134*(4), 2975–2987.

Mielke, J. (2012). A phonetically based metric of sound similarity. *Lingua, 122*(2), 145–163.

Mullennix, J. W., Ross, A., Smith, C., Kuykendall, K., Conard, J., & Barb, S. (2011). Typicality effects on memory for voice: Implications for earwitness testimony. *Applied Cognitive Psychology, 25*(1), 29–34.

Newman, R. S., Clouse, S. A. & Burnham, J. L. (2001). The perceptual consequences of within-talker variability in fricative production. *The Journal of the Acoustical Society of America, 109*(3), 1181–1196.

Nielsen, K. (2011). Specificity and abstractness of VOT imitation. *Journal of Phonetics, 39*(2), 132–142.

Nygaard, L. C., & Pisoni, D. B. (1998). Talker-specific learning in speech perception. *Perception & Psychophysics, 60*(3), 355–376.

Orchard, T. L., & Yarmey, A. D. (1995). The effects of whispers, voice-sample duration, and voice distinctiveness on criminal speaker identification. *Applied Cognitive Psychology, 9*(3), 249–260.

Orena, A. J., Theodore, R. M., & Polka, L. (2015). Language exposure facilitates talker learning prior to language comprehension, even in adults. *Cognition, 143*, 36–40.

O'Toole, A.J., Deffenbacher, K. A., Valentin, D., McKee, K., Huff, D., & Abdi, H. (1998). The perception of face gender: The role of stimulus structure in recognition and classification. *Memory & Cognition, 26*(1), 146–160.

Palmeri, T. J., Goldinger, S. D. & Pisoni, D. B. (1993). Episodic encoding of voice attributes and recognition memory for spoken words. *Journal of Experimental Psychology: Learning, Memory, and Cognition, 19*(2), 309.

Papcun, G., Kreiman, J., & Davis, A. (1989). Long-term memory for unfamiliar voices. *The Journal of the Acoustical Society of America, 85*(2), 913–925.

Perrachione, T. K., Chiao, J. Y., & Wong, P. C. (2010). Asymmetric cultural effects on perceptual expertise underlie an own-race bias for voices. *Cognition, 114*(1), 42–55.

Perrachione, T. K., Del Tufo, S. N. & Gabrieli, J. D. (2011). Human voice recognition depends on language ability. *Science, 333*(6042), 595–595.

Perrachione, T. K., Del Tufo, S. N. & Gabrieli, J. D. Human voice recognition depends on language ability'. In *Applied Psycholinguistics* (in press).

Perrachione, T. Recognizing speakers across languages. In Sascha Frühholz & Pascal Belin *The Oxford handbook of voice perception*. Oxford: Oxford University Press. (in press)

Perrachione, T. K., & Wong, P. C. M. (2007). Learning to recognize speakers of a non-native language: Implications for the functional organization of human auditory cortex. *Neuropsychologia, 45*(8), 1899–1910.

Pierrehumbert, J. B. (2001). Exemplar dynamics: Word frequency, lenition and contrast. *Typological studies in language, 45*, 137–158.

Puts, D. A., Gaulin, S. J. & Verdolini, K. (2006). Dominance and the evolution of sexual dimorphism in human voice pitch. *Evolution and Human Behavior, 27*(4), 283–296.

Riding, D., Lonsdale, D., & Brown, B. (2006). The effects of average fundamental frequency and variance of fundamental frequency on male vocal attractiveness to women. *Journal of Nonverbal Behavior, 30*(2), 55–61.

Saxton, T. K., Caryl, P. G. & Craig Roberts, S. (2006). Vocal and facial attractiveness judgments of children, adolescents and adults: The ontogeny of mate choice. *Ethology*, *112*(12), 1179–1185.

Senior, B., Hui, J., & Babel, M. (2018). Liu vs. Liu vs. Luke? Name influence on voice recall. *Applied Psycholinguistics*, **39**(6), 1117–1146.

Stevenage, S. V., Clarke, G. & McNeill, A. (2012). The "other-accent" effect in voice recognition. *Journal of Cognitive Psychology*, *24*(6), 647–653.

Strand, E. A. (1999). Uncovering the role of gender stereotypes in speech perception. *Journal of Language and Social Psychology*, *18*(1), 86–100.

Sumner, M., & Kataoka, R. (2013). Effects of phonetically-cued talker variation on semantic encoding. In *The Journal of the Acoustical Society of America134*(6), EL485–EL491.

Sumner, M., Kim, S. K., King, E., & McGowan, K. B. (2014). The socially weighted encoding of spoken words: a dual-route approach to speech perception. *Frontiers in Psychology*, *4*, 1015.

Sumner, M., & Samuel, A. G. (2005). Perception and representation of regular variation: The case of final/t. *Journal of Memory and Language*, *52*(3), 322–338.

Theodore, R. M., Myers, E. B. & Lomibao, J. A. (2015). Talker-specific influences on phonetic category structure. *The Journal of the Acoustical Society of America*, *138*(2), 1068–1078.

Theodore, R. M., & Miller, J. L. (2010). Characteristics of listener sensitivity to talker-specific phonetic detail. *The Journal of the Acoustical Society of America*, *128*(4), 2090–2099.

Thompson, C. P. (1987). A language effect in voice identification. *Applied Cognitive Psychology*, *1*(2), 121–131.

Van Bezooijen, R. (1995). Sociocultural aspects of pitch differences between Japanese and Dutch women. *Language and Speech*, *38*(3), 253–265.

Vaughn, Vaughn, C., Baese-Berk, M., & Idemaru, K. Re-examining phonetic variability in native and non-native speech. In *Phonetica*, pp. 1–32 (in press).

Venables, W. N. & Ripley, B. D. (2002). *Modern applied statistics with S*. Fourth. ISBN 0-387-95457-0. New York: Springer. http://www.stats.ox.ac.uk/pub/MASS4.

Wade, T., Jongman, A., & Sereno, J. (2007). Effects of acoustic variability in the perceptual learning of non-native-accented speech sounds. *Phonetica*, *64*(2–3), 122–144.

Winters, S. J., Levi, S. V. & Pisoni, D. B. (2008). Identification and discrimination of bilingual talkers across languages. *The Journal of the Acoustical Society of America*, *123*(6), 4524–4538.

Xie, X., & Myers, E. (2015). The impact of musical training and tone language experience on talker identification. *The Journal of the Acoustical Society of America*, *137*(1), 419–432.

Yarmey, A. D. (1991). Descriptions of distinctive and non-distinctive voices over time. *Journal of the Forensic Science Society*, *31*(4), 421–428.

Zuckerman, M., & Miyake, K. (1993). The attractive voice: What makes it so? *Journal of Nonverbal Behavior*, *17*(2), 119–135.

Chapter 7
Voice, Sexual Selection, and Reproductive Success

Alexandre Suire, Michel Raymond, and Melissa Barkat-Defradas

Abstract Beyond the linguistic content it conveys, voice is one of the fundamental aspects of human communication. It conveys an array of bio-psycho-social information about a speaker and enables the expression of a wide range of emotional and affective states so as to elicit a whole range of auditory impressions. Such aspects are of a great importance in determining the outcomes of competitive and courtship interactions as they influence the access to mating partners and thus reproduction. Sexual selection, the mechanism that promotes biological and social traits that confer a reproductive benefit, provides an interesting theoretical framework to understand the functional role of the human voice from an evolutionary perspective. This chapter aims to provide an overview of the research that lies at the crossroad of the human voice and evolutionary biology.

Keywords Sexual selection · Reproductive success · Mate choice · Contest competition · Voice · Attractiveness

7.1 Evolutionary Background

7.1.1 Sexual Selection

Sexual selection is an evolutionary process by which a specific trait, either biological or social, is selected depending on the advantages it confers to the individual that bears it in order to access sexual partners for reproduction (Darwin, 1871). Reproductive

A. Suire (✉) · M. Raymond · M. Barkat-Defradas
Institut des Sciences de l'Evolution de Montpellier, University of Montpellier,
Centre National de la Recherche Scientifique, Institut pour la Recherche et le Développement,
Ecole Pratique des Hautes Etudes – Place Eugène Bataillon, 34095 Montpellier, France
e-mail: alexandre.suire@umontpellier.fr

M. Raymond
e-mail: michel.raymond@umontpellier.fr

M. Barkat-Defradas
e-mail: melissa.barkat-defradas@umontpellier.fr

B. Weiss et al. (eds.), *Voice Attractiveness*, Prosody, Phonology and Phonetics,
https://doi.org/10.1007/978-981-15-6627-1_7

success is thus a key aspect to assess. It describes an individual's capacity to pass its genes onto the next generation in a way that its descendants can pass it too. It can be estimated, given the situations, by one or several components, such as survival, fertility, or the number of offsprings that are produced in the next generation. Sexual selection can be divided into two distinct selection processes: intra- and intersexual competition (Andersson, 1994).

On one hand, intrasexual competition refers to contest competition that occurs between same-sex individuals. When competition implies a physical confrontation, sexual selection will favor the evolution of any characteristic that strengthens the force and endurance of individuals, or any characteristic that diminishes the physical prowess of competitors. This leads to the evolution of specific "weapons" designed to repel and fight conspecifics. For instance, the antlers of male red deers are important physical attributes in duels during the mating season (Clutton-Brock, Guinness, & Albon, 1982), likewise the impressive body size of male sea lions, a key determinant of male–male fights to access harem of females (Ralls & Mesnick, 2009).

On the other hand, intersexual competition refers to the process of competition that depends on the choice made by opposite sex members, a mechanism commonly termed mate choice. This mechanism depends on sexual attractiveness (Sect. 7.2b deals with it). Evolutionary theory predicts that the sex that invests the more in reproduction (in the form of anisogamy and parental care) should have the scrutiny upon choosing a mate. This type of selection explains the origin of many extravagant characteristics, such as vivid colors, excessive plumage, and complex songs in male bird species (Bennett & Owens, 2002). Such traits are usually termed "ornaments". The most classical example is the tail of the blue peafowl, with its elongated upper tail which bears colorful eyespots.

In humans, many specific traits, such as height, the body size, and the immune system have been well studied under sexual selection theory and have provided a better understanding of their function within human mating systems (Miller, 1998; Puts, 2010). As we will see, contest competition and mate choice are two important evolutionary mechanisms that can also shed light on the evolution of the human voice.

7.1.2 Vocal Dimorphism

Humans display one of the most important vocal acoustic sexual dimorphism across anthropoids (Puts et al., 2016).

Differences in acoustic characteristics between the voices of men and women have long been recognized and studied (Titze, 1989). Men's vocal tract is about 15–20% longer than women because of their larger larynx and lower placement in the neck. Men's vocal chords are also about 50% longer and significantly more massive than those of women. These anatomical differences, which develop during puberty under the influence of the estrogen/testosterone ratio, explain the lower vocal resonant frequencies of male voices (Fitch & Giedd, 1999). Most notably, the fundamental

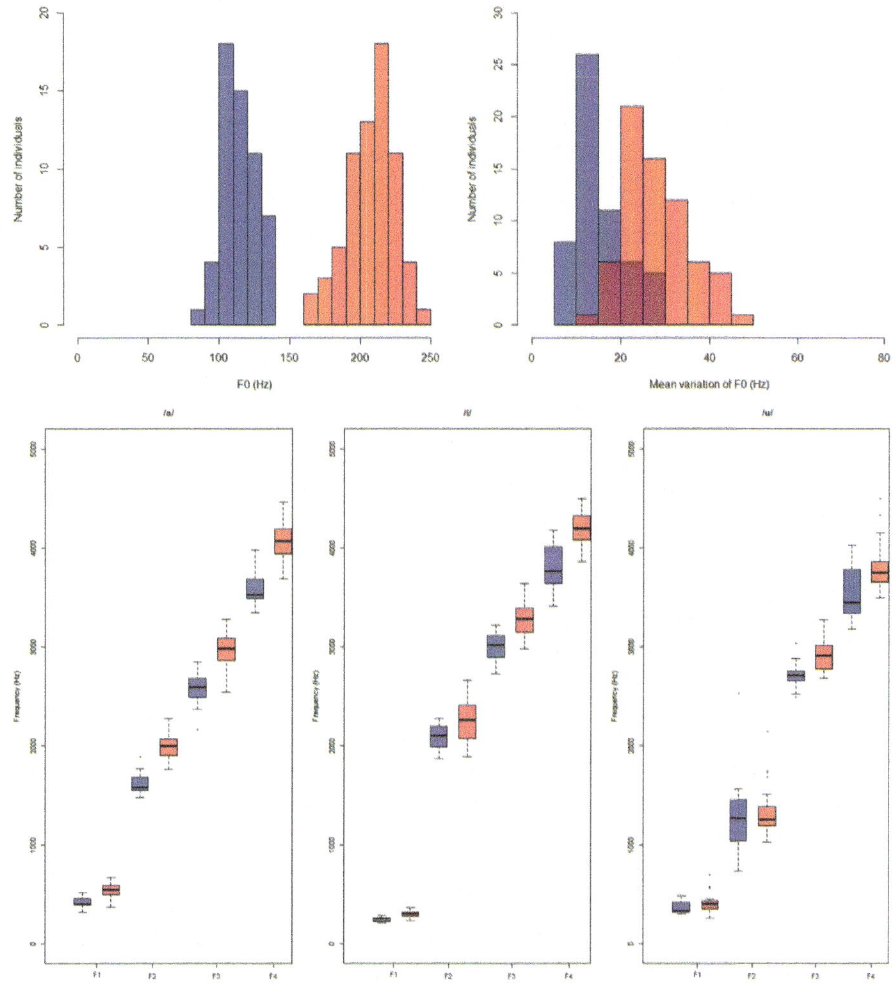

Fig. 7.1 Distribution of F0, F0-SD, and mean values of formant frequencies (F1–F4) for the vowels /a/, /i/ and /u/ for men (blue) and women (red). Purple values represent overlap between sexes. Acoustic data drawn from spontaneous speech; $n_{men} = 60$, $n_{women} = 68$ (Suire, unpublished data)

frequency shows relatively little to no overlap between the two sex: women's fundamental frequency is typically double that of men (Fig. 7.1). Additionally, men display lower formant frequencies from F1 to F4 compared to women, and such differences are consistent across different types of vowels (Simpson, 2009). Although less understood, the variation of F0 (generally noted as F0-SD) also appears to be sexually dimorphic, with men having a more monotonous voice than women (Puts et al., 2012).

Although the proximate mechanisms (i.e., physiology and anatomy) explain the observed difference between men and women, it does not tell which evolutionary factor has led to this phenomenon. When a trait shows a strong dimorphism between the two sex, it is reasonably well grounded to see sexual selection as a potential explaining factor. Although vocal attractiveness and dominance may be less relevant to human mating success in modern life than it has been during most of human evolution, the underlying logic of the following studies is that past contest competition and mate choice would have favored signals of threat potential and mate attraction (Puts, 2010).

7.2 The Functional Role of the Human Voice

7.2.1 Contest Competition and Vocal Dominance

Within same-sex competition, dominance is a key perception to assess. It can be defined as the capacity of one individual to repel competitors. Several studies have highlighted the importance of the fundamental and formant frequencies in the perception of both social and physical dominance, especially in men.

For instance, it has been regularly shown that men with a more masculine voice, i.e., lower F0 and formant frequencies, are perceived as more dominant by same-sex individuals, in both experimental settings (Feinberg, Jones, Little, Burt, & Perrett, 2005; Feinberg et al., 2006; Puts et al. 2006; Puts et al. 2007; Jones, Feinberg, DeBruine, Little, & Vukovic, 2010; Watkins et al. 2010; Wolff & Puts, Wolff & Puts 2010) and correlational studies (Aronovitch, 1976; Hodges-Simeon, Gaulin, & Puts, 2010). Moreover, in a competitive setting, men who perceived themselves as more dominant speak in a lower voice pitch and in a more monotonous manner when speaking to competitors. Conversely, men who feel non-confident or more "submissive" speak in a higher voice pitch (Puts et al., 2006). Interestingly, aggressive and dominant communicative behavior can possibly go beyond simple acoustics, by differentially producing phonetic variants relevant to the perception of masculinity (Kempe, Puts, & Cárdenas, 2013). For instance, taller and more masculine men with higher levels of circulating testosterone levels used less the alveolar stop consonant /t/, as a mean to display threat potential. Effects of side observer or context-dependent displays of aggression may be equally important to signal power and authority to an audience, as it has been reported that observers seeing a man speaking aggressively to other men are perceived as more dominant (Jones, DeBruine, Little, Watkins, & Feinberg, 2011).

Another consequence of having a deeper voice is that it can lead to higher social positions in men. For instance, it has been shown that people prefer to select a leader with a more masculine voice (Anderson & Klofstad, 2012; Klofstad, Anderson, & Peters, 2012), which can also influence voting behaviors (Tigue et al., 2012) and predict actual presidential election outcomes (Klofstad, 2016; Banai, Banai, & Bovan, 2017). Interestingly, voice pitch can be linked to leadership's positions within companies: CEOs with lower pitch voices managed larger companies, earned more money, and enjoy longer tenures (Mayew et al., 2013). More generally, voice pitch and formant frequencies seem to signal potential threat and aggression, higher social status (including social dominance), all of which may have been particularly important in past human environments (Puts, 2010).

For women, there are relatively few studies that have looked at the acoustic correlates of dominance. One study from Borkowska & Pawlowski (2011) showed that men and women perceived women with lower voice pitch as more dominant, with women being more sensitive to this vocal cue than men. Another study showed that feminine voices were perceived as more flirtatious and more attractive to men, and women were most sensitive to formant dispersion (i.e., the relative distance of two adjacent formants) than the fundamental frequency, suggesting that women may track competitors' femininity using this vocal cue (Puts et al., 2011).

The lack of studies for women's vocal dominance can be partly explained by the fact that past research has shown that competition among women, at least during human evolutionary history, relies very little on physical combat or aggression; women are assumed to be more prone to use indirect aggression. Such attempts may include social manipulation, for instance, by spreading false information about one's reputation or interfering with friendships and group inclusion of competitors (Fisher, 2015). Therefore, this kind of competition does not lead to larger, taller, and stronger statures in women, and thus women do not need to convey impressions of dominance or largeness through their vocal features against competitors.

Several authors have recently argued that intrasexual competition has mainly driven the evolution of several morphological traits in men, including voice pitch and its resonant frequencies (Puts, 2010; Hill et al., 2013; Kordsmeyer, Hunt, Puts, Ostner, & Penke, 2018), but mate choice should not be regarded as an insignificant evolutionary force in shaping vocal acoustic features (Suire et al., 2018).

7.2.2 Mate Choice and Vocal Attractiveness

Attractiveness, which can be defined as the capacity of one individual to attract opposite sex members, is an important component of voice perception in seductive and romantic settings. Other perceptions, such as the propensity to fidelity or trustworthiness, are also possibly important indexical cues to assess (Vukovic et al., 2011; O'Connor, Pisanski, Tigue, Fraccaro, & Feinberg, 2014a).

In men, consensus toward the attractiveness of relatively more masculine voices has been well established, that is, a relatively lower voice pitch (Collins, 2000; Feinberg et al., 2005, 2006, 2008; Ridings et al., 2006; Jones et al. 2010, but see Shirazi, Puts, & Escasa-Dorne, 2018). Additionally, simultaneously masculinizing pitch and formant frequencies increases men's vocal attractiveness (Feinberg et al. 2005, 2006; Puts, 2005). However, preferences for vocal monotonicity are contradictory (Ridings et al., 2006; Hodges-Simeon et al., 2010) and further studies are needed. Nonetheless, women's visual object memory seems to increase after hearing masculine male voices, but not after hearing feminine male voices or female voices, suggesting that women may be particularly attuned to masculine voices (Smith, Jones, Feinberg, & Allan, 2012). Voice pitch and formants are well-studied acoustic correlates of voice attractiveness, but multiple components of voice quality have not been studied within an evolutionary context and are known to potentially affect vocal attractiveness, such as vocal roughness and breathiness (Suire et al., 2018). In addition, as for vocal dominance, attractiveness can go beyond the acoustics' limits, as it appears that specific sociolinguistic dialects, combined with a lower voice pitch, are preferentially selected by women (O'Connor et al., 2014b).

Interestingly, women's preferences for vocal masculinity seem to shift during the ovulatory cycle. Given that hormonal profiles (i.e., levels of progesterone and estradiol) vary during the ovulatory cycle, women may prefer less masculinized voices in men during the luteal phase as opposed to preferring masculinized voices in men toward ovulation peak (Puts, 2005; Feinberg et al. 2006). This result can be interpreted by the fact that women observe a trade-off when choosing a partner: a more cooperative and submissive individual during the luteal phase, with relatively lower testosterone levels, and a strong, testosterone-filled masculine men when approaching ovulation. Choosing the former can be understood by the fact that a more cooperative men is preferred when a woman seeks a long-term partner, particularly important so as to provide shelter and resources, and choosing the latter may be important when a woman seeks a short-term partner (i.e., one-night stand) to maximize reproductive success (Buss & Schmitt, 1993). However, recent evidence has found no significant shift of women's preferences over the ovulatory cycle for both vocal and facial masculinity (Jones et al., 2018; Jünger, Kordsmeyer, Gerlach, & Penke, 2018).

Regarding men's preferences for women's voices, both experimental and correlational studies have found a consistent positive relationship between attractiveness and F0, that is, men are attracted in average to relatively higher voice pitch (Collins & Missing, 2003; Feinberg et al., 2008; Jones et al., 2010; Borkowska & Pawlowski, 2011; Puts et al., 2011, however, see Tuomi & Fisher, 1979; Hughes et al., 2010, 2014). However, this relationship might not be linear (Borkowska & Pawlowski, 2011), suggesting a possible optimum for women's vocal attractiveness. Moreover, relatively higher formant dispersion (i.e., Df, the relative distance between two consecutive formants, which correlates to the vocal tract length and perceived timbre) is also perceived as more attractive by men (Puts et al., 2011; Babel et al., 2014). Additionally, the variation of the F0 has also been hypothesized to play upon the perception of indexical cues relevant in human competing and mating contexts (Leongómez et al., 2014; Hogdes-Simeon et al., 2010, 2011) but has so far received scant attention. Although sexually dimorphic, it has only been tested for women's preferences (Bruckert et al., 2006; Hodges-Simeon et al., 2010), but one study suggests that men may be attracted to higher F0-SD profiles in women as it may be a cue of femininity (Leongómez et al., 2014).

Nonetheless, it is possible that vocal preferences for both men and women may not be culturally universal. As a matter of fact, physiological and anatomical differences do not explain the full variation in mean F0 between men and women, as individuals of both sexes exhibit considerable variation from one language to another (Rose, 1991; Traunmüller & Eriksson, 1995; Yamazawa & Hollien, 1992; Keating & Kuo, 2012;

Andreeva et al., 2014; Pépiot, 2014). For instance, even under the same speaking conditions and balanced in age, American women exhibit a lower F0 than Japanese women (mean F0: 211 versus 224 Hz, Yamazawa & Hollien, 1992), while Bulgarian and Polish women exhibit a higher F0 than German and English women (mean F0: 272 and 266 Hz versus 210 and 217 Hz, Andreeva et al., 2014). As males and females vary in mean F0 across various languages, this strongly suggests that some of the differences must be accounted for learned behavior or specific sociocultural practices (Simpson, 2009, e.g., Loveday, 1981). For instance, Dutch women display a lower F0 than Japanese women, and interestingly, Dutch and Japanese men tend to prefer female voices that exhibit culturally congruent vocal heights that is: low female voices versus high female voices for Dutch versus Japanese men, respectively (Van Bezooijen, 1995). Even in men, vocal attractiveness may not be solely predicted by voice pitch. For instance, the harmonics-to-noise ratio (a proxy of vocal breathiness) can predict Namibian men's vocal attractiveness (Šebesta et al., 2017).

7.3 Reproductive and Mating Successes

7.3.1 Its Quantification

Giving such observations, it is interesting to know how much variance can voice explain for an individual's overall reproductive success.

Investigating reproductive success within hunter-gatherer societies is of a particular interest because it is argued that such societies better reflect past human environments, practices, and cultures. However, studies are scarce. Hadza men with relatively lower F0 had higher reproductive success (Apicella et al., 2007). However, it has been recently reported that this relationship does not hold when controlling for reputation (Smith et al., 2017). In women, it has been shown that F0 significantly predicted several measures of reproductive success in a group of Namibian females: higher voice pitch was associated with overall higher reproductive success (Atkinson et al., 2012).

An easier measure of reproductive success is to measure mating success, and mostly the number of past-year sexual partners. Although less powerful, this measure is interesting because it represents a time window over which participants' recollections are expected to be accurate (contrary to asking the lifetime number of sexual partners) and the measured acoustics' characteristics are likely to be stable (Hodges-Simeon et al., 2011). Moreover, human mating success should be an important component of expected reproductive success in past environments, as it represents their potential fertility (Perusse, 1993).

Through a simulated dating game, lower F0 negatively correlated to men's mating success (Puts 2005), but another study found that it was not significant (Puts et al., 2006). Using a similar approach, men who spoke in a more monotonous manner (i.e., lower F0-SD) and faster when confronted to a competitor declared more sexual partners over the past year (Hodges-Simeon et al., 2011; Suire et al., 2018). Lastly, it has been reported that female and male vocal attractiveness (when rated by members of the opposite sex) could predict their mating success, their declared number of extra-pair copulations, and their age at first sexual intercourse (Hughes et al., 2004).

However, methodologies varied concerning speech samples used in previous studies; some studies used the recordings of spoken vowels and read speech without any contextual background (Apicella et al., 2007; Atkinson et al., 2012; Hughes et al., 2004; Smith et al., 2017). This approach of read speech has been also intensively used in perceptual studies when attractiveness and dominance need to be judged. This is problematic as it does not properly reflect how an individual vocally behave in ecological settings. Indeed, it has been regularly shown that studies conducted on read/reciting versus spontaneous speech produce quite different results (Howell & Kadi-Hanifi, 1991; Blaauw, 1992; Daly & Zue, 1992). As spontaneous speech is more difficult to analyze experimentally, it has been little used. Nonetheless, the simulated dating game studies have attempted to use it. These studies also provide an interesting way to quantify the relative contribution of both types of sexual selection in shaping vocal acoustic features (Hodges-Simeon et al., 2011; Suire et al., 2018).

7.3.2 The Underlying Biological Quality of Voice

To understand the ultimate reasons behind the correlations between vocal acoustic features, attractiveness, dominance, and reproductive success, the "honest signaling theory" offers an interesting explanation.

Regarding communication systems (i.e., the exchange of information through different mechanisms involving at least two parties), this theory posits that, giving conflicting interests between and within sexes, an individual should give an honest signal to the receiver rather than cheating. This is due to the fact that cheating will select over time for skeptical individuals who, in turn, have no benefits in "listening." Thus, a communication system cannot emerge if false or manipulative information is exchanged actively. Here, voice has long been considered as an honest signal of overall biological quality, given the physiological and anatomical constraints in speech production (Feinberg et al., 2005; Evans et al., 2006; Puts et al., 2006). This means that voice should reflect another trait particularly relevant in contest and mate choice competitions, which are correlated to the aforementioned perceptions.

It has first been suggested that voice should be a reliable signal of body size, a feature particularly important in physical competitions between same-sex individuals. Correlations between vocalizations' frequencies have been well established in numerous species (Bowling et al., 2017) but surprisingly, such correlations are very weak within the human species. A meta-analysis showed that F0 did not explain more than 2% of the variation in body size, and formant frequencies only explained up to 10% (Pisanski et al., 2014a). This is interesting as both men and women still perceptually associate lower pitch voices to larger and taller individuals, and conversely higher pitch voices to thinner and smaller individuals (Rendall, Vokey, & Nemeth, 2007, but see Pisanski et al., Pisanski et al. 2014b).

An alternative hypothesis is the immuno-handicap principle (Zahavi, 1975). It has been suggested that voice should reflect immuno-competence of individuals. As testosterone is a sexual hormone that is immunosuppressive, individuals with higher testosterone levels could bear the costs of impacting their immune system, and are thus supposedly in a better biological shape. Although lower F0 may be linked to higher testosterone circulating levels (Dabbs & Mallinger, 1999; Evans, Neave, Wakelin, & Hamilton, 2008), the immuno-handicap principle has yielded mixed results in humans (Roberts, Buchanan, & Evans, 2004; Boonekamp et al., 2008). Nonetheless, it has been reported that men plasma testosterone levels were positively correlated with sexual language and the use of swear words in the presence of their partners (Mascaro et al., 2018). Additionally, bioavailable testosterone was also found to be associated with the sound pressure level of the normal speaking voice in men and the softest speaking voice in women (Jost et al. 2018). The most convincing study to date has shown that some masculinized vocal characteristics were correlated to a specific antibody (Arnocky et al., 2018). The authors showed that men with lower voice pitch and formant position had higher concentrations of immunoglobulin A, an antibody produced by the mucus and constituting the first line of immune defense against toxins and infectious agents.

7.4 Conclusion and Future Perspectives

Since the beginning of the 2000s, research has provided a better understanding on the functional role of the human voice from an evolutionary perspective. Although considerable efforts have been dedicated, further studies are needed to understand understudied aspects.

For instance, although acoustic features seem to be heritable (Przybyla, Horii, & Crawford, 1992; Debruyne, 2002) and possibly related to the prenatal and/or pre-pubertal androgen exposure (Fouquet, Pisanski, Mathevon, & Reby, 2016), little is still known of its biological foundations. Other understudied acoustic components part of voice quality, such as roughness and breathiness, have also been little studied and are known to potentially affect attractiveness perceptions (Šebesta et al., 2017; Suire et al., 2018). Sociocultural variation in vocal preferences is also one important avenue for research. Additional efforts should also be devoted to study the interaction between linguistic material and vocal acoustic features to project indexical cues relevant to mating and competing contexts. Lastly, another interesting avenue for further research is to investigate vocal modulation, a capacity described as a volitional control of nonverbal vocal features evolutionarily linked to traits important in the context of sexual selection. However, context-dependent vocal modulation patterns have been little relatively studied so far, but provides evidence that individuals of both sexes alter several acoustic characteristics to signal traits relevant to contest competition and mate choice (see Pisanski et al. 2016 for a review).

References

Anderson, R. C., & Klofstad, C. A. (2012). Preference for leaders with masculine voices holds in the case of feminine leadership roles. *PloS One*, *7*(12), e51216.

Andersson, M. B. (1994). *Sexual selection*. Princeton University Press.

Andreeva, B., Demenko, G., Möbius, B., Zimmerer, F., Jügler, J., & Oleskowicz-Popiel, M. (2014). Differences of pitch profiles in Germanic and Slavic languages. In *Fifteenth Annual Conference of the International Speech Communication Association*.

Apicella, C. L., Feinberg, D. R., & Marlowe, F. W. (2007). Voice pitch predicts reproductive success in male hunter-gatherers. *Biology Letters*, *3*(6), 682–684.

Arnocky, S., Hodges-Simeon, C., Ouellette, D., & Albert, G. (2018). Do men with more masculine voices have better immunocompetence? *Evolution and Human Behavior*.

Aronovitch, C. D. (1976). The voice of personality: Stereotyped judgments and their relation to voice quality and sex of speaker. *The Journal of Social Psychology*, *99*(2), 207–220.

Atkinson, J., Pipitone, R. N., Sorokowska, A., Sorokowski, P., Mberira, M., Bartels, A., et al. (2012). Voice and handgrip strength predict reproductive success in a group of indigenous African females. *PloS One*, *7*(8), e41811.

Babel, M., McGuire, G., & King, J. (2014). Towards a more nuanced view of vocal attractiveness. *PloS one*, *9*(2), e88616.

Banai, I. P., Banai, B., & Bovan, K. (2017). Vocal characteristics of presidential candidates can predict the outcome of actual elections. *Evolution and Human Behavior*, *38*(3), 309–314.

Bennett, P. M., & Owens, I. P. (2002). Evolutionary ecology of birds: Life histories, mating systems and extinction.

Blaauw, E. (1992). Phonetic differences between read and spontaneous speech.

Boonekamp, J. J., Ros, A. H., & Verhulst, S. (2008). Immune activation suppresses plasma testosterone level: A meta-analysis. *Biology Letters*, *4*(6), 741–744.

Borkowska, B., & Pawlowski, B. (2011). Female voice frequency in the context of dominance and attractiveness perception. *Animal Behaviour*, *82*(1), 55–59.

Bowling, D. L., Garcia, M., Dunn, J. C., Ruprecht, R., Stewart, A., Frommolt, K. H., et al. (2017). Body size and vocalization in primates and carnivores. *Scientific Reports*, *7*, 41070.

Bruckert, L., Lienard, J. S., Lacroix, A., Kreutzer, M., & Leboucher, G. (2006). Women use voice parameters to assess men's characteristics. *Proceedings of the Royal Society B: Biological Sciences*, *273*(1582), 83–89.

Buss, D. M., & Schmitt, D. P. (1993). Sexual strategies theory: An evolutionary perspective on human mating. *Psychological Review*, *100*(2), 204.

Charles, D. (1871). *The descent of man and selection in relation to sex*. London: Murray.

Clutton-Brock, T. H., Guinness, F. E., & Albon, S. D. (1982). *Red deer: Behavior and ecology of two sexes*. University of Chicago press.

Collins, S. A. (2000). Men's voices and women's choices. *Animal behaviour*, *60*(6), 773–780.

Collins, S. A., & Missing, C. (2003). Vocal and visual attractiveness are related in women. *Animal Behaviour*, *65*(5), 997–1004.

Dabbs, J. M, Jr., & Mallinger, A. (1999). High testosterone levels predict low voice pitchamong men. *Personality and Individual Differences*, *27*(4), 801–804.

Daly, N. A., & Zue, V. (1992). Statistical and linguistic analyses of F0 in read and spontaneous speech.

Debruyne, F., Decoster, W., Van Gijsel, A., & Vercammen, J. (2002). Speaking fundamental frequency in monozygotic and dizygotic twins. *Journal of Voice*, *16*(4), 466–471.

Evans, S., Neave, N., & Wakelin, D. (2006). Relationships between vocal characteristics and body size and shape in human males: An evolutionary explanation for a deep male voice. *Biological Psychology*, *72*(2), 160–163.

Evans, S., Neave, N., Wakelin, D., & Hamilton, C. (2008). The relationship between testosterone and vocal frequencies in human males. *Physiology & Behavior*, *93*(4–5), 783–788.

Feinberg, D. R., Jones, B. C., Smith, M. L., Moore, F. R., DeBruine, L. M., Cornwell, R. E., et al. (2006). Menstrual cycle, trait estrogen level, and masculinity preferences in the human voice. *Hormones and Behavior*, *49*(2), 215–222.

Feinberg, D. R., Jones, B. C., Little, A. C., Burt, D. M., & Perrett, D. I. (2005). Manipulations of fundamental and formant frequencies influence the attractiveness of human male voices. *Animal Behaviour*, *69*(3), 561–568.

Feinberg, D. R., DeBruine, L. M., Jones, B. C., & Little, A. C. (2008). Correlated preferences for men's facial and vocal masculinity. *Evolution and Human Behavior*, *29*(4), 233–241.

Fisher, M. L. (2015). Women's competition for mates: Experimental findings leading to ethological studies. *Human Ethology Bulletin*, *30*(1), 53–70.

Fitch, W. T., & Giedd, J. (1999). Morphology and development of the human vocal tract: A study using magnetic resonance imaging. *The Journal of the Acoustical Society of America*, *106*(3), 1511–1522.

Fouquet, M., Pisanski, K., Mathevon, N., & Reby, D. (2016). Seven and up: Individual differences in male voice fundamental frequency emerge before puberty and remain stable throughout adulthood. *Royal Society Open Science*, *3*(10), 160395.

Hill, A. K., Hunt, J., Welling, L. L., Cárdenas, R. A., Rotella, M. A., Wheatley, J. R., et al. (2013). Quantifying the strength and form of sexual selection on men's traits. *Evolution and Human Behavior*, *34*(5), 334–341.

Hodges-Simeon, C. R., Gaulin, S. J., & Puts, D. A. (2010). Different vocal parameters predict perceptions of dominance and attractiveness. *Human Nature*, *21*(4), 406–427.

Hodges-Simeon, C. R., Gaulin, S. J., & Puts, D. A. (2011). Voice correlates of mating success in men: examining "contests" versus"mate choice" modes of sexual selection. *Archives of sexual behavior*, *40*(3), 551–557.

Hodges-Simeon, C. R., Gaulin, S. J., & Puts, D. A. (2010). Different vocal parameters predict perceptions of dominance and attractiveness. *Human Nature*, *21*(4), 406–427.

Hodges-Simeon, C. R., Gaulin, S. J., & Puts, D. A. (2011). Voice correlates of mating success in men: Examining "contests" versus "mate choice" modes of sexual selection. *Archives of Sexual Behavior*, *40*(3), 551–557.

Howell, P., & Kadi-Hanifi, K. (1991). Comparison of prosodic properties between read and spontaneous speech material. *Speech Communication*, *10*(2), 163–169.

Hughes, S. M., Farley, S. D., & Rhodes, B. C. (2010). Vocal and physiological changes in response to the physical attractiveness of conversational partners. *Journal of Nonverbal Behavior*, *34*(3), 155–167.

Hughes, S. M., Mogilski, J. K., & Harrison, M. A. (2014). The perception and parameters of intentional voice manipulation. *Journal of Nonverbal Behavior*, *38*(1), 107–127.

Hughes, S. M., Dispenza, F., & Gallup, G. G, Jr. (2004). Ratings of voice attractiveness predict sexual behavior and body configuration. *Evolution and Human Behavior*, *25*(5), 295–304.

Jones, B. C., Hahn, A. C., Fisher, C. I., Wang, H., Kandrik, M., Han, C., et al. (2018). No compelling evidence that preferences for facial masculinity track changes in women's hormonal status. *Psychological Science*, *29*(6), 996–1005.

Jones, B. C., Feinberg, D. R., DeBruine, L. M., Little, A. C., & Vukovic, J. (2010). A domain-specific opposite-sex bias in human preferences for manipulated voice pitch. *Animal Behaviour*, *79*(1), 57–62.

Jones, B. C., DeBruine, L. M., Little, A. C., Watkins, C. D., & Feinberg, D. R. (2011). 'Eavesdropping'and perceived male dominance rank in humans. *Animal Behaviour*, *81*(6), 1203–1208.

Jost, L., Fuchs, M., Loeffler, M., Thiery, J., Kratzsch, J., Berger, T., et al. (2018). Associations of sex hormones and anthropometry with the speaking voice profile in the adult general population. *Journal of Voice*, *32*(3), 261–272.

Jünger, J., Kordsmeyer, T. L., Gerlach, T. M., & Penke, L. (2018). Fertile women evaluate male bodies as more attractive, regardless of masculinity. *Evolution and Human Behavior*.

Keating, P., & Kuo, G. (2012). Comparison of speaking fundamental frequency in English and Mandarin. *The Journal of the Acoustical Society of America*, *132*(2), 1050–1060.

Kempe, V., Puts, D. A., & Cárdenas, R. A. (2013). Masculine men articulate less clearly. *Human Nature, 24*(4), 461–475.

Klofstad, C. A., Anderson, R. C., & Peters, S. (2012). Sounds like a winner: Voice pitch influences perception of leadership capacity in both men and women. *Proceedings of the Royal Society of London B: Biological Sciences*, rspb20120311.

Klofstad, C. A. (2016). Candidate voice pitch influences election outcomes. *Political Psychology, 37*(5), 725–738.

Kordsmeyer, T. L., Hunt, J., Puts, D. A., Ostner, J., & Penke, L. (2018). The relative importance of intra-and intersexual selection on human male sexually dimorphic traits. *Evolution and Human Behavior*.

Leongómez, J. D., Binter, J., Kubicová, L., Stolarová, P., Klapilová, K., Havlícek, J., & Roberts, S. C. (2014). Vocal modulation during courtship increases proceptivity even in naive listeners. *Evolution and Human Behavior, 35*(6), 489–496.

Loveday, L. (1981). Pitch, politeness and sexual role: An exploratory investigation into the pitch correlates of English and Japanese politeness formulae. *Language and Speech, 24*(1), 71–89.

Mascaro, J. S., Rentscher, K. E., Hackett, P. D., Lori, A., Darcher, A., Rilling, J. K., & Mehl, M. R. (2018). Preliminary evidence that androgen signaling is correlated with men's everyday language. *American Journal of Human Biology*, e23136.

Mayew, W. J., Parsons, C. A., & Venkatachalam, M. (2013). Voice pitch and the labor market success of male chief executive officers. *Evolution and Human Behavior, 34*(4), 243–248.

Miller, G. F. (1998). How mate choice shaped human nature: A review of sexual selection and human evolution. *Handbook of evolutionary psychology: Ideas, issues, and applications* (pp. 87–129).

O'Connor, J. J., Pisanski, K., Tigue, C. C., Fraccaro, P. J., & Feinberg, D. R. (2014a). Perceptions of infidelity risk predict women's preferences for low male voice pitch in short-term over long-term relationship contexts. *Personality and Individual differences, 56*, 73–77.

O'Connor, J. J., Fraccaro, P. J., Pisanski, K., Tigue, C. C., O'Donnell, T. J., & Feinberg, D. R. (2014b). Social dialect and men's voice pitch influence women's mate preferences. *Evolution and Human Behavior, 35*(5), 368–375.

Pépiot, E. (2014). Male and female speech: A study of mean f0, f0 range, phonation type and speech rate in Parisian French and American English speakers. In Speech Prosody 7 (pp. 305–309).

Perusse, D. (1993). Cultural and reproductive success in industrial societies: Testing the relationship at the proximate and ultimate levels. *Behavioral and Brain Sciences, 16*(2), 267–283.

Pisanski, K., Fraccaro, P. J., Tigue, C. C., & O'connor, J. J., & Feinberg, D. R., (2014b). Return to Oz: Voice pitch facilitates assessments of men's body size. *Journal of Experimental Psychology: Human Perception and Performance, 40*(4), 1316.

Pisanski, K., Fraccaro, P. J., Tigue, C. C., O'Connor, J. J., Röder, S., Andrews, P. W., et al. (2014a). Vocal indicators of body size in men and women: A meta-analysis. *Animal Behaviour, 95*, 89–99.

Pisanski, K., Cartei, V., McGettigan, C., Raine, J., & Reby, D. (2016). Voice modulation: A window into the origins of human vocal control? *Trends in Cognitive Sciences, 20*(4), 304–318.

Przybyla, B. D., Horii, Y., & Crawford, M. H. (1992). Vocal fundamental frequency in a twin sample: Looking for a genetic effect. *Journal of Voice, 6*(3), 261–266.

Puts, D. A. (2005). Mating context and menstrual phase affect women's preferences for male voice pitch. *Evolution and Human Behavior, 26*(5), 388–397.

Puts, D. A., Hill, A. K., Bailey, D. H., Walker, R. S., Rendall, D., Wheatley, J. R., et al. (2016). Sexual selection on male vocal fundamental frequency in humans and other anthropoids. *Proceedings of the Royal Society of London B: Biological Sciences, 283*(1829), 20152830.

Puts, D. A. (2010). Beauty and the beast: Mechanisms of sexual selection in humans. *Evolution and Human Behavior, 31*(3), 157–175.

Puts, D. A., Gaulin, S. J., & Verdolini, K. (2006). Dominance and the evolution of sexual dimorphism in human voice pitch. *Evolution and Human Behavior, 27*(4), 283–296.

Puts, D. A., Hodges, C. R., Cárdenas, R. A., & Gaulin, S. J. (2007). Men's voices as dominance signals: Vocal fundamental and formant frequencies influence dominance attributions among men. *Evolution and Human Behavior, 28*(5), 340–344.

Puts, D. A., Barndt, J. L., Welling, L. L., Dawood, K., & Burriss, R. P. (2011). Intrasexual competition among women: Vocal femininity affects perceptions of attractiveness and flirtatiousness. *Personality and Individual Differences*, *50*(1), 111–115.

Puts, D. A., Apicella, C. L., & Cárdenas, R. A. (2012). Masculine voices signal men's threat potential in forager and industrial societies. *Proceedings of the Royal Society of London B: Biological Sciences*, *279*(1728), 601–609.

Ralls, K., & Mesnick, S. (2009). Sexual dimorphism. In *Encyclopedia of marine mammals*(Second Edition) (pp. 1005–1011).

Rendall, D., Vokey, J. R., & Nemeth, C. (2007). Lifting the curtain on the Wizard of Oz: Biased voice-based impressions of speaker size. *Journal of Experimental Psychology: Human Perception and Performance*, *33*(5), 1208.

Riding, D., Lonsdale, D., & Brown, B. (2006). The effects of average fundamental frequency and variance of fundamental frequency on male vocal attractiveness to women. *Journal of Nonverbal Behavior*, *30*(2), 55–61.

Roberts, M. L., Buchanan, K. L., & Evans, M. R. (2004). Testing the immunocompetence handicap hypothesis: A review of the evidence. *Animal Behaviour*, *68*(2), 227–239.

Rose, P. (1991). How effective are long term mean and standard deviation as normalisation parameters for tonal fundamental frequency? *Speech Communication, 10*(3), 229–247.

Šebesta, P., Kleisner, K., Tureček, P., Kočnar, T., Akoko, R. M., Třebický, V., et al. (2017). Voices of Africa: Acoustic predictors of human male vocal attractiveness. *Animal Behaviour*, *127*, 205–211.

Shirazi, T. N., Puts, D. A., & Escasa-Dorne, M. J. (2018). Filipino women's preferences for male voice pitch: Intra-individual, life history, and hormonal predictors. *Adaptive Human Behavior and Physiology*, *4*(2), 188–206.

Simpson, A. P. (2009). Phonetic differences between male and female speech. *Language and Linguistics Compass*, *3*(2), 621–640.

Smith, K. M., Olkhov, Y. M., Puts, D. A., & Apicella, C. L. (2017). Hadza men with lower voice pitch have a better hunting reputation. *Evolutionary Psychology, 15*(4), 1474704917740466. Suire, A., Raymond, M., Barkat-Defradas, M. (2018). *Human vocal behavior within competitive and courtship contexts and its relation to mating success*. Accepted: Evolution and Human Behavior (in press).

Smith, D. S., Jones, B. C., Feinberg, D. R., & Allan, K. (2012). A modulatory effect of male voice pitch on long-term memory in women: evidence of adaptation for mate choice? *Memory & Cognition*, *40*(1), 135–144.

Suire, A., Raymond, M., & Barkat-Defradas, M. (2018). Human vocal behavior within competitive and courtship contexts and its relation to mating success. *Evolution and Human Behavior, 39*(6), 684–691.

Tigue, C. C., Borak, D. J., O'Connor, J. J., Schandl, C., & Feinberg, D. R. (2012). Voice pitch influences voting behavior. *Evolution and Human Behavior*, *33*(3), 210–216.

Titze, I. R. (1989). Physiologic and acoustic differences between male and female voices. *The Journal of the Acoustical Society of America*, *85*(4), 1699–1707.

Traunmüller, H., & Eriksson, A. (1995). The frequency range of the voice fundamental in the speech of male and female adults. Unpublished manuscript.

Tuomi, S. K., & Fisher, J. E. (1979). Characteristics of simulated sexy voice. *Folia Phoniatrica et Logopaedica, 31*(4), 242–249.

Van Bezooijen, R. (1995). Sociocultural aspects of pitch differences between Japanese and Dutch women. *Language and Speech*, *38*(3), 253–265.

Vukovic, J., Jones, B. C., Feinberg, D. R., DeBruine, L. M., Smith, F. G., Welling, L. L., et al. (2011). Variation in perceptions of physical dominance and trustworthiness predicts individual differences in the effect of relationship context on women's preferences for masculine pitch in men's voices. *British Journal of Psychology*, *102*(1), 37–48.

Watkins, C. D., Fraccaro, P. J., Smith, F. G., Vukovic, J., Feinberg, D. R., DeBruine, L. M., et al. (2010). Taller men are less sensitive to cues of dominance in other men. *Behavioral Ecology*, *21*(5), 943–947.

Wolff, S. E., & Puts, D. A. (2010). Vocal masculinity is a robust dominance signal in men. *Behavioral Ecology and Sociobiology*, *64*(10), 1673–1683.

Yamazawa, H., & Hollien, H. (1992). Speaking fundamental frequency patterns of Japanese women. *Phonetica*, *49*(2), 128–140.

Zahavi, A. (1975). Mate selection-a selection for a handicap. *Journal of theoretical Biology*, *53*(1), 205–214.

Chapter 8
On Voice Averaging and Attractiveness

Pascal Belin

Abstract Several experiments investigating the perceptual, acoustical and neural bases of the 'voice attractiveness averaging phenomenon' are briefly summarized. We show that synthetic voice composites generated by averaging multiple (same gender) individual voices (short syllables) are perceived as increasingly attractive with the number of voices averaged. This phenomenon, independent of listener or speaker gender and analogous to a similar effect in the visual domain for face attractiveness, is explained in part by two acoustical correlates of averaging: reduced 'Distance-to-Mean', as indexed by the Euclidean distance between a voice and its same-gender population average in f0-F1 space and increased voice 'texture smoothness' as indexed by increased harmonics-to-noise ratio (HNR). These two acoustical parameters co-vary with perceived attractiveness and manipulating them independently of one another also affects attractiveness ratings. The neural correlates of implicitly perceived attractiveness consist in a highly significant negative correlation between attractiveness and fMRI signal in large areas of bilateral auditory cortex, largely overlapping with the Temporal Voice Areas, as well as inferior prefrontal cortex: more attractive voices elicit less activity in these regions. While the correlations in auditory areas were largely explained by distance-to-mean and HNR, inferior prefrontal areas bilaterally were observed even after co-varying out variance explained by these acoustical parameters, suggesting a role as abstract voice attractiveness evaluators.

Keywords Averageness · Aperiodicity · Distance-to-mean · Distinctiveness · Pitch · Formant dispertion

---

P. Belin (✉)
Institut de Neurosciences de La Timone, CNRS et Aix-Marseille
Université Département de Psychologie, Université de Montréal, Montreal, Canada
e-mail: pascal.belin@univ-amu.fr

B. Weiss et al. (eds.), *Voice Attractiveness*, Prosody, Phonology and Phonetics,
https://doi.org/10.1007/978-981-15-6627-1_8

8.1 Introduction

The faces shown in Fig. 8.1a are computer generated: they are the pixel-wise average of a large number of pictures of different faces after conformation to a same configuration (eyes and mouth in the same position). Observers typically find these faces more attractive than most of the individual constituting faces. This so-called 'averaging attractiveness phenomenon' has been observed since the nineteenth century and the beginnings of photography when experimenters such as Sir Francis Galton noticed that by superimposing portraits of different individuals on a same photographic plate one obtained a quite attractive picture (Galton, 1878; Jastrow, 1885). Since those pioneering times the averaging attractiveness phenomenon has been replicated many times with more sophisticated computer graphics techniques such as in Fig. 8.1a (Langlois et al., 2000; Langlois & Roggman, 1990; Perrett, May, & Yoshikawa, 1994; Thornhill & Gangestad, 1999).

There are two main accounts for the face averaging attractiveness phenomenon. One account from evolutionary psychology—the 'good genes' explanation—proposes that we tend to prefer averaged faces because if they were real faces they would signal a potential mate with particularly high fitness. Indeed, facial features such as proximity to the population average, facial symmetry, or face texture smoothness appear to signal high fitness in real faces (Grammer, Fink, Moller, & Thornhill, 2003; Langlois & Roggman, 1990; Thornhill & Gangestad, 1999). The averaging procedure enhances all three of these features, artificially signalling high fitness in a synthetic face, and hence their attractiveness. Another account from cognitive psychology—the 'perceptual fluency' account—proposes that observers prefer averaged faces because they are closer in face space, i.e. more similar to a central face prototype based on which all face identities are coded, and so they are easier to process, and hence more attractive (Winkielman, Halberstadt, Fazendeiro, & Catty, 2006). These two accounts are not mutually exclusive: the 'perceptual fluency' account can be viewed as an explanation at the proximate level, in terms of cognitive mechanisms implementing the effect, while the 'good genes' account is an explanation at a more ultimate level, in terms of the selective evolutionary pressures that gave rise to such a phenomenon in our ancestors.

Crucially, both the cognitive and evolutionary accounts suggest that a similar phenomenon could exist for voices. Thanks to the development of voice morphing technology, and the excellent and generous contribution of Professor Hideki Kawahara at Wakayama University, we were able to test that hypothesis for the first time in Bruckert et al. (2010).

8.2 Voice Attractiveness Increases with Averaging

To start addressing the complex problem of voice averaging, we decided to focus on the simpler problem, more manageable in an experimental setting, of averaging of brief, quasi-stationary vocalizations, and opted to use short syllables as stimuli.

We reasoned that such stimuli, for which time plays minimal role, would be easier to process through averaging than longer, more complex and variable utterances. Quasi-static syllables are also analogous to the static photographs with which most face attractiveness research has been performed so far.

We selected from a database of high-quality recordings of English syllables (Hillenbrand, Getty, Clark, & Wheeler, 1995) as set of recordings of the syllable /had/ spoken in isolation by 32 different male and 32 female American speakers (duration: mean ± s.d.: female voices: 320 ± 51 ms; male voices: 267 ± 42 ms). We then identified in each stimulus a set of spectro-temporal landmarks to be put in correspondence across speakers during averaging. As shown by the black dots in Fig. 8.1b, these landmarks consisted of first three formant frequencies at onset and offset of phonation, and at the beginning of the formant transition of the final /d/. We then used the Straight software (Kawahara & Matsui, 2003) to generate voice composites consisting of an interpolation of the aperiodic and spectral temporal density components of varying numbers of individual voices of the same gender (arbitrarily chosen, such that not all possible composites have been generated). For each speaker

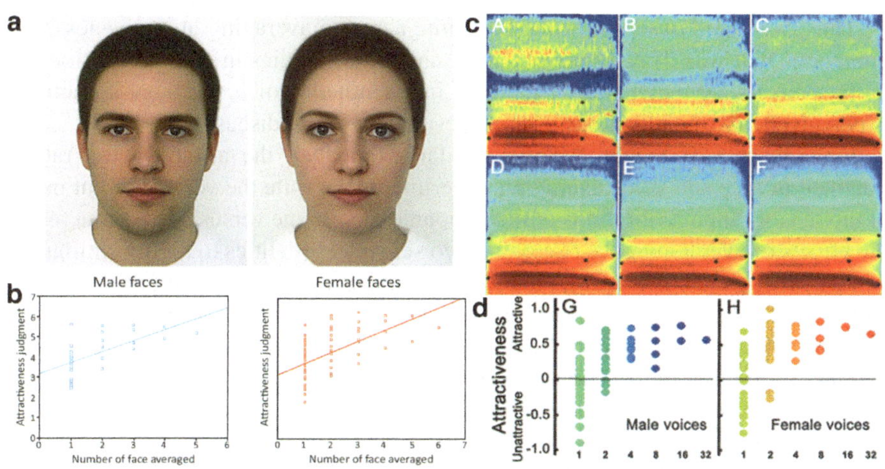

Fig. 8.1 Face and voice attractiveness judgments as a function of averaging. **a** Face composites generated by averaging 32 male faces (left) and 64 female faces (right). **b** Attractiveness ratings as a function of number of faces averaged. Note the steady increase in attractiveness ratings with increasing number of averaged faces, for both male (left) and female (right) faces. Reproduced with permission from Braun et al. (2001). **c** Spectrograms of voice composites generated by averaging an increasing number of voices of the same gender (different speakers uttering the syllable 'had'). Top left panel: 1-voice composite; middle top panel: 2-voice composite; right top panel: 4-voice composite; bottom left panel: 8-voice composite; bottom middle panel: 16-voice composite; bottom right panel: 32-voice composite. **d** Attractiveness ratings as a function of number of voices averaged in the composites (individual points). Note the steady increase in attractiveness ratings with increasing number of averaged voices, for both male (left) and female (right) voices. Reproduced with permission from Bruckert et al. (2010)

gender, this procedure resulted in thirty two 1-voice composites (the individual voices resynthesized), sixteen 2-voice composites, eight 4-voice composites, four 8-voice composites, two 16-voice composites, and a single average of all voices of the same gender, the 32-voice composite. Example composite stimuli are shown in Fig. 8.1c.

We then played these stimuli in a pseudorandom order to 25 listeners (13 females) who were asked to rate the perceived attractiveness of each stimulus using a visual analogue scale ranging from 'not at all' to 'extremely' attractive. Analysis of the data provided striking results (Bruckert et al., 2010). As shown in Fig. 8.1d, for the 1-voice composites (the resynthesized original voices) we found as expected a normal distribution of attractiveness ratings around the average: for both male and female speakers, most of the voices were rated with average attractiveness while a few voices were perceived as more attractive than average and others as less. But as soon as two or more voices where averaged together, we witnessed a marked progressive increase of average attractiveness ratings, similar for the male and female voices. 4-voice composites were already perceived as markedly above average and 16- and 32-voice composites all resulted in very high ratings. The correlation between attractiveness z-scores and number of voices in the composite was highly significant ($p < 0.001$) for both male and female voices (Bruckert et al., 2010).

Thus, we could observe for the first time a 'voice averaging attractiveness phenomenon' that was directly predicted by analogous studies in face perception: the more speakers are included by averaging in a synthetic voice, the more attractive it is perceived. Two implications of these results are worth discussing.

First, there is a highly striking similarity between the attractiveness ratings obtained in face and voice averaging experiments. Despite the very different nature of the sensory input (vibrations of the tympanic membrane versus light on the retina) the effects of averaging gave rise in the two sensory modalities to highly similar and gender-independent averaging-induced attractiveness increases (compare Fig. 8.1b and d). This beautifully illustrates the notion of similar functional architectures for face and voice processing in the human brain. Indeed, many sources of evidence from patient observation to neuroimaging studies converge to the notion that the computational problems posed by face and voice processing, being of very similar nature, and subjected in our ancestors to comparable evolutionary pressures, are addressed by the brain using similar neurophysiological solutions (Yovel & Belin, 2013).

Second, and more relevant to voice attractiveness, the voice averaging attractiveness phenomenon opens an exciting window onto the acoustical underpinnings of this complex percept. Indeed, the averaging procedure had at least two independent acoustic effects on the synthesized composites. Including an increasing number of different voices in the composite's resulted in: (i) a progressive decrease in the distance-to-mean (increased similarity to the average) and (ii) a progressive decrease in the amount of aperiodicity (increased harmonicity or voice texture smoothness).

8.3 Effects of Distance-to-Mean

Because of the linear combination of individual spectral temporal landmarks in the composites during averaging, at each successive averaging step the resulting composites mathematically became closer in acoustical space to the average: their fundamental frequency and formant frequency values became increasingly similar to those of the 32-voice average, resulting at each step in decreasing average 'distance-to-mean', as defined by the Euclidean distance between a voice and the same-gender average in f0-F1 (first formant frequency) space. In other words, the more voices are averaged together, the more the resulting composite sounds like the population average. This suggests that distance-to-mean could potentially provide an acoustical parameter relevant for voice attractiveness. We tested that hypothesis in two different ways: (i) by examining the relationship between distance-to-mean and perceived attractiveness in our set of natural and synthetic voices and (ii) by explicitly manipulating distance-to-mean (but not other parameters such as aperiodicity) in synthetic voices.

We first tested, independently for male and female voices, whether distance-to-mean in our set of voice composites would correlate with their average perceived attractiveness. For both voice genders, we found highly significant negative correlations between distance-to-mean and attractiveness: the higher the distance-to-mean, the lower the perceived attractiveness. As including composites of all levels in this analysis, known to be both closer to the average and more attractive involves some level of circularity, we repeated the analysis by only including the 1-voice composites, resynthesized versions of the original recordings (indistinguishable by ear): the results remained strongly significant, for both male and female voices. Thus, in our set of 32 male and 32 female voices, those that were naturally closer to the same-gender average were also perceived as more attractive (Bruckert et al., 2010)—a result that should be tested on larger samples (Fig. 8.2).

Does modifying distance-to-mean also modify perceived attractiveness? We tested the hypothesis by using morphing to generate, for each of the 32 individual voices of each gender, a pair of synthetic voices that differed from the original by having been moved either towards the average or away from the average by the exact similar amount of acoustical change (50% of the natural distance-to-mean). We predicted that although the new synthetic voices were acoustically equally dissimilar to the original, the one closer to the average would be perceived as more attractive. Results confirmed that prediction for both voice genders (Bruckert et al., 2010).

Thus, not only are voices naturally closer to the same-gender average perceived as more attractive, but acoustically modifying voices to move them closer to the average also makes them more attractive than moving them away. Distance-to-mean thus appears as one important acoustical correlate of voice attractiveness. Interestingly, distance-to-mean can be consciously modified, if not by altering formant frequencies (largely dependent on vocal tract size) but by consciously modifying one's pitch of voice so that our average fundamental frequency is closer to the gender-typical value

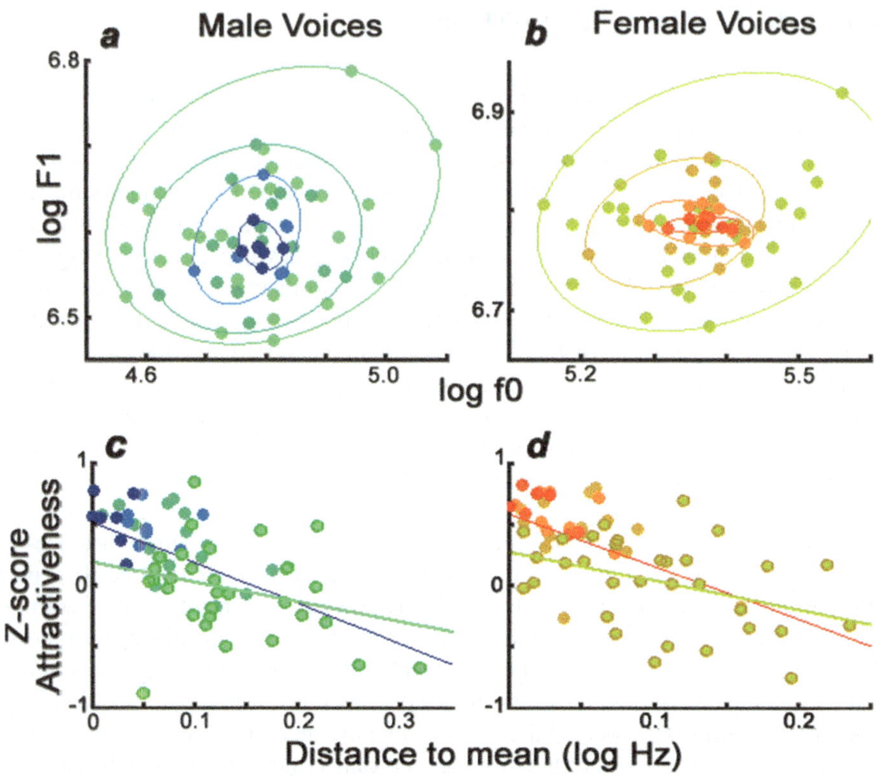

Fig. 8.2 Effects of distance-to-mean. **a.** Male voice composites are represented as coloured dots in logf0-logF1 space. Colour indicates degree of averaging with darker colours indicating more voices in the composite. Lines indicate the smallest elliptic contours containing all composites of a same degree of averaging. Note how composites progressively become closer to the average with increasing number of constituting voices. **b.** Female voice composites, legend as in (**a**). **c.** Relation between distance-to-mean and attractiveness ratings for each male voice composite (coloured dots). Lines indicate the regression line when all composites are considered (blue line) or when only the 1-voice composites are considered (green line). **d.** Relation between distance-to-mean and attractiveness for female voice composites. Legend as in (**c**)

(about 125 Hz for men and 215 Hz for women Hillenbrand et al., 1995), not too low and not too high, as a means of 'vocal make-up' to enhance one's perceived attractiveness.

8.4 Effects of Voice Texture Smoothness

Another important effect of averaging on the acoustical structure of voices, largely independent from the effect of distance-to-mean in f0-formant frequency space, is a progressive decrease in the amount of aperiodicity with the number of voices averaged, as the morphing procedure averages out aperiodic noise in the signal. This effect can be plainly seen in Fig. 8.1c as the spectrograms become progressively smoother with the increasing number of voices in the composite from the top left panel (1-voice composite, showing much spectro-temporal irregularities) to the bottom right panel (32-voice composite) with a very smooth structure. This effect is analogous to the increase in face texture smoothness (see Fig. 8.1a) caused by averaging as individual local variations in luminance and reflection (the 'villainous irregularities' of Galton, 1878) are averaged out across individual faces. This effect of smoothing of the 'voice texture' can be quantified using measures such as the harmonics-to-noise ratio (HNR) that captures the amount of regularity in the sound. When the harmonic-to-noise ratio of each composite is plotted as a function of its number of constituent voices (Bruckert et al., 2010), there is a clear and highly significant progressive increase in HNR along with number of voices in the composite that nearly mirrors the increase in attractiveness ratings. Thus, the amount of energy in the aperiodic component of voice could constitute another acoustical correlate of voice attractiveness.

We tested this hypothesis by generating for each of the 32 male and 32 female voices of our sample, a 'smoother' and 'rougher' version of each voices. Those were generated by moving stimuli away or closer to the average by equal amounts of acoustic change, as for the manipulation of distance-to-mean above, but this time modifying only the aperiodic component of voice. We verified that the 'smoother' synthetic voices had greater harmonics-to-noise ratio than the 'rougher' for both voice genders. We then presented listeners with voice pairs made of the smoother and rougher version of a same original voice and asked them to decide the one they found the more attractive. Subjects overwhelmingly preferred the smoother version with reduced periodicity and increased HNR to the rougher version (Bruckert et al., 2010).

Overall, the increase in voice attractiveness induced by averaging highlights distance-to-mean and voice textures smoothness as two largely independent and important acoustical correlates of voice attractiveness. They can potentially be used to predict listeners' ratings and can be manipulated in synthetic, but also in natural voices, to artificially increase perceived attractiveness. Note, however, that while distance-to-mean already correlated with attractiveness ratings in natural, unaveraged voices, this was not the case for HNR that showed essentially no relation with attractiveness ratings for the natural voices. This suggest that, while both parameters contribute to the attractiveness averaging effect, distance-to-mean is more important than HNR in determining the attractiveness of natural voices.

8.5 Neural Correlates of Perceived Voice Attractiveness

We then turned to the question of the neural correlates of voice attractiveness. Indeed neuroimaging studies have shown linear or quadratic relations between perceived facial attractiveness and neural activity in orbitofrontal cortex as well as in amygdala

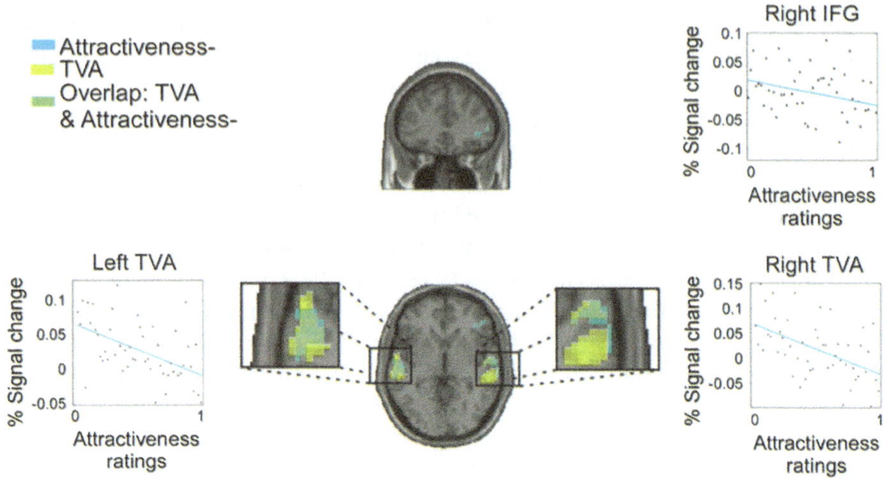

Fig. 8.3 Neural correlates of perceived attractiveness. Cerebral regions modulated by implicitly perceived attractiveness during passive listening to voices. Cortical areas in blue showed significant negative correlation between BOLD signal and attractiveness (graphs in insets showing regression lines for three regions of interest): more attractive voices elicited less neural activity in those regions. They largely overlap with the voice-sensitive temporal voice areas (in yellow) but also involve right inferior prefrontal cortex (top central panel)

(Winston, O'Doherty, Kilner, Perrett, & Dolan, 2007). To address this question in the domain of voice perception, we performed a functional magnetic resonance imaging (fMRI) study in normal participants (Bestelmeyer et al., 2012). They were scanned while passively listening to our set of voice composites presented in a pseudorandom order. We used a so-called 'cluster volume acquisition' fMRI protocol with brief silent intervals during fMRI volume acquisitions allowing the presentation of voice stimuli during silent periods for optimal stimulation. Subjects were not informed of our focus on voice attractiveness and were simply instructed to listen to the voices and press the button when they would hear an infrequent pure-tone stimulus.

In the fMRI analyses, we first asked whether there would be regions of the brain in which stimulus-induced activity would co-vary with the average attractiveness rating obtained offline for each voice. Indeed a well-defined network of cortical region showed significant correlations between fMRI signal and attractiveness ratings (Fig. 8.3). Most prominently, negative correlations were observed in large areas of bilateral superior temporal gyrus and sulci, overlapping with the voice-selective temporal voice areas (TVA) of auditory cortex (Belin, Zatorre, Lafaille, Ahad, & Pike, 2000; Pernet, Charest, Belizaire, Zatorre, & Belin, 2007) of secondary auditory cortex. But such negative correlation was also observed in the inferior frontal gyrus (IFG) of the right hemisphere, outside of voice-sensitive regions (Bestelmeyer et al., 2012).

We asked whether part of these strong negative correlations could be partly explained by one or the other acoustic parameters highlighted above—distance-to-mean or texture smoothness (as measured by HNR). We ran new analyses in which we searched for correlations between the brain activity elicited by each voice and their distance-to-mean on one hand and their HNR on the other. Both analyses revealed auditory cortical regions in which voice-elicited fMRI signal significantly correlated with these measures, although in different locations. Areas of secondary auditory cortex overlapping with the TVAs bilaterally showed a positive correlation with distance-to-mean (voices farther away from the mean—also less attractive on average—eliciting greater signal). This phenomenon has since been replicated and extended in subsequent work (Latinus & Belin, 2011; Latinus, McAleer, Bestelmeyer, & Belin, 2013). The positive correlation between distance-to-mean and neural activity constitutes a hallmark of 'norm-based coding' as evidenced in visual cortex for face identity processing (Leopold, Bondar, & Giese, 2006): individual voices appear to be coded in the TVAs as a function of their difference with the average voice: whether the negative correlation with attractiveness in those areas is a consequence of, or drives, the positive correlation with distance-to-mean remains to be established. Other, more anterior parts of the TVAs instead showed a negative correlation with HNR, with more aperiodic voices eliciting higher activity. Thus, the large negative correlation between attractiveness and fMRI signal is in part explained by a sensitivity of auditory cortex to the two underlying acoustical features shown as determinant for perceived attractiveness.

But could we detect attractiveness-related changes that would be independent of the underlying acoustics? We addressed that question by performing another analysis in which measures of distance-to-mean HNR were included in the model and regressed out to examine variance not accounted for by these parameters. Results showed that the large negative correlation in the auditory cortex had disappeared, confirming that it was largely explained by the HNR and distance-to-mean of the voices. However, two bilateral regions of inferior prefrontal cortex, pars triangularis, survived after removing variance accounted for by acoustics: these regions still showed the negative relation with attractiveness. This region is part of Broca's area (Anwander, Tittgemeyer, von Cramon, Friederici, & Knosche, 2007) and is strongly connected to sensory cortex (Petrides & Pandya, 2009). In addition to its involvement in language perception, bilateral activity in Broca's area has been linked to auditory working memory in which increased task demands correlate with increased activity (Martinkauppi, Rama, Aronen, Korvenoja, & Carlson, 2000; Arnott, Grady, Hevenor, Graham, & Alain, 2005). Our results thus may suggest that increasingly unattractive voices demand larger processing resources and may point towards the role of the IFG pars triangularis as being involved not only in the processing of language and affective prosody but also in integrating acoustic information received from bilateral TVA into a unified percept of attractiveness.

A clear limitation of the above findings is that they were obtained with the use of brief vowels and hence cannot be easily generalized to realistic speaking situations in which a number of additional cues are present, including intonation, speaking rate, etc. Therefore, our results concern only one component that contributes to perceived voice attractiveness in realistic settings. Nonetheless, these findings have important potential implications for voice-based technology, suggesting simple ways of enhancing the attractiveness of synthetic voices at a time when automated voice production systems become ubiquitous.

Acknowledgments I gratefully acknowledge my co-authors on publications discussed above: Laetita Bruckert, Patricia E.G. Bestelmeyer, Marianne Latinus, Julien Rouger, Ian Charest, Guillaume A. Rousselet, Hideki Kawahara and Frances Crabbe. P.B. was supported by the French Fondation pour la Recherche Médicale (AJE201214) and Agence Nationale de la Recherche (PRIMAVOICE), and by grants ANR-16-CONV-0002 (ILCB), ANR-11-LABX-0036 (BLRI).

References

Anwander, A., Tittgemeyer, M., von Cramon, D. Y., Friederici, A. D., & Knosche, T. R. (2007). Connectivity-based parcellation of Broca's area. *Cerebral Cortex, 17*(4), 816–825.

Arnott, S. R., Grady, C. L., Hevenor, S. J., Graham, S., & Alain, C. (2005). The functional organization of auditory working memory as revealed by fMRI. *Journal of Cognitive Neuroscience, 17*, 819–831.

Belin, P., Zatorre, R. J., Lafaille, P., Ahad, P., & Pike, B. (2000). Voice-selective areas in human auditory cortex. *Nature, 403*, 309–312.

Bestelmeyer, P. E., Latinus, M., Bruckert, L., Rouger, J., Crabbe, F. & Belin, P. (2012). Implicitly perceived vocal attractiveness modulates prefrontal cortex activity. *Cereb Cortex* **22**, 1263–1270, https://doi.org/10.1093/cercor/bhr204

Braun, C., Gruendl, M., Marberger, C., & Scherber, C. (2001). Beautycheck—Ursachen und Folgen von Attraktivitae.

Bruckert, L., Bestelmeyer, P., Latinus, M., Rouger, J., Charest, I., Rousselet, G. A., et al. (2010). Vocal attractiveness increases by averaging. *Current Biology, 20*(2), 116–120.

Galton, F. (1878). Composite portraits. *Journal of the Anthropological Institute, 8*, 132–144.

Grammer, K., Fink, B., Moller, A. P., & Thornhill, R. (2003). Darwinian aesthetics: Sexual selection and the biology of beauty. *Biological Reviews, 78*(3), 385–407.

Hillenbrand, J. M., Getty, L. A., Clark, M. J., & Wheeler, K. (1995). Acoustic characteristics of American English vowels. *Journal of the Acoustical Society of America, 97*, 3099–3111.

Jastrow, J. (1885). Composite portraiture. *Science, 6*, 165.

Kawahara, H., & Matsui, H. (2003). Auditory morphing based on an elastic perceptual distance metric in an interference-free time-frequency representation. In*Proceedings of the International Conference on Acoustics, Speech, and Signal Processing* (pp. 256–259).

Langlois, J. H., & Roggman, L. A. (1990). Attractive faces are only average. *Psychological Science, 1*(2), 115–121.

Langlois, J. H., Kalakanis, L., Rubenstein, A. J., Larson, A., Hallam, M., & Smoot, M. (2000). Maxims or myths of beauty? A meta-analytic and theoretical review. *Psychological Bulletin, 126*(3), 390–423.

Latinus, M., & Belin, P. (2011). Anti-voice adaptation suggests prototype-based coding of voice identity. *Frontiers in Psychology, 2*, 175. https://doi.org/10.3389/fpsyg.2011.00175.

Latinus, M., McAleer, P., Bestelmeyer, P. E., & Belin, P. (2013). Norm-based coding of voice identity in human auditory cortex. *Current Biology, 23*(12), 1075–1080.

Leopold, D. A., Bondar, I. V., & Giese, M. A. (2006). Norm-based face encoding by single neurons in the monkey inferotemporal cortex. *Nature, 442*(7102), 572–575.

Martinkauppi, S., Rama, P., Aronen, H. J., Korvenoja, A., & Carlson, S. (2000). Working memory of auditory localization. *Cereb Cortex, 10*, 889–898.

Pernet, C., Charest, I., Belizaire, G., Zatorre, R. J., & Belin, P. (2007). The temporal voice area (TVA): Spatial characterization and variability. *Neuroimage, 36*, S1–S168.

Perrett, D. I., May, K. A., & Yoshikawa, S. (1994). Facial shape and judgements of female attractiveness. *Nature, 368*, 239–242.

Petrides, M., & Pandya, D. N. (2009). Distinct parietal and temporal pathways to the homologues of Broca's area in the monkey. *PLOS Biology, 7*, e1000170.

Thornhill, R., & Gangestad, S. W. (1999). Facial attractiveness. *Trends in Cognitive Sciences, 3*(12), 452–460.

Winkielman, P., Halberstadt, J., Fazendeiro, T., & Catty, S. (2006). Prototypes are attractive because they are easy on the mind. *Psychological Science, 17*(9), 799–806.

Winston, J. S., O'Doherty, J., Kilner, J. M., Perrett, D. I., & Dolan, R. J. (2007). Brain systems for assessing facial attractiveness. *Neuropsychologia, 45*(1), 195–206. https://doi.org/10.1016/j.neuropsychologia.2006.05.009.

Yovel, G., & Belin, P. (2013). A unified coding strategy for processing faces and voices. *Trends in Cognitive Sciences, 17*(6), 263–271.

Part III
Prosody

Chapter 9
Attractiveness of Male Speakers: Effects of Pitch and Tempo

Hugo Quené, Geke Boomsma, and Romée van Erning

Abstract Men with lower pitched voices tend to be rated as more attractive by female listeners; this tendency has been attributed to female sexual selection. Males not only speak with a lower pitch than females, however, but they also tend to speak at a faster tempo. Therefore, this study investigates whether speech tempo also affects the subjective attractiveness of male speakers for female listeners. To this end, sentences read by 24 male speakers were changed in relative tempo (factors 0.85, 1.00, and 1.15) and in overall pitch (-1.5, 0, $+1.5$ semitone), and were presented with and without fictitious portraits of the speakers. Ratings of speakers' attractiveness by female heterosexual listeners show significant effects of both tempo and pitch, in that voices with increased pitch and with decreased tempo are rated as significantly less attractive. In conclusion, female listeners rate a male speaker as less attractive if his voice pitch is increased (higher) and if his speech tempo is decreased (slower). Therefore, both tempo and pitch may be relevant for speech-based sexual selection of males by females.

Keywords Sexual selection · Voice pitch · Speech tempo · Speaking rate · Attractiveness · Experiment · Proportional odds model

9.1 Introduction

Male and female speakers differ in their average fundamental frequency (F0, perceived as pitch), viz., typically about 110 Hz for males and 205 Hz for females (Holmberg, Hillman, & Perkell, 1988; Simpson, 2009, Puts, Apicella, & Cárdenas, 2012).

H. Quené (✉) · G. Boomsma · R. van Erning
Utrecht institute of Linguistics, Utrecht University, Trans 10, 3512 JK
Utrecht, The Netherlands
e-mail: h.quene@uu.nl

G. Boomsma
e-mail: gekeboomsma@hotmail.com

R. van Erning
e-mail: romeevanerning@gmail.com

B. Weiss et al. (eds.), *Voice Attractiveness*, Prosody, Phonology and Phonetics,
https://doi.org/10.1007/978-981-15-6627-1_9

This large and significant difference in F0 develops in conjunction with primary and secondary sexual characteristics, during puberty. This suggests that the pitch difference may be related to some sexual function. The voice pitch of an adult male speaker is indeed reportedly related to the speaker's level of testosterone (Dabbs & Mallinger, 1999; Puts et al., 2012) and to the speaker's self-reported number of children (Apicella, Feinberg, & Marlowe, 2007) (but see Smith, Olkhov, Puts, & Apicella, 2017 for the mediating effect of hunting reputation). Thus, a male speaker's pitch may indicate his health and physical dominance, by virtue of the intercorrelations between male speakers' pitch and testosterone level (Dabbs & Mallinger, 1999; Puts et al., 2012), body height (Pisanski et al., 2014), physical strength (Puts et al., 2012), and masculinity (Clark & Henderson, 2003; Archer, 2006). Female listeners may, therefore, use voice pitch to assess the male speaker's physical suitability for producing and protecting offspring, i.e., in sexual selection via female choice of mate (Andersson, 1994). Indeed, ratings of attractiveness by female listeners are (negatively) correlated with the male speaker's F0 (Collins, 2000; Bruckert, Liénard, Lacroix, Kreutzer, & Leboucher, 2006), and experiments have confirmed that manipulations of F0 influence these attractiveness ratings (Feinberg, Jones, Little, Burt, & Perrett, 2005). In addition, voice pitch may be used to indicate health and dominance among male competitors, i.e., in sexual selection via male–male competition (Puts, Gaulin, & Verdolini, 2006), a mechanism which may be more important than female choice (Hill et al., 2013; Kordsmeyer, Hunt, Puts, Ostner, & Penke, 2018).

Males not only speak with a lower F0 than females, however, but they also tend to speak at a faster speech rate or tempo than females (about 5% faster) (Quené, 2008; Jacewicz, Fox, & Wei, 2010). This difference too may be related to male dominance, as the faster tempo presumably indicates the speaker's cognitive abilities and motor skills through his speaking. The faster tempo requires more physical energy (Moon & Lindblom, 2003), even more so because the male speech organs have somewhat more mass than the females', and it also requires more cognitive effort in linguistic planning and motor control. Indeed, faster speakers tend to be rated as more convincing, reliable, empathic, serious, active, and competent (Apple, Streeter, & Krauss, 1979; Smith, Brown, Strong, & Rencher, 1975). Presumably, then, female listeners also use a male speaker's tempo, to assess his motor skills and cognitive suitability as a potential mate.

This study aims primarily to replicate previous findings on female preference for male voices with lower *pitch*, and secondly to extend that work by investigating the presumed female preference for male speakers speaking at a faster *tempo*. Thirdly, we are interested in the interaction between the two factors. From a sexual selection perspective, a speaker who combines a low pitch with a fast tempo may be most attractive (and vice versa), because this combination would suggest a healthy physique as well as good motor and cognitive capabilities, a combination which is presumably more rare in potential male partners than the separate capabilities and characteristics.

The experiment reported below addresses these questions by manipulating Dutch sentences in tempo and pitch, and then asking Dutch female listeners to rate the attractiveness of the speaker. This attractiveness rating is regarded here as a proxy for the female listener's degree of preference for that male speaker in sexual selection, although vocal attractiveness also affects other social attributions (Babel, McGuire, & King, 2014).

In addition, this study also investigates whether these hypothesized effects of pitch and tempo are moderated by the presence of visual cues about the speaker in a portrait photo (see details below). On the one hand, humans have evolved to assess speakers not only by ear but also by eye,[1] so the task of rating a speakers' attractiveness may be more ecologically valid when a portrait is available. On the other hand, the presence of visual cues may well dampen the effects of prosodic cues. The fourth aim of this study was, therefore, to establish whether and how the presence of a portrait photo would affect a listeners' ratings of attractiveness of the speaker.

9.2 Methods

The experiment consisted of two sessions, in which the same speech stimuli were used. In the first session, speaker's voices were presented without a simultaneous portrait photo. In the second session, which included listeners who participated in the first session as well as new listeners, the same speech stimuli were presented *with* a portrait photo, in order to assess the effects of the portrait on listeners' responses. Listeners' task was to rate the attractiveness of the speaker.

During each session, a listener rated two different sentences spoken by the same speaker. One sentence was unchanged from the original, and the other sentence was manipulated orthogonally in pitch and/or in tempo, as described below. (This single-interval rating paradigm was chosen, instead of a two-interval forced-choice paradigm, because the latter would have highlighted the phonetic manipulations in one of the two speech intervals, and thus would have introduced biases in the responses subsequent to a listener noticing the manipulations).

The within-listener and within-speaker design allows for testing our primary predictions regarding the hypothesized effects of manipulated pitch and manipulated tempo on the subjective voice attractiveness of male speakers. Listeners' judgements are predicted to be affected by the phonetic manipulations, with higher ratings for lowered pitch and faster tempo, and with lower ratings for higher pitch and slower tempo, as argued above. The effects of phonetic manipulations may interact, and may be moderated by the photo conditions.

---

[1] Although present-day listeners may be used to hear speakers without seeing them, this is presumably not how speech has evolved in humans.

9.2.1 Participants

Listeners were 208 students or employees at Utrecht University, from 8 different undergraduate course groups taught in Dutch. In order to conceal the research topic (knowledge of which might have biased responses), targeted participants as well as other persons were tested and subsequently presented with a questionnaire asking about gender, sexual orientation, age, speech/hearing problems, and guess about the purpose of the experiment. Data from 58 persons were excluded for various reasons listed in Table 9.1.

Subsequent analysis was based on data from 150 remaining targeted participants: all female, self-identified other than lesbian, within age range 17–29 (median age 20, median absolute deviation 1.5, at second session; this was done to select participants from approximately the same age range as the speakers, to improve ecological validity), without self-identified speech/hearing problems, and not aware of the purpose of the study. All participants were highly proficient in Dutch, as their native language or as a non-native language attested at an advanced academic level (B2 or higher).

9.2.2 Materials

Stimulus sentences were taken from Dutch spontaneous monologues by 24 male speakers (age $M = 18.0$, $s = 0.7$, range 16–19 years), who spoke about an informal topic of their own choice. These monologues had been previously recorded for a different study at 44.1 kHz (for further details, see Quené & Orr, 2014; Quené, Orr, & van Leeuwen, 2017). Two sentences were selected from each speaker's interview. Selected sentences were between 2.5 and 3.5 s in duration, which were spoken fluently and without a long pause, with neutral content, comprehensible without context, and not elliptic (i.e., contained both a subject and an inflected verb). Thus the sentences

Table 9.1 Numbers of participants, with reasons for exclusion from data analysis. Multiple reasons may apply to a single participant

Description	Female	Male	Total
All participants tested	$\geq$155	$\leq$53	208
Aborted prematurely	$\leq$3	$\leq$1	3
Already knew purpose of study	6	3	9
Speech/hearing problems	7	4	11
No valid responses	0	1	1
Gender male or unspecified	0	$\leq$35	35
Orientation lesbian	3	0	3
Age $<$ 16 or $>$ 30	3	0	3
Participants remaining	150	0	150

should provide listeners with enough speech material to rate voice attractiveness, without requiring listeners' inference of context or grammar.

For each of the 24 × 2 selected stimulus sentences, average syllable duration (excluding pauses, Quené, 2008 and average F0 (over voiced portions) were measured using `Praat` (Boersma & Weenink, 2015). These measurements were then analyzed by means of linear mixed models (Quené & van den Bergh, 2004; 2008; Bates et al., 2015; R Core Team, 2018) with only the intercept as a fixed predictor, and with speakers as random intercepts. The estimated average syllable duration was 0.188 s ($s_u = 0.015$, $s_e = 0.026$, ICC $= 0.25$, i.e., with most variance between sentences within speakers), and the estimated average F0 was 116 Hz ($s_u = 16$, $s_e = 7$, ICC $=$ 0.82, i.e., with most variance between speakers).

In order to once again conceal the research topic, similar filler stimuli, but spoken by female speakers, were also included in the experiment. These filler sentences were taken from recorded monologues of 24 female speakers (each contributing one sentence) from the same corpus and using the same selection criteria as for male speakers. Neither the filler sentences themselves nor any responses to these fillers were further analyzed.

For the second session, each individual speaker (male or female voice) was matched to an individual portrait photo. These photos were taken from 3 public databases of facial portraits (Hancock, 2008; Nefian, 1999; Spacek, 2008) and did *not* portray the actual speakers. The selected photos of 24 males and 24 females each showed one person in the target age range (18–25 years) with a neutral facial expression. All selected photos were cropped and/or resized to the same size.

9.2.3 Speech Manipulations

One of the two sentences of each male speaker was retained as a baseline stimulus with unchanged tempo and unchanged pitch. The other sentence of each male speaker was varied in tempo (factors 0.85, 1.00, 1.15) and in overall pitch ($-1.5, 0, +1.5$ semitone), yielding 8 manipulated versions of each sentence. The changes are well above the respective just noticeable differences (Quené, 2006; 'THart, Collier, & Cohen, 1990) and they correspond to approximately $\pm 1 s_e$ for both manipulations, while the resulting sentences still sound very natural to us. Filler sentences by female speakers were not varied. Tempo and pitch were manipulated by means of `sox` (Bagwell, 2013). Finally, stimulus and filler sentences were all scaled to −0.5 dB relative to the maximum amplitude.

9.2.4 Procedure

The 8 manipulated versions of each sentence were distributed over 8 experimental lists, counterbalanced over the 24 male speakers. The 24 unchanged male-spoken sentences and 24 female-spoken filler sentences were added to each experimental list. Hence, the unchanged sentences of all speakers were presented to all listeners, whereas the changed sentences were partitioned over lists so that each listener heard only a single changed version of a particular sentence. This design allowed subsequent within-speaker and within-listener comparisons of baseline and changed versions. The 72 sentences were presented in quasi-random order[2] (which was however the same across the 8 lists).

The experiment was conducted in a classroom setting, with each experimental list presented to a separate undergraduate course group. In the first session, speech stimuli were presented (using PowerPoint) over the classroom sound system. In the second session, typically a few days later, the same speech stimuli were presented with simultaneous portraits visible, using the same sound system and the classroom computer projector. The inter-stimulus interval was 3 s in both sessions, as determined in pilot tests.

Of the remaining 150 participants, 76 participated only in the first session (absent from the second session), 20 only in the second session (absent from the first session), and 54 participated in both sessions, the latter group allowing within-subject comparisons.

Participants were instructed to rate the attractiveness of the speaker on a 7-point Likert scale (1 extremely unattractive, 7 extremely attractive) on a printed response sheet. For the first session, their instruction was as follows (in translation):

> … In a moment you will hear 72 sound fragments of people saying something. We'd like to ask you to indicate for every sound fragment how attractive you find the speaker. You have about 3 s to respond for each person.

For the second session, participants' instruction was as follows (in translation):

> … In a moment you will see 72 photos of people. With every face you will also hear a sound fragment. We'd like to ask you to indicate for every person how attractive [Dutch: "hoe aantrekkelijk"] you find that person. You have about 3 s to respond for each person.

After the rating sessions, participants were invited to answer a brief questionnaire about their gender, age, native language(s), hearing problems, speech problems, dexterity, and sexual orientation as heterosexual or homosexual or bisexual or unknown (including unwilling to answer); see Sect. 9.2.1.

9.3 Results

The average ratings by the targeted listeners observed in the listening experiment are summarized in Table 9.2. The lower standard error in the baseline condition is due to the larger number of responses in this condition, because all listeners have judged the unchanged sentences of all speakers (see Sect. 9.2.4).

[2]Between stimuli involving the same speaker, at least 5 different test or filler sentences were presented.

Table 9.2 Mean responses (by targeted listeners only) of subjective attractiveness on a 7-point scale, broken down by manipulations of tempo and pitch, with standard errors in parentheses

		Pitch		
		Lower	Unchanged	Higher
Tempo	Slower	2.78 (0.06)	2.94 (0.06)	2.47 (0.05)
	Unchanged	3.28 (0.06)	3.30 (0.02)	2.55 (0.05)
	Faster	3.15 (0.06)	3.39 (0.06)	2.55 (0.06)

The separate responses given by each of the 150 remaining listeners to each of the 24 unchanged and 24 manipulated speech stimuli were analyzed by means of a cumulative-link mixed-effects model (CLMM) (Quené & van den Bergh, 2004; Christensen, 2015). This family of models (also known as proportional odds models) regards the dependent variable as ordinal, and coefficients represent the changes in log odds of a response falling in the jth category or higher. In other words, a CLMM as used here is somewhat similar to a GLMM (Quené & Van den Bergh, 2008), but with multiple ordered response categories. Fixed predictors in the CLMM were the 8 manipulated conditions of tempo and pitch (using dummy coding, with the unchanged condition as baseline), the centered trial number,[3] and the absence (baseline code 0) or presence (contrast code 1) of a portrait photo. Two-way interactions between photo and manipulation conditions were also included as fixed predictors. Random predictors in the CLMM were listeners ($n = 150$), speakers ($n = 24$), and sentences ($n = 48$) as three crossed random intercepts. The main effect of the photo condition was also included as a random slope at the speaker level, thus allowing for nonuniform effects of the portrait photo across speakers.

The fixed regression coefficients, random variances and correlations, and category thresholds estimated by the CLMM described above are listed in Table 9.3.

The **fixed** part of the CLMM shows several interesting effects. In the conditions without a photo (first session), the conditions with slower tempo, as well as the conditions with higher pitch, all yield a significant negative effect: slower tempo is *less* attractive than the unchanged baseline, and so is higher pitch. However, none of the opposite conditions with faster tempo (conditions FU and FL), and none of the conditions with lower pitch, yields a positive effect: faster tempo is equally attractive as the unchanged baseline, and so is lower pitch.

Second, the photo condition yielded a large and significant negative main effect, with considerably lower ratings in the second session (with photo) as compared to the first session (without photo). The interactions suggest that the negative effect of adding a photo is significantly mitigated, in particular, in those phonetic conditions yielding the most negative ratings without a photo. As discussed below, this interaction pattern may suggest a floor effect.

[3]This centered trial number was scaled by factor 0.1 for computational reasons.

Table 9.3 Estimated coefficients of the CLMM, for intercepts and effects of conditions of tempo (S: slower, U: unchanged, F: faster) and pitch (L: lower, U: unchanged, H: higher), trial number (centered and scaled), and photo condition. Random effects are reported in units of variance of log odds (logit), with standardized correlation among random effects; a significant correlation is marked with an asterisk ($p < 0.05$ according to bootstrapped 95% confidence interval of the correlation, over 200 bootstrap replications). Fixed effects are reported in log odds (logit) units; significant coefficients are marked with an asterisk ($p < 0.05$)

Random: listeners	Variance			
(Intercept)	0.9569			
Random: speakers	Variance	Correlation		
(Intercept)	0.8032			
Photo	0.4856	−0.57*		
Random: sentences	Variance			
(Intercept)	0.4076			
Fixed	Estimate	Std. Error	*z* value	*p* value
cond.SH	−1.75	0.22	−8.11	<0.0001*
cond.SU	−0.78	0.21	−3.65	0.0003*
cond.SL	−0.97	0.21	−4.54	<0.0001*
cond.UH	−1.29	0.21	−6.05	<0.0001*
cond.UL	−0.10	0.21	−0.49	0.6224
cond.FH	−1.84	0.22	−8.55	<0.0001*
cond.FU	0.15	0.21	0.71	0.4806
cond.FL	−0.07	0.21	−0.32	0.7457
photo	−1.38	0.16	−8.84	<0.0001*
cond.SH:photo	0.59	0.16	3.36	0.0008*
cond.SU:photo	0.26	0.17	1.53	0.1257
cond.SL:photo	0.45	0.17	2.63	0.0087*
cond.UH:photo	0.17	0.18	0.98	0.3261
cond.UL:photo	0.12	0.17	0.71	0.4789
cond.FH:photo	1.14	0.17	6.53	<0.0001*
cond.FU:photo	−0.20	0.17	−1.22	0.2225
cond.FL:photo	−0.01	0.17	−0.04	0.9700
trial	−0.09	0.06	−1.59	0.1120
Category thresholds	Estimate	Std. Error	*z* value	
1|2	−3.25	0.24	−13.36	
2|3	−1.43	0.24	−5.91	
3|4	−0.04	0.24	−0.17	
4|5	1.19	0.24	4.94	
5|6	2.75	0.24	11.28	
6|7	4.82	0.26	18.55	

Finally, the coefficients in the fixed part of the CLMM did not show an effect of the trial number on listeners' judgements: listeners did not tend to increase or decrease their ratings during a session.

The **random** part of the CLMM shows that speakers' intercepts correlate with speakers' slope of the photo condition ($r = -0.57$): speakers whose voices were judged as more attractive tended to "lose" less when combined with an alleged portrait, or in other words, the negative main effect of the photo portrait was relatively stronger (more negative) for less-attractive voices.

9.4 Discussion

First, the results confirm previously reported effects of **pitch** manipulations on attractiveness (Collins, 2000; Feinberg et al., 2005): male voices with increased pitch are rated as less attractive by heterosexual female listeners. While previous studies used only short vowel stimuli, these findings are partially replicated here with sentence-length stimuli. This result further corroborates the evidence for the role of male voice pitch in sexual selection through female choice of mate. In spite of this effect in the manipulated stimuli, however, the corresponding effect was not observed for voices with decreased pitch.

Second, the results confirm our prediction that manipulations of **tempo** also affect the speaker's attractiveness, with slower speech being less attractive. Slower speakers may be regarded as less attractive because speech tempo may indicate the speaker's (relatively poor) motor skills and cognitive capabilities. Again, the corresponding effect was not observed for voices with increased tempo.

In comparison, the detrimental effect of slower tempo appears to be somewhat smaller than that of higher pitch (cf. Table 9.3). This difference in effect size for pitch and tempo may be explained in three ways. One explanation could be that pitch constitutes a more salient cue in sexual selection than tempo, because pitch varies more between speakers (and less within speakers) than tempo does (cf. Sect. 9.2.2 for variations in our stimuli), so that pitch may be a more reliable indicator of the speaker's individual characteristics than tempo. Another plausible explanation could be that our pitch manipulations were perceptually larger than our tempo manipulations, relative to the individual differences between speakers. The prosodic measurements and manipulations described above (Sects. 9.2.2–9.2.3), however, do not support this latter explanation: the pitch manipulations are about $\pm\frac{1}{2}s_u$ whereas the tempo manipulations are relatively larger, about $\pm 2s_u$ (for comparison, both manipulations were about $\pm 1s_e$ in magnitude). A third explanation was proposed by Babel et al. (2014) who argue that attractive voice properties may not be universal, but dependent on cultural preferences; the weights of tempo and pitch properties on voice attractiveness may thus be culturally constrained. Further research, with different sizes of phonetic manipulations and with listeners sampled from different cultures, would be required to rule out one or more of these explanations.

Third, the results do not support the hypothesized interaction between pitch and tempo cues on speakers' attractiveness. In the first session (without photo), neither lower pitch, nor faster tempo, nor the combination of these two manipulations yielded a positive effect on voice attractiveness. Moreover, lowest ratings were obtained in conditions with increased pitch, irrespective of the tempo manipulations. This suggests that the combined traits of physical and cognitive capabilities are somehow assessed independently, contrary to the expectations outlined in Sect. 9.1.

Finally, the results suggest that the **photo** portraits may have introduced floor effects in this experiment. Coefficients in the fixed part of the CLMM suggest that conditions yielding the lowest ratings without a photo (session 1) also decrease less with a photo (session 2), which may be because the conditions involving less-attractive speech cannot "lose" as many points when combined with a photo. In addition, speakers who are rated as more attractive tend to "lose" more when accompanied by a photo ($r = -0.57$, Table 9.3), which may again be because less-attractive speakers cannot be rated below the floor of the Likert scale. The photos were included in the experimental design in order to investigate the effects of (ecologically valid) visual cues on voice attractiveness ratings. However, the unexpected negative effect of adding a portrait photo may have resulted in ratings that were too low to show the effects of phonetic properties. One possible explanation is that the photos were taken from relatively old sources (portraits were at least 8 years old at the time of testing) and may have contained outdated visual cues regarding style, hairdress, etc., for the target listeners in our study. More speculatively, there may have been some unknown mismatch between (non-Dutch) portraits (Hancock, 2008; Nefian, 1999; Spacek, 2008) and (Dutch) voices, yielding a negative effect on the ratings in the with-portrait condition. For further phonetic research into listeners' attractiveness judgements, we recommend to refrain from randomly matched portraits accompanying the voice stimuli.

9.5 Conclusions

Female listeners rate a male speaker as less attractive if his voice pitch is increased and if his speech tempo is decreased, relative to a baseline sentence with unchanged pitch and tempo. These effects suggest that both pitch and tempo play a role in speech-based sexual selection of males by females, although our results suggest that the underlying mechanisms for pitch and tempo may well be different. Voice pitch indicates the speaker's health and physical dominance (Dabbs & Mallinger, 1999; Puts et al., 2012; Collins, 2000; Feinberg et al., 2005), while speech tempo may indicate the speaker's motor skills and cognitive competence (Apple et al., 1979; Smith et al., 1975). The effect of voice pitch on attractiveness is larger than that of speech tempo, perhaps because pitch varies relatively more between speakers than within speakers, in contrast to tempo, so that pitch may constitute a more reliable cue to a speaker's individual characteristics.

Acknowledgments Results from a different, related study (using the same audio stimuli, always presented with photos, with different participants) were reported at the Speech Prosody 2016 conference (Boston, U.S.A.). We thank Nivja de Jong, Gerrit Bloothooft, Huub van den Bergh, the audience at Speech Prosody 2016, and three anonymous reviewers, for helpful comments and suggestions.

References

Andersson, M. B. (1994). *Sexual selection*. Princeton: Princeton Univesity Press.

Apicella, C. L., Feinberg, D. R., & Marlowe, F. W. (2007). Voice pitch predicts reproductive success in male hunter-gatherers. *Biology Letters, 3*, 682–684.

Apple, W., Streeter, L. A., & Krauss, R. M. (1979). Effects of pitch and speech rate on personal attributions. *Journal of Personality and Social Psychology, 37*, 715–727.

Archer, J. (2006). Testosterone and human aggression: An evaluation of the challenge hypothesis. *Neuroscience and Biobehavioral Reviews, 30*, 319–345.

Babel, M., McGuire, G., & King, J. (2014). Towards a more nuanced view of vocal attractiveness, *PLoS ONE, 9*(2), e88616. https://doi.org/10.1371/journal.pone.0088616

Bagwell, C. (2013). Sound eXchange (SOX), version 14-4-1. http://sourceforge.net/projects/sox/.

Bates, D., Maechler, M., Bolker, B., & Walker, S. (2015). lme4: Linear mixed-effects models using Eigen and S4. R package version 1.1-9. http://CRAN.R-project.org/package=lme4

Boersma, P., & Weenink, D. (2015) Praat: Doing phonetics by computer, version 6.0 http://www.praat.org

Bruckert, L., Liénard, J.-S., Lacroix, A., Kreutzer, M., & Leboucher, G. (2006). Women use voice parameters to assess men's characteristics. *Proceedings of the Royal Society B: Biological Sciences, 273*(1582), 83–89.

Christensen, R. H. B. (2015). ordinal—Regression Models for Ordinal Data. R package version 2015.6-28. http://www.cran.r-project.org/package=ordinal/.

Clark, A. S., & Henderson, L. P. (2003). Behavioral and physiological responses to anabolic-androgenic steroids. *Neuroscience and Biobehavioral Reviews, 27*, 413–436.

Collins, S. A. (2000). Men's voices and women's choices. *Animal Behaviour, 60*, 773–780.

Dabbs, J. M., & Mallinger, A. (1999). High testosterone levels predict low voice pitch among men. *Personality and Individual Differences, 27*, 801–804.

Feinberg, D. R., Jones, B. C., Little, A. C., Burt, D. M., & Perrett, D. I. (2005). Manipulations of fundamental and formant frequencies influence the attractiveness of human male voices. *Animal Behaviour, 69*, 561–568.

Hancock, P. (2008). Utrecht ECVP: Psychological Image Collection at Stirling. http://pics.stir.ac.uk/.

't Hart, J. T., Collier, R., & Cohen, A. (1990). *A perceptual study of intonation: An experimental-phonetic Approach to speech perception*. Cambridge: Cambridge University Press.

Hill, A. K., Hunt, J., Welling, L. L. M., Cárdenas, R. A., Rotella, M. A., Wheatley, J. R., et al. (2013). Quantifying the strength and form of sexual selection on men's traits. *Evolution and Human Behavior, 34*, 334–341.

Holmberg, E. B., Hillman, R. E., & Perkell, J. S. (1988). Glottal airflow and transglottal air pressure measurements for male and female speakers in soft, normal, and loud voice. *The Journal of the Acoustical Society of America, 84*, 511–529.

Kordsmeyer, T. L., Hunt, J., Puts, D. A., Ostner, J., & Penke, L. (2018). The relative importance of intra- and intersexual selection on human male sexually dimorphic traits. *Evolution and Human Behavior, 39*, 424–436.

Jacewicz, E., Fox, R. A., & Wei, L. (2010). Between-speaker and within-speaker variation in speech tempo of American English. *The Journal of the Acoustical Society of America, 128*, 839–850.

Moon, S.-J., & Lindblom, B. (2003). Two experiments on oxygen consumption during speech production: Vocal effort and speaking tempo. In *Proceedings of XVth International Congress of Phonetic Sciences* (pp. 3129–3132), Barcelona, Spain.

Nefian, A. V. (1999). Georgia Tech face database. http://www.anefian.com/research/face_reco.htm

Orr, R., Quené, H., van Beek, R., Diefenbach, T., van Leeuwen, D. A., Huijbregts, M. (2011). An International English speech corpus for longitudinal study of accent development. In *InterSpeech 2011, 27–31 Aug, Florence, Italy, Proceedings* (pp. 1889–1892).

Puts, D. A., Gaulin, S. J. C., & Verdolini, K. (2006). Dominance and the evolution of sexual dimorphism in human voice pitch. *Evolution and Human Behavior*, *27*, 283–296.

Puts, D. A., Apicella, C. L., & Cárdenas, R. A. (2012). Masculine voices signal men's threat potential in forager and industrial societies. *Proceedings of the Royal Society B: Biological Sciences*, *279*, 601–609.

Pisanski, K., Fraccaro, P. J., Tigue, C. C., O'Connor, J. J. M., Röder, S., Andrews, P. W., et al. (2014). Vocal indicators of body size in men and women: A meta-analysis. *Animal Behaviour*, *95*, 89–99.

Quené, H. (2006). On the just noticeable difference for tempo in speech. *Journal of Phonetics*, *35*, 353–362.

Quené, H. (2008). Multilevel modeling of between-speaker and within-speaker variation in spontaneous speech tempo. *The Journal of the Acoustical Society of America*, *123*, 1104–1113.

Quené, H., Orr, R., Long-term convergence of speech rhythm in L1 and L2 English. In *Speech Prosody 2014, 20–23 May, Dublin, Ireland, Proceedings* (pp. 342–345).

Quené, H., Orr, R., & van Leeuwen, D. (2017). Phonetic similarity of /s/ in native and second language: Individual differences in learning curves. *Journal of the Acoustical Society of America*, *142*(6), EL519–EL524. https://doi.org/10.1121/1.5013149.

Quené, H., & van den Bergh, H. (2004). On multi-level modeling of data from repeated measures designs: A tutorial. *Speech Communication*, *43*(1–2), 103–121.

Quené, H., & Van den Bergh, H. (2008). Examples of mixed-effects modeling with crossed random effects and with binomial data. *Journal of Memory and Language*, *59*(4), 413–425.

R Core Team (2018) R: A language and environment for statistical computing. R Foundation for Statistical Computing, Vienna, Austria, version 3.4.4. http://www.R-project.org.

Simpson, A. P. (2009). Phonetic differences between male and female speech. *Language and Linguistics Compass*, *3*, 621–640.

Smith, B. L., Brown, B. L., Strong, W. J., & Rencher, A. C. (1975). Effects of speech rate on personality perception. *Language & Speech*, *18*, 145–152.

Smith, K. M., Olkhov, Y. M., Puts, D. A., & Apicella, C. L. (2017). Hadza men with lower voice pitch have a better hunting reputation. *Evolutionary Psychology*. https://doi.org/10.1177/1474704917740466.

Spacek, L. (2008). Collection of facial images (faces94, faces95). http://cswww.essex.ac.uk/mv/allfaces/.

Chapter 10
The Contribution of Amplitude Modulations in Speech to Perceived Charisma

Hans Rutger Bosker

Abstract Speech contains pronounced amplitude modulations in the 1–9 Hz range, correlating with the syllabic rate of speech. Recent models of speech perception propose that this rhythmic nature of speech is central to speech recognition and has beneficial effects on language processing. Here, we investigated the contribution of amplitude modulations to the subjective impression listeners have of public speakers. The speech from US presidential candidates Hillary Clinton and Donald Trump in the three TV debates of 2016 was acoustically analyzed by means of modulation spectra. These indicated that Clinton's speech had more pronounced amplitude modulations than Trump's speech, particularly in the 1–9 Hz range. A subsequent perception experiment, with listeners rating the perceived charisma of (low-pass filtered versions of) Clinton's and Trump's speech, showed that more pronounced amplitude modulations (i.e., more 'rhythmic' speech) increased perceived charisma ratings. These outcomes highlight the important contribution of speech rhythm to charisma perception.

Keywords Amplitude modulations · Speech rhythm · Modulation spectrum · Charisma perception · Temporal envelope · Political debates

10.1 Introduction

Any spoken utterance, regardless of talker, language, or linguistic content, contains fast-changing spectral information (e.g., vowel formants, consonantal frication, etc.) as well as slower changing temporal information. The temporal information in speech is particularly apparent in the temporal envelope of speech, which includes the fluctuations in amplitude from consonants (constricted vocal tract, lower amplitude) to vowels (unconstricted vocal tract, higher amplitude), from stressed (prominent) to

H. R. Bosker (✉)
Max Planck Institute for Psycholinguistics, P. O. Box 310, 6500 AH Nijmegen, The Netherlands
e-mail: HansRutger.Bosker@mpi.nl

Psychology of Language Department, Donders Institute for Brain Cognition and Behaviour, Radboud University, Kapittelweg 29, 6525 EN Nijmegen, The Netherlands

B. Weiss et al. (eds.), *Voice Attractiveness*, Prosody, Phonology and Phonetics,
https://doi.org/10.1007/978-981-15-6627-1_10

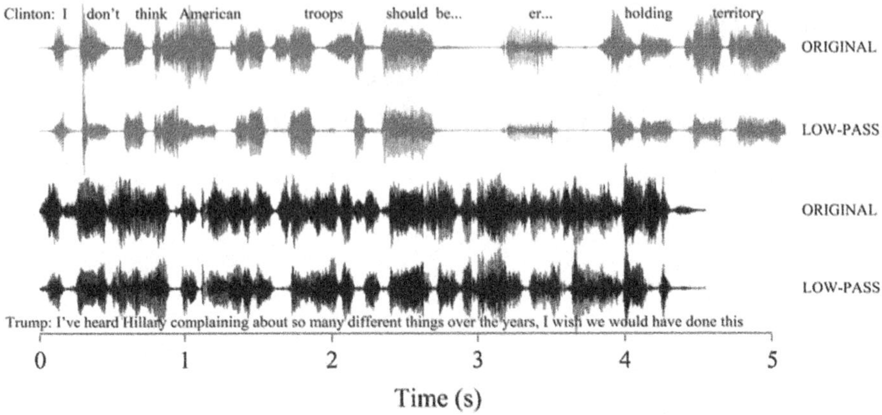

Fig. 10.1 Excerpts of Clinton's speech (in gray) with a notable syllabic rhythm around 3 Hz and Trump's speech (in black) with a notable lack of consistent slow-amplitude modulations. Below each waveform are the low-pass filtered versions of the excerpts, demonstrating that the original slow-amplitude modulations are maintained to a large degree

unstressed syllables (less prominent), etc. For instance, the top example in Fig. 10.1 has pronounced fluctuations in amplitude (also known as amplitude modulations) occurring at around 3 Hz, related to the syllabic rate of the utterance (i.e., roughly three syllables per second).

The temporal dynamics of speech (e.g., energy patterns and syllable durations in speech) are semi-regular at multiple (segmental, syllabic, sentential) timescales (Poeppel, 2003; Rosen, 1992). Hence, speech is an intrinsically rhythmic signal, with 'rhythmic' referring to the semi-regular recurrence over time of waxing and waning prominence profiles in the amplitude signature of speech (for other conceptualizations of speech rhythm, see Kohler, 2009; Nolan & Jeon, 2014). Naturally produced syllable rates typically do not exceed a rate of 9 Hz (Ghitza, 2014; Jacewicz, Fox, & Wei, 2010; Pellegrino, Coupé, & Marsico, 2011; Quené, 2008; Varnet, Ortiz-Barajas, Erra, Gervain, & Lorenzi, 2004). As such, most of the energy in the amplitude modulations in the speech signal is found below 9 Hz (Ghitza & Greenberg, 2009; Greenberg & Arai, 1999, 2004), across a range of typologically distant languages (Ding et al., 2017; Varnet, Ortiz-Barajas, Erra, Gervain, & Lorenzi, 2017), with the most prominent modulation frequencies near the average syllable rate of 3–4 Hz (Delgutte 1998).

In recent models of speech perception (Ghitza 2011; Giraud & Poeppel, 2012; Peelle & Davis, 2012), this rhythmic nature of speech is said to play a central role in speech recognition. For instance, speakers who are intrinsically more intelligible than others show more pronounced low-frequency modulations in the amplitude envelope (Bradlow, Torretta, & Pisoni, 1996). In fact, when the slow amplitude fluctuations in speech are degraded or filtered out, intelligibility drops dramatically (Drullman, Festen, & Plomp, 1994; Ghitza, 2012; Houtgast & Steeneken, 1973), while speech with only minimal spectral information remains intelligible as long as low-frequency temporal modulations are preserved (Shannon, Zeng, Kamath, Wygonski, & Ekelid, 1995). Similarly, speech stream segregation (understanding speech in noise; Aikawa & Ishizuka, 2002), word segmentation (resolving continuous speech into words; Cutler, 1994; Cutler & Butterfield, 1992; Cutler & Norris, 1988), and phoneme perception (Bosker, 2017a; Bosker & Ghitza, 2018; Quené, 2005) are all influenced by regular energy fluctuations in speech.

A powerful demonstration of the contribution of regular amplitude modulations to speech comprehension is the finding that otherwise unintelligible speech can be made intelligible by imposing an artificial rhythm (Bosker & Ghitza, 2018; Doelling, Arnal, Ghitza, & Poeppel, 2014; Ghitza, 2012, 2014). For instance, Bosker and Ghitza (2018) took Dutch recordings of seven-digit telephone numbers (e.g., "215–4653") and compressed these by a factor of 5 (i.e., make the speech five times as fast while preserving spectral properties such as pitch and formants). This heavy compression manipulation made the intelligibility of the telephone numbers drop from the original 99% to about 39% digits correct. However, Bosker and Ghitza then imposed an artificial rhythm onto the heavily compressed speech, by taking 66 ms windows of compressed speech and spacing these apart by 100 ms of silence (i.e., inserting 100-ms silent intervals). This 'repackaged' condition did not contain any additional linguistic or phonetic information compared to the heavily compressed speech; it only differed in having a very pronounced amplitude modulation around 6 Hz. The authors found that imposing this artificial rhythm onto the compressed speech boosted intelligibility (from 39 to 71%) digits correct, demonstrating that regular amplitude modulations play a central role in speech perception.

Rhythmic amplitude modulations in speech not only affect speech intelligibility but they also play a role in spoken communication more generally. For instance, syntactic processing (Roncaglia-Denissen, Schmidt-Kassow, & Kotz, 2013), semantic processing (Rothermich, Schmidt-Kassow, & Kotz, 2012), and recognition memory (Essens & Povel 1985) are all facilitated by regular meter. Moreover, there are even suggestions in the literature that listeners explicitly prefer listening to speech with a clear rhythmic structure. For instance, Obermeier et al. (2013) took four-verse stanzas from old German poetry and independently manipulated the rhyme and meter of these poetry fragments. Rhyme was manipulated by substituting rhyming sentence-final words with non-rhyming words with the same metrical structure (maintaining meter), while meter was manipulated by substituting a sentence-medial word with a word with mismatching metrical structure (e.g., "Nacht" > "Dunkelheit"; maintaining rhyme in sentence-final words). Native German participants rated the original and manipulated fragments of poetry on liking and perceived intensity. Results indicated that non-rhyming and non-metrical stanzas received lower ratings on both the liking and perceived intensity scales, suggesting that the presence of rhythmical structure induces greater esthetic liking and more intense emotional processing (Obermeier et al., 2013, 2016).

Here, we examined the contribution of rhythmic amplitude modulations to the perception of charisma in public speakers' voices. Charisma and charismatic leadership are intensively studied topics, with clear implications for public speakers, politics, religion, and society at large. There seems to be a consensus in the literature that being a charismatic speaker is a necessary precondition for being a charismatic leader. In fact, how one speaks (i.e., performance characteristics, such as pitch, loudness, prosody, etc.) has been argued to contribute to charisma perception more than what one says (i.e., the linguistically formulated communicative message; Awamleh & Gardner, 1999; Rosenberg & Hirschberg, 2009). Several studies have, therefore, attempted to find acoustic correlates of charisma in public speakers' voices (see also in this volume; Rosenberg & Hirschberg, this volume; Brem & Niebuhr, this volume). For instance, pausing behavior (D'Errico, Signorello, Demolin & Poggi, 2013), speech rate (D'Errico et al., & Poggi, 2013), overall intensity (Niebuhr, Voße & Brem, 2016), number and type of disfluencies (Novák-Tót, Niebuhr, & Chen, 2017), and timbre (Weiss and Burkhardt, 2010) have all been identified as contributing to perceived charisma and personality. However, although there are suggestions in the literature that greater variability in pitch and intensity contours increases perceived charisma (D'Errico et al., 2013; Niebuhr et al., 2016; Rosenberg & Hirschberg, 2009), it is unclear what the role of the rhythm of speech is in charisma perception. Therefore, the present research goal was to investigate how political debaters make use of variation in the amplitude envelope in speech production and how this variation, in turn, may affect speech perception.

Regarding rhythm in speech production, we report an acoustic comparison of the temporal amplitude modulations in the speech produced by two presidential candidates in the American elections of 2016: Hillary Clinton and Donald Trump. Recordings from three national presidential debates were collected and the speech produced by both candidates was first matched for overall intensity. Thereafter, their speech was analyzed by means of modulation spectra (Bosker & Cooke, 2018; Ding et al., 2017; Krause & Braida, 2004). These modulation spectra quantify the power of individual modulation frequency components present in a given signal (e.g., see Fig. 10.2), with power on the y-axis and modulation frequency on the x-axis. They can be used to assess which modulation frequencies are most prominent in different signals (e.g., speech and music show well-separated peaks around 5 and 2 Hz, respectively; Ding et al. 2017) but also to compare the overall power (in different frequency bands) across talkers or speech registers (Krause & Braida, 2004). For instance, Bosker and Ghitza 2018 calculated modulation spectra of spoken sentences produced in quiet (plain speech) and the same sentences produced in noise (Lombard speech). Results showed greater power in Lombard speech compared to plain speech, particularly in the 1–4 Hz range, demonstrating that talkers produce more pronounced amplitude modulations when talking in noise, presumably to aid speech comprehension.

Similarly, the present acoustic analysis compared the power of different modulation frequency bands across the two talkers. Greater power in the modulation spectrum of one speaker over another would reveal a more pronounced temporal envelope in that particular candidate's speech (i.e., greater amplitude modulations). Specifically, we expect power differences to occur within the frequency range of typical speech rates, namely below 9 Hz because (1) this modulation range is most characteristic of spontaneous speech (Ding et al., 2017); and (2) previous research indicates that differences between speech registers (plain vs. Lombard speech) are apparent in the lower modulation range (Bosker and Ghitza 2018). Power differences in this 1–9 Hz modulation range would be indicative of a more regular syllabic rhythm. Moreover, the locations of peaks in the modulation spectrum would reveal which modulation frequencies are most pronounced in that speaker's amplitude envelope, being indicative of a specific rhythm preference. By contrast, differences in the power of modulation frequencies between 9–15 Hz are expected to be smaller (if present at all) since this modulation range is less pronounced in speech and is not straightforwardly related to particular acoustic or perceptual units in speech.

When it comes to quantifying rhythm in speech, modulation spectra have several advantages over other rhythm metrics that have been introduced in the literature, such as %V (percentage over which speech is vocalic; Ramus et al. (1999)), $Theta$C (standard deviation of consonantal intervals; Ramus et al. (1999)), PVI (pairwise variability index; Grabe and Low (2002)), or normalized metrics such as VarcoV and VarcoC (Dellwo, 2006; White and Mattys, 2007). These metrics assess durational variability (Loukina et al., 2011), not necessarily periodicity. That is, both isochronous and anisochronous distributions of vowels and consonants can have the same %V. Moreover, such measures are influenced by between-language differences, whereas modulation spectra are not (Ding et al., 2017).

Going beyond merely identifying differences in the use of rhythm between speakers in speech production, we also tested the contribution of pronounced amplitude modulations to speech perception. Specifically, a rating experiment was carried out with low-pass filtered versions of (a subset of) the speech from both speakers. Filtering was applied to reduce the contribution of lexical-semantic information to participants' judgments while maintaining the temporal structure of the acoustic signal (see Fig. 10.1), forcing listeners to base their judgments primarily on temporal characteristics. In line with the introduced beneficial effects of rhythmic regularity on speech intelligibility and esthetic liking, we hypothesized that the perceived charisma ratings would correlate with the speech rhythm in the signals. That is, speech fragments with more pronounced amplitude modulations in the 1–9 Hz range would be expected to be rated as more charismatic than speech fragments with less pronounced amplitude modulations. If corroborated, this would indicate that speech rhythm not only contributes to intelligibility and the qualitative appreciation of the linguistic message but also to the subjective impression listeners have of a (public) speaker.

10.2 Acoustic Analysis

10.2.1 Method

10.2.1.1 Materials

Recordings of all three presidential debates between Hillary Clinton and Donald Trump were retrieved from Youtube. The first debate (NBC News 2016) took place at Hofstra University, Hempstead, NY, USA, on September 26, 2016, and had the form of a traditional debate: the two candidates responded to questions posed by a moderator. The second debate (ABC News, 2016a) was broadcasted from Washington University in St. Louis, St. Louis, MO, USA, on October 9, 2016. This debate was structured as a 'town hall discussion' with the candidates responding mostly to audience member questions. To illustrate, Fig. 10.1 shows two excerpts of Clinton's and Trump's speech in the second debate. The presence of a 3 Hz syllabic 'beat' is clearly visible in Clinton's waveform, whereas Trump's speech notably lacks slow-amplitude modulations. Finally, the third debate (ABC News, 2016b) took place at the University of Nevada, Las Vegas, Las Vegas, NV, USA, on October 19, 2016, and had the form of a traditional debate again.

All monologue speech from either candidate was manually annotated. That is, only those speech fragments in which one talker and one talker alone was speaking (uninterrupted monologue including all pauses, corrections, hesitations, etc.) was analyzed. Speech fragments that included crosstalk, laughter, applause, questions posed by the moderator, etc., were excluded from analyses. Monologues longer than approximately 35 s were cut into smaller fragments of <35 s at sentence boundaries. For the first debate, these annotations resulted in 93 speech fragments produced by Clinton (duration: $M = 24s$; $SD = 7$ s; $range = $ 5–36 s; $total = 2263$ s) and 98 speech fragments produced by Trump (duration: $M = 25$ s; $SD = 7$ s; $range = $ 6–35 s; $total = 2514$ s). For the second debate, these annotations resulted in 77 speech fragments produced by Clinton (duration: $M = 29$ s; $SD = 5$ s; $range = $ 8–36 s; $total = 2243$ s) and 82 speech fragments produced by Trump (duration: $M = 27$ s; $SD = 6$ s; $range = $ 7–35 s; $total = 2241$ s). For the third debate, these annotations resulted in 93 speech fragments produced by Clinton (duration: $M = 24$ s; $SD = 7$ s; $range = $ 5–35 s; $total = 2245$ s) and 76 speech fragments produced by Trump (duration: $M = 23$ s; $SD = 8$ s; $range = $ 5–34 s; $total = 1779$ s).

10.2.1.2 Procedure

Before analysis of the speech fragments, the overall power (root mean square; RMS) in each fragment was normalized (set to an arbitrary fixed value), thus matching the overall power of the speech from both speakers. Following this normalization procedure, the speech fragments from each debate were analyzed separately.

First, the modulation spectrum of each individual speech fragment produced by Clinton was calculated, using a method adapted from (Bosker and Cooke 2018). It involved filtering the speech fragment by a band-pass filter spanning the 500–4000 Hz range and deriving the envelope of the filter's bandlimited output (i.e., Hilbert envelope). The envelope signal was zero-padded to the next power of 2 higher than the length of the longest fragment of that particular speaker to achieve the same frequency resolution across recordings. This signal was then submitted to a Fast Fourier Transform (FFT), resulting in the modulation spectrum of that particular speech fragment. Finally, the average power in two frequency bands was calculated: average power in the 1–9 Hz range and average power in the 9–15 Hz range, resulting in two different observations for each of the speech fragments. Note that natural speech rates typically fall below 9 Hz. The same steps were then repeated for Trump's speech fragments.

This analysis procedure was followed for each of the three debates and formed the two dependent variables (average power below and above 9 Hz) for statistical analyses reported below. In order to visualize the average rhythmicity in the speech of one speaker in one debate, all individual modulation spectra of one speaker in one debate were downsampled by a factor of 25 and thereafter averaged.

10.2.2 Results

Data from the three debates are reported separately to allow for comparison across debates. Note, however, that follow-up analyses did not reveal large qualitative differences between the outcomes of the three debates.

10.2.2.1 First Debate

The average modulation spectra of the speech produced by both speakers in each of the three debates is given in Fig. 10.2.

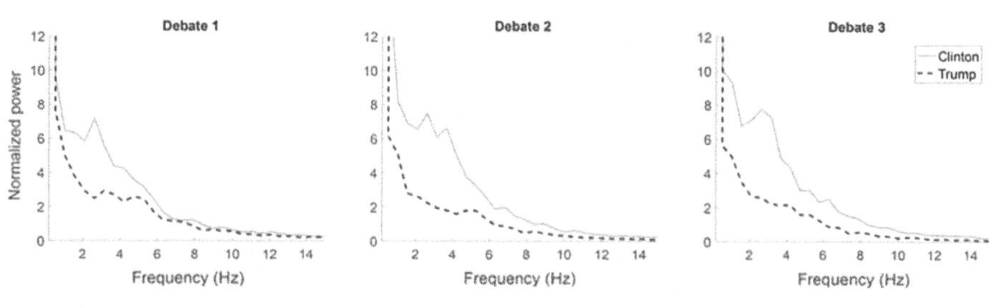

Fig. 10.2 Average modulation spectra of the speech produced by Hillary Clinton (gray solid lines) and Donald Trump (black dashed lines), separately for the three presidential debates

A simple linear model was built in R (R Development Core Team, 2012) separately for each of the two frequency bands (1–9 and 9–15 Hz), predicting the average power for each of the two speakers. The first model, predicting power in the 1–9 Hz range, showed a significant effect of Speaker ($b = 1.265$, $F(1, 189) = 90.91$, $p < 0.001$, $adjusted R^2 = 0.321$), indicating that Clinton's speech contained more power in the lower frequencies compared to Trump's speech. The other model, predicting power in the 9–15 Hz range, also showed a significant difference between the two speakers, only with a much smaller effect size ($b = 0.164$, $F(1, 189) = 42.75$, $p < 0.001$, *adjusted* $R^2 = 0.180$). These findings reveal that, in the first presidential debate, Clinton's speech contained more power in the 1–9 Hz range, and also slightly more power in the frequency range above 9 Hz.

10.2.2.2 Second Debate

The average modulation spectra of all speech produced by the two speakers in the second debate are given in Fig. 10.2.

Again, simple linear models were built separately for each of the two frequency bands (1–9 Hz and 9–15 Hz). The first model, predicting power in the 1–9 Hz range, showed a significant effect of Speaker ($b = 2.322$, $F(1, 157) = 434.5$, $p < 0.001$, $adjusted R^2 = 0.733$), as did the second model, predicting power in the 9–15 Hz range, only with a considerably smaller effect size ($b = 0.263$, $F(1, 157) = 250.9$, $p < 0.001$, *adjusted* $R2 = 0.613$). These findings reveal that, in the second presidential debate, Clinton's speech contained considerably more power in the 1–9 Hz range, and also somewhat more power in the frequency range above 9 Hz.

Note that, similar to the first debate, there is a clear peak in the modulation spectrum of Clinton around 3 Hz. This peak indicates a pronounced syllabic rhythm around 3 Hz in the amplitude envelope of Clinton's speech (cf. Fig. 10.1).

10.2.2.3 Third Debate

The average modulation spectra of the speech produced by both speakers in the third debate are given in Fig. 10.2.

Once more, simple linear models were built separately for each of the two frequency bands (1–9 Hz and 9–15 Hz). The first model, predicting power in the 1–9 Hz range, showed a significant effect of Speaker ($b = 2.427$, $F(1, 167) = 207.5$, $p < 0.001$, *adjusted* $R^2 = 0.551$), as did the second model, predicting power in the 9–15 Hz range, only with a considerably smaller effect size ($b = 0.350$, $F(1, 167) = 197.6$, $p < 0.001$, *adjusted* $R^2 = 0.539$). These findings from the third debate mirror those from the second debate: Clinton's speech contained considerably more power in the 1–9 Hz range, and also slightly more power in the frequency range above 9 Hz.

10.3 Perception Experiment

10.3.1 Participants

Native Dutch participants ($N = 20$; 17 females, 3 males; $M_{age} = 25$) with normal hearing were recruited from the Max Planck Institute's participant pool. Participants in all experiments reported here gave informed consent as approved by the Ethics Committee of the Social Sciences department of Radboud University (project code: ECSW2014-1003-196).

10.3.2 Material

Only speech fragments from the third debate were included in the perception experiment because (1) it was impossible to include the speech from all debates in a single rating experiment for reasons of length and (2) the third debate showed the largest difference between the two talkers in the power of amplitude modulations in the 1–9 Hz range.

Speech fragments from the third debate were first scaled to 70 dB using Praat (Boersma & Boersma, 2016). We did not want raters to base their judgments on the linguistic content of the speech since this was not controlled across the two speakers. Therefore, all speech was low-pass filtered (450 Hz cutoff, using a Hann window with a roll-off width of 25 Hz as implemented in Praat) to avoid lexical-semantic interference, while preserving sufficient ecological validity (being like naturally filtered speech, as if overhearing a person in another room). This manipulation crucially leaves the amplitude fluctuations present in the original speech signals relatively intact (cf. Fig. 10.1). After low-pass filtering, the speech was scaled to 70 dB.

10.3.3 Procedure

Participants in the experiment listened to the low-pass filtered speech fragments from either Clinton or Trump (counter-balanced across participants) in random order. Participants were instructed to rate the items for charisma, basing their judgments on the sound of the speech. They were explicitly pointed to the speaker's identity (but remained unaware that ratings of the other speaker were also collected). Nevertheless, they were told not to let any potential political or personal preferences influence their ratings. The use of a between-participants design reduced the contrast between the two speakers, thus further minimizing potential biases due to speaker sex, pitch, political stance, etc. Participants were instructed to rate the items for charisma using an Equal Appearing Interval Scale (Thurstone, 1928), including seven stars with labeled extremes (not charismatic on the left; very charismatic on the right).

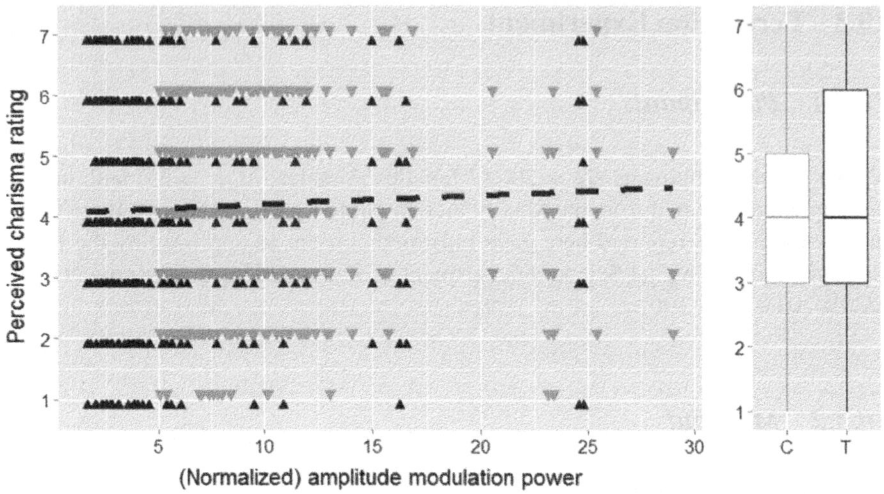

Fig. 10.3 *Left panel*: Individual perceived charisma ratings (on a scale from 1 "not charismatic" to 7 "very charismatic") of each speech fragment as a function of the (normalized) average power of amplitude modulations in the 1–9 Hz range. Gray triangles indicate speech fragments from Clinton and black triangles those from Trump. The black dashed line shows a (simple) linear regression line across all data points. *Right panel:* Boxplots showing the charisma ratings split for the two speakers (C = Clinton; T = Trump)

10.3.4 Results

The average perceived charisma rating of the speech of Clinton was 4.1, while Trump received an average rating of 4.3. Speech fragments with outlier values for the average power of amplitude modulations in the 1–9 Hz range (i.e., $> 2 * SD$; $n = 8$) were excluded to avoid the heavy weight of these outliers on the correlation analyses reported below. Figure 10.3 shows the individual perceived charisma ratings of speech fragments as a function of the average power of amplitude modulations in the 1–9 Hz range.

The right panel of Fig. 10.3 suggests that, on average, Trump (black) received higher charisma ratings than Clinton (gray). The left panel suggests that the charisma ratings seem to be a function of the average power of amplitude modulations in the 1–9 Hz range, with greater power of the amplitude modulations leading to higher charisma ratings.

Perceived charisma ratings were entered into a simple linear model, including the predictor's Speaker (categorical predictor; deviation coding, with Trump coded as −0.5 and Clinton as +0.5), Modulation Power Below 9 Hz (continuous predictor; z-scored), Modulation Power Above 9 Hz (continuous predictor; z-scored), and interactions between Speaker and the two Modulation Power predictors. This model, first, revealed a significant effect of Modulation Power Below 9 Hz ($b = 0.318$, $F(5, 1664) = 2.245$, $p = 0.041$). This indicates that, across the two talkers, speech with greater power in the 1–9 Hz range led to higher charisma ratings. Second, we found a main effect of Speaker ($b = -0.209$, $F(5, 1664) = 2.245$, $p = 0.014$), suggesting that Trump's speech was rated as more charismatic overall than Clinton's speech. No effect of Modulation Power Above 9 Hz was observed ($p = 0.151$), nor was their statistical evidence for either interaction term.

10.3.5 General Discussion

The present research goal was to investigate the role of temporal amplitude modulations in charisma perception in political debates. An acoustic analysis of the speech from two presidential candidates, Hillary Clinton and Donald Trump, in three different debates was carried out by means of modulation spectra, revealing the spectral content of the amplitude envelopes. Also, a perception experiment investigated whether judgments of perceived charisma would be sensitive to the speech rhythm in the acoustic signal.

Comparison of the amplitude spectra of Hillary Clinton's and Donald Trump's speech revealed considerably greater power in the modulation spectra of Clinton's speech than in those of Trump's speech. This power difference cannot be due to overall intensity differences between the two speakers since all speech was normalized in overall power prior to analysis, matching the overall intensity of Clinton's and Trump's speech fragments. Also, the power difference cannot be attributed to differences in habitual speech rate since such differences would be expected to lead to peaks at different frequencies in the modulation spectra, rather than differences in overall power. Instead, this finding indicates that there was a more pronounced temporal envelope in Clinton's speech (compared to Trump's speech).

Note that this power difference was concentrated (i.e., largest) in the 1–9 Hz range, the range of typical syllable rates (Ding et al., 2017; Ghitza & Greenberg, 2009; Greenberg & Arai, 1999, 2004). This suggests that the power difference between Clinton and Trump is driven by more pronounced syllabic amplitude fluctuations in the speech of Clinton. Moreover, across the three debates, there seems to be a relatively consistent peak around 3 Hz in Clinton's modulation spectra, suggesting a preferred syllabic rate. In contrast, Trump's modulation spectra lack pronounced peaks, indicating particularly flat, that is, unmodulated amplitude envelope contours.

Whether or not Clinton used this particular speaking style (with regular amplitude modulations) purposefully and strategically remains unknown. In this regard, one may note that speakers, in general, tend to produce greater amplitude modulations when instructed to produce clear speech (Krause & Braida, 2004) or when talking in noise (Bosker & Cooke, 2018), presumably for reasons of achieving greater intelligibility. As such, Clinton's speaking style during the three debates examined here may be the result of her extensive experience with making herself understood during public addresses. We may speculate that the influence of the enhanced modulation signature of Clinton's speech did not influence charisma perception alone. Regular energy fluctuations have been shown to benefit speech recognition (Doelling et al.,

2014; Ghitza, 2012, 2014), particularly in noisy listening conditions (Aikawa and Ishizuka, 2002), and, as such, may have improved Clinton's intelligibility in the noisy environment of a live debate. This seems particularly relevant considering the large number of interruptions (i.e., overlapping speech) that Clinton encountered during the three debates (Trump: $N = 106$ vs. Clinton : $N = 27$). Also, rhythmic amplitude modulations facilitate recognition memory (Essens & Povel 1985), potentially serving Clinton's political aims at the time.

One may also speculate about the absence of amplitude modulations in Trump's speech. Tian's recent analysis (Tian, 2017) of Trump's disfluency patterns during these presidential debates indicated that Trump was considerably more disfluent than Clinton. Trump was found to use particularly many repetitions, repairs, and abandoned utterances (Tian, 2017); all types of disfluencies that signal less extensive utterance planning and self-monitoring. As such, Tian suggested that Trump used less rehearsed utterances compared to Clinton. This difference in utterance planning can well be thought to underlie the difference in rhythmic structure between the two speakers: putting more effort in cognitive planning would also allow the speaker to better temporally organize the syllabic structure of the utterance, and especially so with increased public-speaking experience.

The outcomes of the perception experiment supported two conclusions. First, more pronounced amplitude modulations biased raters toward higher perceived charisma ratings. Across all speech fragments from both talkers, we observed that those items with a higher power of amplitude modulations in the 1–9 Hz range also received higher perceived charisma ratings—independent from the main speaker effect. This suggests that the rhythm of speech contributes to perceived charisma, with implications for public speakers in general.

The second conclusion is that Trump's speech was, on the whole, rated as more charismatic than Clinton's. Although this may seem at odds with the observation that less pronounced amplitude modulations result in lower perceived charisma ratings, it is important to realize that listeners could base their judgments on a larger set of acoustic characteristics than just rhythm. It is unlikely that participants in the study based their perceived charisma ratings solely on the amplitude modulation signatures of the speech signals. Many other (acoustic) characteristics are likely to have contributed to participants' judgments—even in the case of low-pass filtered speech (i.e., without access to linguistic content). One potential acoustic cue that was available to listeners and that may account for the main effect of Speaker is pitch. The low-pass filter applied to the speech only filtered out spectral information above 450 Hz, leaving fundamental frequencies relatively intact. As such, the low-pass filtered stimuli still contained acoustic cues to talker gender (distinction male vs. female cued by pitch). Indeed, talker gender is known to bias charisma ratings (and the perception of other personality traits), with male talkers generally being perceived as more charismatic than female talkers (Brooks, Huang, Kearney, & Murray, 2014; Niebuhr, Skarnitzl, & Tylecková, 2018; Novák-Tát, 2017). Therefore, the main effect of Speaker is likely driven by a range of acoustic and social factors that were not controlled for. Still, it is important to note that the correlation between more pronounced amplitude modulations and higher perceived charisma ratings held across talkers (no interaction between modulation power and speaker). This means that, despite an overall difference between the male and female voice, enhanced amplitude modulations in speech equally affected the ratings of Trump's and Clinton's speech.

Another possible explanation for the overall effect of Speaker could be related to the concept of 'effectiveness windows' in charisma perception (Niebuhr, Tegtmeier, & Brem, 2017). It has been proposed that public speakers, in attempting to persuade their audiences, should use charisma-relevant acoustic cues within particular functional ranges, avoiding, for instance, exaggerated vocal characteristics. Maybe Clinton's consistent use of regular amplitude modulations was perceived as an "overdose" of charismatic vocal cues, thus at some point hurting, rather than serving, the subjective impression listeners had of her. However, such an interpretation would also predict an inverse U-curve in the relationship between modulation power and charisma perception, such that greater rhythmicity would be beneficial only up to a certain point. However, follow-up statistical analyses (i.e., testing for a quadratic effect of Modulation Power Below 9 Hz) and visual inspection of Fig. 10.3 do not support the presence of such a U-shaped relationship, arguing against this particular explanation.

The fact that we used low-pass filtered speech may be seen as both a strength as well as a limitation of the current study. It is a strength of the methodology of the experiment because this allowed us to isolate the (temporal) acoustics of the speech from the linguistic content. In this fashion, potential interference from the linguistic message was reduced. At the same time, one may argue that it limits the generalizability of the present findings since in most natural communicative situations we hear unfiltered speech. For our current purposes, we valued experimental control higher than ecological validity and future studies may investigate whether the rhythm of speech also influences charisma perception in more natural settings.

Another limitation of this study is that we only performed correlational analyses. Even though we are unaware of possible confounds, we acknowledge that the present empirical evidence does not necessarily warrant the conclusion that more pronounced amplitude modulations causally influence perceived charisma. Future investigations may, for instance, examine this causal relationship by directly manipulating the modulation depth of speech fragments—while keeping all other (acoustic, linguistic, social) cues present in the signal constant.

Finally, one further highly relevant issue in the field of charisma research is the role of listener variation in charisma perception. Most empirical studies of charisma perception have used subjective ratings collected from young university students. In fact, some studies, like the present one, recruited non-native speakers of the language under study (e.g., Brem & Niebuhr, this volume). It remains unclear how variation among raters might impact charisma perception and the perceptual weight assigned to various vocal characteristics. Is charisma perception language- or culture-dependent (cf. D'Errico, 2013)? Do non-native speakers of a language weight the acoustic cues to charisma differently from native speakers, possibly through influences from their L1? Do male and female raters differ in how they judge male versus female public speakers (cf. Brem & Niebuhr, this volume)? What is the role of one's own speech production patterns on the perception of others (cf. Bosker, 2017b)? For instance, do fast talkers find fast speech more attractive or persuasive than others? These questions regarding inter-individual variation in charisma perception are promising avenues for future research.

10.4 Conclusion

The present outcomes shed light on the use and function of speech rhythm in political debates, specifically comparing the speech produced by Hillary Clinton and Donald Trump in three presidential debates in 2016. Clinton's speech was observed to contain more power in the modulation spectra, particularly in the 1–9 Hz range, suggesting more pronounced amplitude modulations in her speech (compared to Trump). This may be argued to indicate that Clinton planned her utterances more extensively, allowing more opportunity to temporally organize the syllabic structure of her utterances. At the same time, the lack of rhythmic amplitude modulations in Trump's speech may indicate a level of spontaneity in his speech production, with little attempt to pre-plan certain utterances.

Perceptual data revealed a positive correlation between the strength of amplitude modulations in the syllabic range (1–9 Hz), on the one hand, and perceived charisma ratings, on the other hand. This suggests that greater rhythm in the speech of a public speaker positively influences listeners' impressions of the speaker charisma. Thus, it highlights the important contribution of speech rhythm to charisma perception.

Acknowledgments The author was supported by a Gravitation grant from the Dutch Government to the Language in Interaction Consortium. Parts of the acoustic analysis have been presented at Interspeech 2017, Stockholm, Sweden. Thanks go to YouTube and the various news agencies for making the digital recordings of the presidential debates freely available. Thanks also to Annelies van Wijngaarden for coordinating the perception experiment, to Joe Rodd for help with visualizing the data, and to the student-assistants in the Psychology of Language Department of the Max Planck Institute for Psycholinguistics for help with annotating the speech recordings.

References

ABC News. (2016). *FULL VIDEO: Donald Trump vs Hillary Clinton—2nd Presidential Debate.* Retrieved October 9, 2016, https://www.youtube.com/watch?v=h-gkBUbU_F4.

ABC News (2016). *FULL VIDEO: Donald Trump vs Hillary Clinton—3rd Presidential Debate.* Retrieved October 9, 2016, from https://www.youtube.com/watch?v=LsA6Gj8y8rU.

Aikawa, K., & Ishizuka, K. (2002). Noise-robust speech recognition using a new spectral estimation method "PHASOR". In *Proceedings of Acoustics, Speech, and Signal Processing (ICASSP)* (pp. 397–400).

Awamleh, R., & Gardner, W. L. (1999). Perceptions of leader charisma and effectiveness: The effects of vision content, delivery, and organizational performance. *The Leadership Quarterly, 10*(3), 345–373.

Boersma, P., & Weenink, D. (2016). Praat: Doing phonetics by computer. Computer program.

Bosker, H. R. & Cooke, M. (2018). Talkers produce more pronounced amplitude modulations when speaking in noise. *Journal of the Acoustical Society of America, 143*(2), EL121-EL126.

Bosker, H. R. (2017a). Accounting for rate-dependent category boundary shifts in speech perception. *Perception & Psychophysics*, *79*(1), 333–343.

Bosker, H. R. (2017b). How our own speech rate influences our perception of others. *Journal of Experimental Psychology: Learning, Memory, and Cognition*, *43*(8), 1225–1238.

Bosker, H. R., & Ghitza, O. (2018). Entrained theta oscillations guide perception of subsequent speech: Behavioural evidence from rate normalisation. *Language, Cognition and Neuroscience*, *33*(8), 955–967.

Bradlow, A. R., Torretta, G. M., & Pisoni, D. B. (1996). Intelligibility of normal speech I: Global and fine-grained acoustic-phonetic talker characteristics. *Speech Communication*, *20*(3–4), 255–272.

Brooks, A. W., Huang, L., Kearney, S. W., & Murray, F. E. (2014). Investors prefer entrepreneurial ventures pitched by attractive men. *Proceedings of the National Academy of Sciences*, *111*(12), 4427–4431.

Cutler, A. (1994). Segmentation problems, rhythmic solutions. *Lingua*, *92*, 81–104.

Cutler, A., & Butterfield, S. (1992). Rhythmic cues to speech segmentation: Evidence from juncture misperception. *Journal of Memory and Language*, *31*(2), 218–236.

Cutler, A., & Norris, D. (1988). The role of strong syllables in segmentation for lexical access. *Journal of Experimental Psychology: Human Perception and Performance*, *14*(1), 113–121.

Delgutte, B., Hammond, B., & Cariani, P. (1998). Neural coding of the temporal envelope of speech: relation to modulation transfer functions. Psychophysical and physiological advances in hearing, 595–603.

Dellwo, V. (2006). Rhythm and speech rate: A variation coefficient for Δ C. *Language and language-processing* (pp. 231–241). Frankfurt a. M.: Peter Lang.

D'Errico, F., Signorello, R., Demolin, D., & Poggi, I. (2013). The perception of charisma from voice: A cross-cultural study. In *Proceedings of Affective Computing and Intelligent Interaction (ACII)* (552–557).

Ding, N., Patel, A., Chen, L., Butler, H., Luo, C., & Poeppel, D. (2017). Temporal modulations in speech and music. *Neuroscience and Biobehavioral Reviews*, *14*(1), 113–121.

Doelling, K. B., Arnal, L. H., Ghitza, O., & Poeppel, D. (2014). Acoustic landmarks drive delta-theta oscillations to enable speech comprehension by facilitating perceptual parsing. *NeuroImage*, *85*, 761–768.

Drullman, R., Festen, J. M., & Plomp, R. (1994). Effect of reducing slow temporal modulations on speech recognition. *The Journal of the Acoustical Society of America*, *95*(5), 2670–2680.

Essens, P. J., & Povel, D.-J. (1985). Metrical and nonmetrical representations of temporal patterns. Perception & Psychophysics, 37(1), 1–7. https://doi.org/10.3758/bf03207132

Ghitza, O. (2011). Linking speech perception and neurophysiology: Speech decoding guided by cascaded oscillators locked to the input rhythm. *Frontiers in Psychology, 2*, 130.

Ghitza, O. (2012). On the role of theta-driven syllabic parsing in decoding speech: Intelligibility of speech with a manipulated modulation spectrum. *Frontiers in Psychology, 3*, 238.

Ghitza, O. (2014). Behavioral evidence for the role of cortical Θ oscillations in determining auditory channel capacity for speech. *Frontiers in Psychology, 5*, 652.

Ghitza, O., & Greenberg, S. (2009). On the possible role of brain rhythms in speech perception: Intelligibility of time-compressed speech with periodic and aperiodic insertions of silence. *Phonetica*, *66*(1–2), 113–126.

Giraud, A.-L., & Poeppel, D. (2012). Cortical oscillations and speech processing: Emerging computational principles and operations. *Nature Neuroscience*, *15*(4), 511–517.

Grabe, E., & Low, E. L. (2002). Durational variability in speech and the rhythm class hypothesis. *Laboratory Phonology*, *7*, 515–546.

Greenberg, S., & Arai, T. (2004). What are the essential cues for understanding spoken language? *IEICE Transactions on Information and Systems, E87-D*(5), 1059–1070.

Greenberg, S., & Arai, T. (1999). Speaking in shorthand—A syllable-centric perspective for understanding pronunciation variation. *Speech Communication*, *29*(2), 159–176.

Houtgast, T., & Steeneken, H. J. (1973). Modulation transfer-function in room acoustics as a predictor of speech intelligibility. *Acustica, 28*(1), 66–73.
Jacewicz, E., Fox, R. A., & Wei, L. (2010). Between-speaker and within-speaker variation in speech tempo of American English. *The Journal of the Acoustical Society of America, 128*(2), 839–850.
Kohler, K. J. (2009). Rhythm in speech and language. *Phonetica, 66*(1–2), 29–45.
Krause, J. C., & Braida, L. D. (2004). Acoustic properties of naturally produced clear speech at normal speaking rates. *The Journal of the Acoustical Society of America, 115*(1), 362–378.
Loukina, A., Kochanski, G., Rosner, B., Keane, E., & Shih, C. (2011). Rhythm measures and dimensions of durational variation in speech. *The Journal of the Acoustical Society of America, 129*(15), 3258–3270.
NBC News. (2016). *FULL VIDEO: The First Presidential Debate: Hillary Clinton and Donald Trump (Full Debate)*. Retrieved September 28, 2016, from https://www.youtube.com/watch?v=855Am6ovK7s.
Niebuhr, O., Skarnitzl, R., & Tylecková, L. (2018). The acoustic fingerprint of a charismatic voice - Initial evidence from correlations between long-term spectral features and listener ratings. In *Proceedings of Speech Prosody* (pp. 359–363).
Niebuhr, O., Voße, J., & Brem, A. (2016). What makes a charismatic speaker? A computer-based acoustic-prosodic analysis of Steve Jobs tone of voice. *Computers in Human Behavior, 64*, 366–382.
Niebuhr, O., Tegtmeier, S., & Brem, A. (2017). Advancing research and practice in entrepreneurship through speech analysis—from descriptive rhetorical terms to phonetically informed acoustic charisma metrics. *Journal of Speech Sciences, 6*(3), 3–26.
Nolan, F., & Jeon, H.-S. (2014). Speech rhythm: A metaphor? *Philosophical Transactions of the Royal Society B-Biological Sciences, 369*(1658).
Novák-Tót, E., Niebuhr, O., & Chen, A. (2017). A gender bias in the acoustic-melodic features of charismatic speech? In *Proceedings of Interspeech* (pp. 2248–2252).
Obermeier, C., Menninghaus, W., von Koppenfels, M., Raettig, T., Schmidt-Kassow, M., Otterbein, S., & Kotz, S. A. (2013). Aesthetic and emotional effects of meter and rhyme in poetry. *Frontiers in Psychology, 4*(10).
Obermeier, C., Kotz, S. A., Jessen, S., Raettig, T., von Koppenfels, M., & Menninghaus, W. (2016). Aesthetic appreciation of poetry correlates with ease of processing in event-related potentials. *Cognitive, Affective, & Behavioral Neuroscience, 16*(2), 362–373.
Peelle, J. E., & Davis, M. H. (2012). Neural oscillations carry speech rhythm through to comprehension. *Frontiers in Psychology, 3*(10).
Pellegrino, F., Coupé, C., & Marsico, E. (2011). Across-language perspective on speech information rate. *Language, 87*(3), 539–558.
Peter, J., & Povel, D. -J. (1985). Metrical and nonmetrical representations of temporal patterns. *Perception & Psychophysics, 37*(1), 1–7.
Poeppel, D. (2003). The analysis of speech in different temporal integration windows: Cerebral lateralization as 'asymmetric sampling in time'. *Speech Communication, 41*(1), 245–255.
Quené, H., & Port, R. (2005). Effects of timing regularity and metrical expectancy on spoken-word perception. *Phonetica, 62*(1), 1–13.
Quené, H. (2008). Multilevel modeling of between-speaker and within-speaker variation in spontaneous speech tempo. *The Journal of the Acoustical Society of America, 132*(2), 1104–1113.
R Development Core Team. (2012). R: A Language and Environment for Statistical Computing. Computer program.
Ramus, F., Nespor, M., & Mehler, J. (1999). Correlates of linguistic rhythm in the speech signal. *Cognition, 73*, 265–292.
Roncaglia-Denissen, M. P., Schmidt-Kassow, M., & Kotz, S. A. (2013). Speech rhythm facilitates syntactic ambiguity resolution: ERP evidence. *PloS One, 8*(2), e56000.
Rosen, S. (1992). Temporal information in speech—acoustic, auditory and linguistic aspects. *Philosophical Transactions of the Royal Society of London Series B-Biological Sciences, 336*(1278), 367–373.

Rosenberg, A., & Hirschberg, J. (2009). Charisma perception from text and speech. *Speech Communication, 51*(7), 640–655.

Rothermich, K., Schmidt-Kassow, M., & Kotz, S. A. (2012). Rhythm's gonna get you: Regular meter facilitates semantic sentence processing. *Neuropsychologia, 50*(2), 232–244.

Shannon, R. V., Zeng, F.-G., Kamath, V., Wygonski, J., & Ekelid, M. (1995). Speech recognition with primarily temporal cues. *Science, 270*(5234), 303.

Thurstone, L. L. (1928). Attitudes can be measured. *American Journal of Sociology, 33*, 529–554.

Tian, Y. (2017). Disfluencies in Trump and Clinton first presidential debate. In *Proceedings of the conference Fluency and Disfluency Across Languages and Language Varieties* (pp. 106–109).

Varnet, L., Ortiz-Barajas, M. C., Erra, R. G., Gervain, J., & Lorenzi, C. (2017). A cross-linguistic study of speech modulation spectra. *The Journal of the Acoustical Society of America, 142*(4), 1976–1989.

Verhoeven, J., De Pauw, G., & Kloots, H. (2004). Speech rate in a pluricentric language: A comparison between Dutch in Belgium and the Netherlands. *Language and Speech, 47*(3), 279–308.

Weiss, B., & Burkhardt, F. (2010). Voice attributes affecting likability perception. In *Proceedings of Interspeech* (pp. 2014–2017).

White, L., & Mattys, S. L. (2007). Calibrating rhythm: First language and second language studies. *Journal of Phonetics, 35*(4), 501–522.

Chapter 11
Dress to Impress? On the Interaction of Attire with Prosody and Gender in the Perception of Speaker Charisma

Alexander Brem and Oliver Niebuhr

Abstract Understanding charismatic speech becomes a highly relevant issue in times of globalized markets and mobile on-demand mass media that strengthen the influence of individuals. Pushing phonetic research further into the realm of non-lexical charisma triggers, the present study is the first to investigate the combined effects of variation in attire and prosody on the perception of male and female speaker charisma. A perception experiment was carried out with Attire and Prosody as independent variables, each with two manipulation steps and embedded in a 2 × 2 orthogonal design. A total of 53 participants took part in the experiment and rated eight senior business leaders of well-known US American companies, four males and four females, on three approved charisma-related scales: convincing, passionate, charming. The audio-visual stimuli consisted of a keynote-speech excerpt of a speaker in combination with a matching photograph. Results clearly show that both Attire and Prosody had significant effects on the speakers' perceived charisma. The charisma effects of Attire and Prosody are additive, but in gender-specific ways and with gender-specific effect sizes. A bipartite results pattern among the female speakers further suggests that it depends on their physical attractiveness whether Attire and Prosody conditions have a charisma-supporting or charisma-reducing effect. The results are discussed in terms of their practical implications for the daily business life of men and women.

Keywords Charisma · Passion · Charm · Persuasion · Attire · Public speaking · prosody · English speech · Perception · Expressive · Speech

A. Brem
Innovation and Technology Management, Friedrich-Alexander-Universität Erlangen-Nürnberg, Erlangen, Germany
e-mail: alexander.brem@fau.de

O. Niebuhr (✉)
Mads Clausen Institute, Centre for Electrical Engineering, University of Southern, Odense, Denmark
e-mail: olni@sdu.dk

B. Weiss et al. (eds.), *Voice Attractiveness*, Prosody, Phonology and Phonetics,
https://doi.org/10.1007/978-981-15-6627-1_11

11.1 Introduction

11.1.1 Charisma and Delivery

We live in times in which individual politicians are increasingly able to determine people's opinion and voting behavior (whether for better or for worse), in which managers become an integral part of a company's brand image (like Steve Jobs was for Apple and Elon Musk is for Tesla), and in which entrepreneurship, i.e., the motivation, passion, and persuasive power of individuals, becomes a mainstay of national prosperity in international competition networks. In these times, it is of high societal and economical importance to understand in detail what good speakers actually do and how, in which way, and to what degree they influence listeners. Good speakers draw us under their spell. We cannot help but listen to them, we believe in what they tell us, and we are willing to adopt their opinions, attitudes, and/or agendas. Attracting attention as well as gaining and persuading followers without having to use force or referring to formal authority is the essence of charisma. In the more speaker- than listener-oriented words of Antonakis, Fenley & Liechti (2016: 304), charisma is defined as "values-based, symbolic, and emotion-laden leader signaling".

Charisma leads to more successful brainstorming outputs and salary negotiations (Pentland, 2008), results in better learning outcomes of students and, generally, in more satisfied subordinates (Towler, 2003; Lee, 2014), helps raise more start-up funding (Davis, Hmieleski, Webb, & Coombs, 2017), changes people's opinions and decisions (Brilman, 2015), and makes a product or service appear more credible and likable to customers (Gélinas-Chebat, Chebat, & Vaninsky, 1996). Previous studies also demonstrated that charisma is not a mysterious talent of a few gifted people, as was originally claimed by Weber (1947), but a tangible skill that anyone can learn and improve (Antonakis et al., 2011, 2012).

However, this learning and improving requires that we understand how the mechanisms work that makes a speaker sound charismatic, in particular the mechanisms of the so-called "delivery" that consists of everything a speaker conveys beyond the words themselves. Delivery includes auditory components like the speaker's speech prosody as well as visual components like body language and attire, and cross-modal components like age and gender.[1] Results of experimental studies repeatedly suggested that these components of delivery are—alone or in combination—more important than words for a speaker's charismatic impact (Holladay & Coombs, 1994; Awamleh & Gardner, 1999; Chen et al., 2014; Brilman, 2015).

[1]Note that "gender" is very often used also to refer to the biological concept of "sex", even in the scientific literature and across disciplines (cf. Brooks, Huang, Kearney, & Murray, 2014). Therefore, in order to be easily understandable for a broad, interdisciplinary readership, we decided to use "gender" in the sense of "sex" in the present paper.

11.1.2 The Roles of Prosody and Attire

As the transdisciplinary science "whose goal is the description, modeling and explanation of speech communication in the languages of the world" (Kohler, 2000: 1) and whose areas of instrumental-experimental research range from physiology through acoustics to cognition and perception, phonetics is perhaps in the best position of all scientific disciplines to decipher, objectively quantify, and ultimately understand how and by means of which signal cues charisma is created in the perceiver's brain. In fact, the intensive exploration of charismatic speech in phonetic production and perception experiments has already greatly expanded our knowledge of the acoustic-phonetic indicators of perceived speaker charisma. We know today that acoustic parameters such as the level, range, and dynamics of pitch[2] and intensity patterns, the durations of pauses and utterances, the number of emphatically emphasized words, the vocal tract's resonance frequencies (lower levels of the first three formants), and the timbre of the voice (e.g., in terms of HNR or the Hammarberg index) are all involved in the signaling of speaker charisma (Touati, 1993; Rosenberg & Hirschberg, 2009; Signorello, D'Errico, Poggi, & Demolin, 2012; Scherer, Layher, Kane, Neumann, & Campbell, 2012; D'Errico, Signorello, Demolin, & Poggi, 2013; Chen et al., 2014; Brilman, 2015; Shim et al., 2015; Hiroyuki & Rathcke, 2016; Bosker, 2007; Niebuhr, Thumm, & Michalsky, 2018a, b). In addition, we know that the general relevance of these parameters for speaker charisma does not differ between politics and business (Niebuhr, Brem, Novák-Tót, & Voße 2016; Novák-Tót, Niebuhr, & Chen 2017); maybe not even across cultures.

However, what does differ is which parameter level is appropriate and how strongly each parameter contributes to making a speaker sound charismatic. Not only culture, situation, industry sector, and listener age are relevant factors in this connection (Biadsy, Rosenberg, Carlson, Hirschberg, & Strangert, 2008; Abidi & Gumpert, 2018; Jokisch, Iaroshenko, Maruschke, & Ding, 2018), but also speaker gender. Most prosodic parameters have an identical effect on the charisma of male and female speakers and differ solely in the magnitude of this effect. Two parameters are different, though. These two parameters are pitch level and speaking rate. While men need to raise their pitch levels to sound more charismatic, women need to lower the pitch level (Berger, Niebuhr, & Peters, 2017; Niebuhr et al., 2018b); and while it is beneficial for the charisma effect of male speakers to increase the speaking rate, women must reduce their speaking rate to sound more charismatic (Bachsleitner & Popp, 2018). The gender-specific effect of speaking rate may be due to the fact that women already sound subjectively faster than men at the same objectively measured speaking rate (Weirich & Simpson, 2014).

For the visual components of charismatic delivery, and attire in particular, there are far less solid empirical findings from controlled experimental studies. For male

[2]Similarly as for "gender", we use the term "pitch" here as it is easily understandable to a broad, interdisciplinary (and non-expert) readership. What we actually mean is the acoustic fundamental frequency (F0), from which pitch is derived in the perception of speech signals, see Terhardt (1974) for further information.

speakers, things seem pretty straightforward, though. Compared to any form of casual or smart-casual attire, formal business attire supports the perception of charisma in terms of charisma-related attributes such as competence, credibility, and assertiveness. In addition, the formal business attire of men is quite clearly and narrowly defined as a dark-colored suit, see Furnham and Petrova (2010), Furnham, Chan, and Wilson (2014). In contrast, for women things are not that straightforward. On the one hand, women "have less freedom to wear more comfortable or casual attire" in the workplace (Franz & Norton, 2001: 88, see also Behling & Williams, 1991; Furnham et al., 2014). That is, wearing casual attire is less harmful for men than for women. On the other hand, the appropriate standard for the formal business attire of women is less clearly defined than for men. While every little detail counts for the perception of male charisma (up to the pattern of the tie and the garment of the suit, cf. Howlett, Pine, Orakçıoğlu, & Fletcher, 2013), even somewhat salient differences in female attire like that between a skirt suit and a pantsuit seem to play a lesser role in the perception of charisma-related attributes of women (Morris, Gorham, Cohen, & Huffman, 1996), perhaps because female attire is subject to much greater and faster fashion variation than male attire (Auty & Elliot, 1998; Molloy, 1996).

Furthermore, contradicting the traditional dress-code instructions for women in brochures and guidebooks (cf. Molloy, 1977; McEwan & Agno, 2011; Hoover, 2013), recent papers advise female leaders to "think color" (Karabell, 2016). More specifically, concerning the color range of a proper female business attire, these papers recommend wearing "all shades of red" (Karabell, 2016), i.e., all colors from blue-red to pink, as they are supposed to represent signals of power and charismatic qualities like "confidence and leadership" (Silverberg, 2017). The experimental study of Radeloff (1990) showed that red can compete with traditional business colors like (dark) blue and black when it comes to proper female business attire. Molloy's (1996) more practical research agrees with Radeloff's experimental data. Additionally, he points out that things have changed since the 1980s and that today "using color correctly can give businesswomen an advantage over men" (p. 157). In this context, Radeloff (1990) especially highlights the value of red for businesswomen, whereas for businessmen, the range of wearable colors is typically restricted to black or dark blue and, beyond that, hardly addressed in the literature; or, as in the case of grey and earth tones, associated in brochures and guidebooks with very different statements and recommendations that clearly reflect the lack of a solid empirical basis.

So, while everything from red to pink seems to be more effective than dark blue or black for female speakers' charisma, this color range certainly is a charisma killer for male speakers (e.g., the Financial Post[3] regards red as one of the three worst colors for men to wear in the office). In the opposite direction, while male leaders can at least dare to speak to an audience in jeans, T-shirt or hoody, this kind of casual business attire seems to be an absolute no-go for female leaders.

[3]https://business.financialpost.com/business-insider/the-best-and-worst-colours-to-wear-to-the-office.

11.1.3 Aims and Assumptions

In summary, with respect to the key charisma factor of delivery, intensive phonetic research provided us with a fairly detailed empirical picture of the charisma-relevant parameters of speech prosody and their context-specific phonetic variation. However, this does not apply to the same degree to the visual communication signals of speaker charisma, especially not to the factor attire. Moreover, we still know practically nothing about the interplay of attire and prosody in the perception of speaker charisma, which is interesting not only because both factors make a major contribution to speaker charisma, but also because of the gender-specific differences in each factor.

Therefore, our goal is to expand the empirical knowledge of the non-verbal ingredients of speaker charisma beyond prosody into the visual components of delivery. Continuing our previous studies (see Niebuhr et al., 2017), the focus of this line of research is not on political leaders but on business leaders. The first step presented here addresses the attire of speakers and their interaction with prosody. We report the results of a perception experiment with special emphasis on the gender-specific aspects of prosody and attire. The factor prosody was represented by a two-step manipulation of speaking rate and pitch level in male and female speech stimuli. The factor of attire was also represented by a two-step variation. However, unlike for prosody, this variation was not carried out analogously for male and female speakers, but took into account the fact that for men it is the style of attire that is most relevant in everyday business life, while for women it is primarily the color of attire.

Our study is able to test three basic assumptions. The present experiment:

(1) replicates the known gender-specific effects of pitch level and speaking rate on perceived speaker charisma;
(2) finds an additional gender-specific effect of attire on perceived speaker charisma, with male and female speakers being supported by a dark-colored suit or a red attire, respectively;
(3) finds the gender-specific effects of attire and prosody to be additive in the perception of speaker charisma.

11.2 Method

11.2.1 Speakers

Instead of using specifically designed and staged laboratory data, we opted for an approach with genuine, ecologically valid field data. This was for two reasons. First, for complex concepts like charisma whose multi-faceted perceptual nature is still too poorly understood to replicate it properly and consistently in the laboratory, the practical relevance of research findings critically relies on the authenticity of the analyzed data. We wanted our results to have as much practical value as possible, and one obvious way to enhance the relevance of our findings *for* practitioners was to take real data *from* practitioners. Second, both pilot tests and our own experience from previous studies indicated that subjects in perception experiments take the assessment of speaker charisma the more seriously (i.e., they make more reflective, elaborate judgments) the more well-known and influential the speakers are (cf. also Pearce & Brommel, 1972). This can only be credibly achieved with data of real speakers. All our speakers (or the companies they represent) can be considered similarly well known and influential. A high degree of popularity and influence also had the positive side effect that enough speech and image material of our speakers was available on the internet.

Since we worked with publicly available materials (i.e., field data), we had to choose our speakers such as to minimize the influence of potentially confounding between-speaker variables. On this basis, we chose the following eight speakers, four females (F1–F4) and four males (M1–M4):

(F1) Margret Whitman, born August 4, 1956, in Cold Spring Harbor, New York, USA; CEO and President of Hewlett Packard Enterprise (until January 31, 2018).
(F2) Virgina Marie Rometty, born July 29, 1957, in Chicago, Illinois, USA; CEO and President of IBM.
(F3) Sara Blakely, born February 27, 1971, in Clearwater, Florida, USA; Founder and CEO of Spanx Inc.
(F4) Sheryl Kara Sandberg, born August 28, 1969, in Washington D.C., USA; COO of Facebook Inc.
(M1) Reid Hoffman, born August 5, 1967, in Stanford, California, USA; Co-Founder of LinkedIn, former manager of PayPal.
(M2) Satya Nadella, born August 19, 1967, in Hyderabad, India; CEO of Microsoft.
(M3) Sundar Pichai, born 1972, in Madurai, India; CEO of Google LLC.
(M4) Mark Zuckerberg, born May 14, 1984, in White Plains, New York, USA; CEO of Facebook Inc.

All selected speakers are leading senior managers (CEOs or COOs) of well-known US American companies and were either born in the US or came from other English-speaking countries and then lived in the US for decades. Accordingly, all selected managers were native speakers of English and fluent speakers of American English, albeit with different regional and dialectal characteristics. However, these characteristics can be considered irrelevant to the questions of the present study, not least because—as became apparent from the metadata and participant feedback collected after the perception experiment—our participants were unable to either detect these characteristics or to associate them consistently with a specific geographical origin. Thus, it is unlikely that dialectal or regional stereotypes, their related socio-economic associations, or similar socio-phonetic effects were able to bias the participants' judgments of speaker charisma in a systematic and consistent way, cf. Ladegaard (1998), Bayard, Weatherall, Gallois, and Pittam (2001), Bailey (2003), and Andersson (2009) for the relationships between varieties of English and the judgment of their speakers.

All speakers are from the educated upper social class of the USA; and all were between 30 and 60 years old when they gave those speeches whose excerpts we used to create our stimuli. In this middle biological age range, we can assume all speakers to have the same basic physiological prerequisites with regard to the production of speech prosody (e.g., Schötz, 2006), except for some gender-specific differences, of course (Xu & Sun, 2002; Pépiot, 2013). Similarly, our speakers' age range was chosen small enough to prevent any potential age-related charisma differences from masking the actually investigated main effects of Attire and Prosody. Results of empirical studies suggest that perceived age has a separate influence on speaker charisma (e.g., Jokisch et al., 2018). Speaker charisma increases with age, but not linearly.

All speakers are leading IT executives. This restriction was added because initial results from another study (Abidi & Gumpert, 2018) suggest that speaker charisma is produced and assessed in an industry-specific fashion. For example, it seems that different ideas of charismatic presentations exist in the automotive sector as compared to the IT sector, which, in turn, seems to have different expectations concerning charismatic speeches than the financial sector. Although these results are still very preliminary, we nevertheless wanted to control this factor by keeping our dataset homogenous by focussing on the IT sector. A further advantage of this decision is that the IT sector is the same sector from which also the participants of the perception experiment were recruited. This had the advantage that our participants had already dealt with the eight selected speakers in one way or another, for example, by reading or writing about them in their course of studies or in related journals or newspapers. That is, all participants were similarly familiar with the speakers and well aware of their top positions in market-leading companies (see Rosenberg & Hirschberg, 2009 for the charisma-increasing effect of speaker familiarity and Pearce & Brommel, 1972 for the charisma-increasing effect of a higher social status).

11.2.2 Image Material and the Independent Variable Attire

The independent variable Attire is represented in the experiment by presenting the eight male and female speakers on different photographs. Two photographs were selected for each speaker. One shows the speaker in a more formal or conservative attire. The other one shows the speaker in a more casual or expressive attire. The full set of photographs can be made available to interested persons upon individual request.

Like in the selection of speakers (2.1), a maximum of comparability and control of potentially confounding factors was a major criterion for choosing suitable photographs. This was true within and across each speaker's pair of photographs. For example, all selected photographs showed the eight speakers from a similar angle (frontal view), in a similar posture (standing upright), and against the similar background of a large exhibition hall. Furthermore, all photographs showed the eight speakers with open and slightly rounded lip positions, which indicated that the photograph was taken while giving a speech. Head postures and hand and arm gestures on each photograph additionally characterized their speech as animated and passionate. In addition, we made sure that the two points in time at which the photographs of a speaker had been taken were less than 24 months apart (so as to prevent differences in a speaker's visual age across attire conditions, cf. Grd, 2013) and that the two photographs showed the speaker similarly large, i.e., up to the hips with the legs not being visible. The latter was important as the size of a person on a photograph (or screen) influences the subjective spatial distance of this person to the viewer. This distance, in turn, determines the level of social and emotional connection that the viewer feels for the person on the photograph (Reeves & Nass, 1996). As this connection is obviously related to concepts like perceived charisma, we had to control for this factor in the experiment.

Figure 11.1 shows, as an example, pairs of photographs for one female speaker (Sheryl Sandberg) and for one male speaker (Mark Zuckerberg) similar to those used in the actual experiment. As can be seen in Fig. 11.1, and as was mentioned Sect. 11.1.2, the biggest difference between the pairs of photographs was that, in the case of the male speakers, the independent variable Attire was operationalized

Fig. 11.1 Examples of photographs showing female and male speakers giving a keynote speech in conservative (left) and expressive (right) business attire. Top left photo taken by Pete Souza (2015), top right photo taken by Anthony Quintano (2018), bottom left photo taken by Moritz Hager (2012, photo edited by 1st author), bottom right photo taken by Remy Steinegger (2016). All photo are under CC-BY license

as the difference between a dark-colored business suit and a light-blue T-shirt. In the case of the female speakers, it was operationalized as the difference between a dark-colored and a red or pink pantsuit. Thus, in the case of the male speakers, attire concerns the *style* of clothing, whereas, in the case of female speakers, it concerns the *color* of clothing. For lack of better generic terms that equally apply to both types of attire variation (formal vs. casual is considered inappropriate), we refer the attire variation in both gender conditions as *conservative* versus *expressive*. Note that, based on the literature summarized in the introduction, it is the *conservative* Attire condition that is assumed to support the charisma perception of *male* speakers, whereas the *expressive* Attire condition is assumed to make *female* speakers more charismatic.

Mixing up style and color in the Attire variable follows the charisma-related statements in the literature on gender and attire. However, we also had no other option. It was no problem to find photographs of the male speakers wearing a T-shirt, even for similar public-speaking situations as in the business-suit condition. The same was not possible for the female speakers, though. In fact, for none of our female speakers, we were able to find one single photograph on which the speaker does not wear a formal dress or pantsuit. That is, photographs showing our female executive leaders in a T-shirt, sweater, hoody, jeans, or a similar casual clothing do not exist on the internet, no matter which occasion or which monologue or dialogue situation. We think that this fact resonates well with the literature in Sect. 11.1.2, in that it tells a lot about the different socio-cultural demands on male and female business attire, and about the leeway that male and female executive leaders have for choosing their attire in the workplace (or for public speeches as in the present experiment). Thus, although the two *expressive* Attire conditions of men and women obviously differ at the surface level (style vs. color), the variable Attire is nevertheless appropriately and homogeneously implemented in the experiment, because the two variable levels *conservative* and *expressive* equally cover for both genders the real full range of possible attire variation in the workplace. Yet, an obvious task of subsequent studies is, of course, to repeat the present experiment with staged photographs of fake executive leaders in order to implement the variable Attire in a consistent way across both genders, i.e., as the difference between business suit and T-shirt.

11.2.3 Speech Material and the Independent Variable Prosody

We chose one YouTube clip per speaker from one of his/her major public keynotes held in front of a large audience. Since the durations of speech stimuli are known to correlate positively with the perception of speaker charisma (i.e., the longer the stimulus the higher the speaker charisma, see Biadsy et al., 2008; Rosenberg & Hischberg, 2009), a similarly long speech section of 19–20 s was extracted from all eight YouTube clips. The onsets and offsets of these speech excerpts coincided in all cases with major intonational phrase boundaries (see AE-ToBI, Beckman, Hirschberg, & Shattuck-Hufnagel, 2005) at the beginnings and ends of syntactically complete utterance. Furthermore, all eight speech excerpts were free from disfluencies like hesitational lengthening, hesitation particles, overlong silent pauses (for turn-internal standards, Ten Bosch, Oostdijk, & de Ruiter, 2004), etc. The speech excerpts also contained no applause, music inserts, and other background noises.

Studies by Antonakis et al. (2011, 2012) showed on an experimental basis that, in addition to prosody, traditional morphosyntactic and lexical instruments of rhetoric have an influence on the perceived charisma of a speaker as well (an effect that manifests itself in both speaker ratings and participant behavior). Antonakis et al. summarized these effective rhetorical instruments under the umbrella term "Charismatic Leadership Tactics" (CLTs). These CLTs include, for example, metaphors, analogies, contrasts, rhetorical questions, and three-part lists (marked either explicitly/verbally or implicitly/prosodically). Also, the use of the 1st person (instead of the 3rd person) singular or plural contributes to perceived speaker charisma (Biadsy et al., 2008).

We controlled our speech excerpts such that they all contained a similar total number of CLT items and were dominated by verbs of the 1st person singular or plural. There were 3–4 CLT items within the 19–20 s excerpt of each speaker. The range of CLT items ranged from rhetorical questions ("How do you communicate authentically?", Sheryl Sandberg) through metaphors and analogies ("… we will unlock new platforms", Mark Zuckerberg; "We could not think of our users as wallets", Margret Whitman) or syntagmatic contrast constructions ("We have talked about machine learning […], but it also important to think about …", Sundar Pichai) to three-part lists ("it's black, it's invisible, it's not understood—sight, sound, music …", Virgina Marie Rometty; "It's the same amount of blood, sweat, and tears when you start a company", Reid Hoffman). In addition, all eight speech excerpts are similar in that they outline an inspiring new idea in the context of a visionary future perspective ("You actually do not know inside of it, what it is—and that's what's changing in this new era", Virgina Marie Rometty; "Aiming for something large is really important", Reid Hoffman; "but it's also important to think about how to do this technology can have an immediate impact on people's lives", Sundar Pichai).

Using an online script, the selected waveform signal was extracted from each YouTube clip and stored as an uncompressed audio file (.wav) in mono with a sound quality of 48 kHz and 24-bit. Each speech excerpt was characterized by a moderate speaking rate of on average about (5 syllables per second (syll/s) and a moderate pitch level of on average about 140 Hz (male speakers) or 205 Hz (female speakers). These moderate levels are suitable for performing a parameter manipulation without creating audible artifacts or extreme values of speaking rate and pitch.

The manipulation was done by means of the PSOLA resynthesis algorithm implemented in PRAAT (Mouliner & Charpentier, 1990; Boersma, 2001). For each speech excerpt, two combined PSOLA manipulations were performed and presented to the participants in the perception experiment instead of the original speech excerpt. That is, the perception experiment included only resynthesized audio stimuli. In this way, we ensured that all audio stimuli had the same sound quality.

The first stimulus condition of the independent variable Prosody was created by decreasing the speaking rate by 10% and the pitch level by 2 semitones (st) for each speaker. The pitch level was manipulated in st (i.e., along a logarithmic scale) so that the changes in acoustic F0 were perceptually equal for men and women. The size of the manipulation (2 st) was above the Just Noticeable Difference (JND) and hence audible for participants (Jongman, Qin, Zhang, & Sereno, 2017), but still small enough not to affect the naturalness of the speech. The speaking-rate manipulation was performed linearly across consonants and vowels. This is a simplification. In actual speech, vowel durations would change more as a function of speaking rate than consonant durations (van Santen, 1994); rate changes would also be paralleled by changes in speech reduction that cannot be imitated in resynthesized speech (see Ernestus & Smith, 2018). However, the resulting PSOLA output still sounded natural; also because 10% was, like for pitch, above the JND for speaking-rate changes at the utterance level (Quené, 2004), but small enough for the simplification and other manipulation artifacts to not become salient.

The second stimulus condition of the independent variable Prosody was created exactly inversely to the first one. That is, the speaking rate was increased by 10%, and the pitch level by 2 st compared to the original parameter values.

Note that we manipulated speaking rate and pitch level in combination rather than independently of each other because our focus was not on the interplay of the two prosodic parameters in charisma perception. Both parameters are well investigated in this respect already (Berger et al., 2017). Our aim was to create a strong and reliable variation in prosody-induced charisma and determine its interplay with a variation in attire-induced charisma. To that end, it was an advantage to co-vary two prosodic parameters, especially those whose effects on charisma are consistent and well investigated, also with respect to speaker gender.

At the end of the manipulation procedure, we had two versions of the same 19–20 s speech excerpt for each speaker: one with higher parameter values (+10% speaking rate, +2 st pitch level) and one with lower parameter values (−10% speaking rate, −2 st pitch level). In connection with the independent variable Prosody, the former version is henceforth called the *high* condition; the latter version is referred to as the *low* condition. Note that, like for Attire, the two variable levels have a gender-specific implication for charisma perception. Based on previous findings, male speakers should sound more charismatic in the *high* Prosody condition, whereas female speakers should sound more charismatic in the *low* Prosody condition.

11.2.4 Experiment Design

The more and less charismatic speech excerpts (audio stimuli) of a speaker were combined with the conservatively and expressively dressed photographs (visual stimuli) of that speaker. Thus, all stimuli of the perception experiment were multi-modal. Per speaker, there were $2 \times 2 = 4$ different audio-visual stimulus conditions:

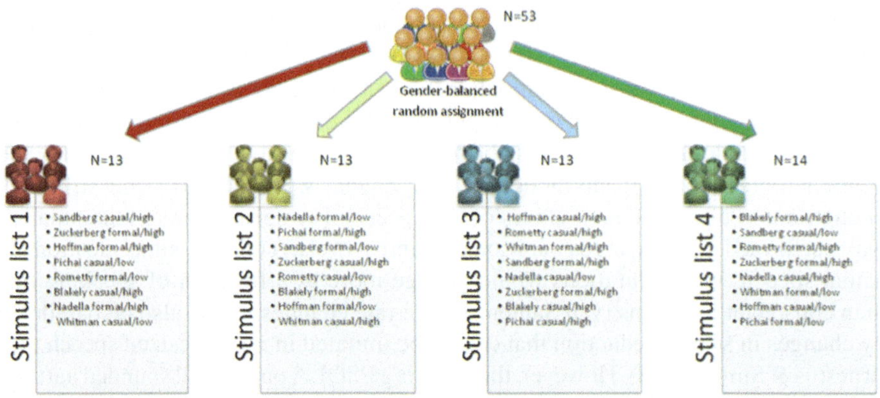

Fig. 11.2 Assignment of the 53 participants to four stimulus lists. Note that the order of the stimuli in each list is an example. Stimulus orders were individually randomized in the experiment

conservative/high, *conservative/low*, *expressive/high*, and *expressive/low*. For eight speakers, this gave a total of 32 stimuli.

In order to keep the experiment short and interesting, the 32 stimuli were not all presented to each participant. Rather, four different stimulus lists were compiled. The four audio-visual stimulus conditions of each speaker were distributed across these four lists such that each participant saw, heard, and rated each speaker only once, see Fig. 11.2. This made it impossible for individual participants to uncover the independent variables and their manipulations and infer from them the actual goal of the experiment. Participants only received eight differently dressed and differently speaking leading managers of US American companies, men and women, whose speaker charisma was to be assessed by them. Note that due to distributing the four audio-visual stimulus conditions across the four lists, both Attire and Prosody became between-subject factors in the experiment design.

Charisma is a complex, multi-faceted concept. Accordingly, our experience from pilot testing suggests in agreement with previous studies that participants respond insecurely and/or heterogeneously when being asked to rate the charisma of a speaker directly on a scale. For this reason, we decomposed charisma into three attributes that participants could rate separately for each audio-visual stimulus: "The speaker is …" (1) convincing (German: überzeugend); (2) passionate (German: enthusiastisch); (3) charming (German: ansprechend). This decomposition creates a clear frame of reference and provides participants with a concrete idea of what they are supposed to rate. In this way, the ratings become simpler and more consistent. The three attributes were chosen, because they are known from previous studies to be highly correlated with perceived speaker charisma (Rosenberg & Hirschberg, 2009), and because they are equally applicable to both attire and speech prosody.

11.2.5 Participant Sample and Experimental Procedure

The experiment was conducted as an online experiment (based on SoSci Survey). A total of 53 participants took part in the experiment; 23 men and 30 women who were between 21 and 48 years old. The average age of the participant sample was 24.7 years. All participants were native speakers of German and undergraduate or graduate students of social-science disciplines ("Innovation and Business" or "Innovation and Technology Management"). All had a very good command of English, i.e., either level B2 or C1 according to university-internal student entry tests. Nevertheless, their skills as non-native speakers were not sufficient to consistently identify regional or dialectal differences between speakers and associate them with positive or negative stereotypes or speaker attributes (Bailey, 2003). The 53 participants were distributed almost equally over the four stimulus lists. The 13 or 14 participants per list included about equal numbers of men and women. Except for controlling these basic factors, the participant-to-list assignment was entirely random.

Each online session of SoSci Survey started with the information that the experiment would be about the assessment of perceived speaker charisma. The compositional concept of charisma was briefly outlined to the participants with reference to Antonakis et al. (2016) who defined charisma as "values-based, symbolic, and emotion-laden leader signaling" (p. 304). In addition, the participants were given some names of particularly charismatic speakers for further illustration. These names included, for example, Steve Jobs, Barak Obama, and Martin Luther King Jr. In order to increase the spontaneity and impartiality of assessments, it was emphasized to the participants that assessments of perceived speaker charisma are inevitably subjective and that there is no right or wrong in subjective assessments.

Subsequently, participants were told that they would successively see and hear eight fairly popular and influential male and female managers (CEOs or COOs) of leading US companies. Each audio-visual stimulus would consist of a photograph of one of these eight managers at an important keynote speech and an approximately 20-second audio clip from that keynote speech. On this basis, their task would simply be to listen to each of the eight audio-visual stimuli separately, i.e., without drawing comparisons between the speakers, and each time as if being part of the speaker's keynote audience. Then, ratings were to be made about how the speaker was experienced in terms of perceived charisma on three scales

- Convincing,
- Passionate,
- Charming.

Participants received the 6-level system of German school grades from "1" (=very good) to "6" (=not good at all) for their assessments, as this is a system that all participants were well familiar with. Judgments were made by clicking, for each charisma attribute, the respective button of a 6-point Likert scale whose endpoints, "very good" and "not good at all", were displayed above the three scales. An example of one judgment trial of the experiment is shown in Fig. 11.3.

Fig. 11.3 Screenshot of one judgment trial of the male speaker Sundar Pichai in the experiment. Photo of Sundar Pichai taken by Maurizio Pesce (2015, edited under CC-BY license by 1st author)

Following the initial instructions, the participants were presented with the eight audio-visual stimuli of their respective stimulus list. The experiment was performed in a self-paced fashion. Each participant received the eight audio-visual stimuli of his/her list in an individually randomized order.

After the experiment was over, a few metadata of the participants were queried. These included age, gender, level of English, familiarity with the eight managers, and further speaker-oriented judgments on estimated age, perceived physical attractiveness, and estimated leadership experience. Furthermore, the participants were asked to specify their foreign or second language skills (besides English) and to give some feedback on the difficulty and the assumed purpose of the experiment as well as on the applicability of the rating scales. Together with this final metadata questionnaire, the entire experiment took about 10–12 min.

11.3 Results

The statistical processing of the data was performed separately for the two quadruplets of male and female speakers, taking into account that we expect the Attire and Prosody manipulations to affect speaker charisma in diametrically opposed ways depending on speaker gender. The gender-specific results are presented in Sects. 11.3.1 and 11.3.2. One of the three charisma-related scales, charming, did not yield conclusive results, and in the feedback questionnaire after the experiment participants also reported problems with applying this scale to the stimuli (we will address these problems in more detail in the discussion). For this reason, only the other two scales—convincing and passionate—were taken into account in the analysis and presentation of the results.

In accord with the use of convincingness and passion scales in previous studies, we found that the two scales are good representatives of charisma and suitable for asking participants to assess perceived speaker charisma. First, the participants rated the application of the scales to the stimuli and the general concept of charisma as simple and intuitive. Second, matching with the participants' report, we found no contradicting ratings in our results data, i.e., no cases in which the convincingness and passion ratings of a single stimulus go in opposite directions. On the contrary, the convincingness and passion ratings are correlated with each other in an order of magnitude that matches with how strongly they correlated with charisma itself in previous studies ($r[200] = 0.55$, $p < 0.001$ for the male speakers' stimuli and $r[200] = 0.69$, $p < 0.001$ for the female speakers' stimuli). That is, convincingness and passion both represent perceived speaker charisma equally well, but are nevertheless related to different facets of the phenomenon. Reflecting this fact, the results section presents the convincingness and passion ratings separately, but at the same time interprets them coherently in terms of perceived speaker charisma.

For the statistical analysis, a three-way General Linear Model (GLM) was used, with the two independent variables Attire and Prosody being fixed factors. As the third fixed factor, Speaker was additionally included in the model (four levels for the four male or female speakers). For supplementary t-tests and multiple comparisons between factor levels (e.g., of the fixed factor Speaker), alpha-error levels were adjusted using the Sidak method. Dependent variable was the rating score 1–6 on the respective grading scale per participant. The participant him/herself was taken into account as a random factor in the GLM. Participant as a random factor was appropriate here for two reasons. First, the participants were randomly selected, and, secondly, we were not interested in identifying possible differences between participants as a previous inspection of the data already indicated no separate systematic effects of participant age, gender, and international/linguistic background. In contrast, in the case of Speaker, we were interested in possible differences among the male or female speakers. For this reason, we made Speaker a fixed factor. However, note that we would arrive at the same conclusions with (male or female) Speaker being a second random factor. Further aspects of the generalization of the findings are addressed in Sect. 11.4.5.

Separate statistical analyses (GLMs) were conducted for the two assessment scales convincing and passionate. Each of these analyses was based on 212 participant ratings, 106 for the variable Attire, and 106 for the variable Prosody. All individual t-tests comparisons were conducted with 52–56 participant ratings in each sample.

We use bar plots below for illustrating the statistical patterns and summarizing the results descriptively. As it would be confusing for many readers that higher rising bars mean worse and lower rising bars better ratings of speaker charisma (as 6 = best and 1 = worst), we plotted the bars upside down. So, the lower a bar reaches the more negative is the charisma-related rating.

11.3.1 Male Speakers

The bar plot in Fig. 11.4 shows the effects of the variable Attire on the rating of the four male speakers. The individual bars display, in a different color for each speaker, the cumulative mean value of the difference between the two Prosody conditions *low* and *high* across all 53 participants. So, for example, if the mean rating of a speaker on the convincingness scale were 3.4 in the Prosody condition *low* and 2.4 in the Prosody condition *high*, then Fig. 11.4 would show a value of +1 for this speaker (recall that higher numbers in the German school grading system mean a worse performance).

The results shown in Fig. 11.4 can be summarized as follows. In terms of the two attributes convincing and passionate, speaker charisma is perceived to be higher for the *conservative* Attire condition than for the *expressive* Attire condition. In other words, wearing a conservative attire supports the speakers to the extent that it doubles their perceived charisma The scale values halved accordingly: For perceived convincingness, we can see a decrease in the overall assessment across the four

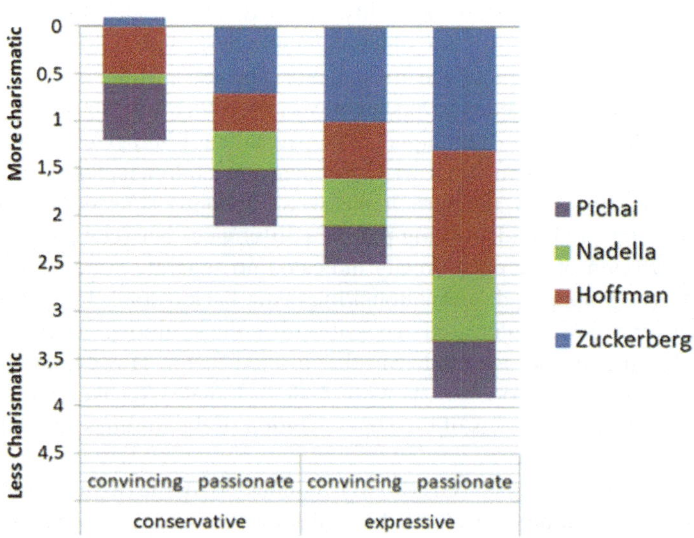

Fig. 11.4 Results of the Attire conditions conservative and expressive on the male-speaker assessments

speakers from 2.5 in the *expressive* Attire condition to about 1.2 in the *conservative* Attire condition. For perceived passion, the cumulative mean value of just under 4.0 in the *expressive* condition is halved to only 2.1 in the *conservative* condition.

With respect to Prosody, Fig. 11.4 shows further that, with one exception, all the mean differences between the two Prosody conditions *low* and *high* give a positive value. This means that each speaker was judged to be more convincing and passionate—i.e., overall more charismatic—for the higher than for the lower parameter values of speaking rate and pitch level.

In the corresponding GLMs the results of Fig. 11.4 manifest themselves in significant main effects of Attire (convincing: $F[1,196] = 440.70$, $p < 0.001$, $\eta_p^2 = 0.69$; passionate: $F[1,196] = 687.20$, $p < 0.001$, $\eta_p^2 = 0.78$) as well as in similar, but in terms of partial Eta-squared slightly weaker significant main effects of Prosody (convincing: $F[1,196] = 219.68$, $p < 0.001$, $\eta_p^2 = 0.53$; passionate: $F[1,196] = 350.75$, $p < 0.001$, $\eta_p^2 = 0.64$). The fixed factor Speaker had significant main effects as well (convincing: $F[3,196] = 307.48$, $p < 0.001$, $\eta_p^2 = 0.83$; passionate: $F[3,196] = 629.17$, $p < 0.001$, $\eta_p^2 = 0.91$). Moreover, there were, for both assessment scales, significant interactions between Speaker and Attire (convincing: $F[3,196] = 33.52$, $p < 0.001$, $\eta_p^2 = 0.34$; passionate: $F[3,196] = 11.85$, $p < 0.001$, $\eta_p^2 = 0.15$). The three-way interaction was not significant.

Figure 11.5 shows in more detail how the speaker rating changes as a result of the Prosody variable, pooled across the two scales convincing and passionate. There is a significant interaction of the variable Prosody with the variable Attire (convincing: $F[1,196] = 6.79$, $p < 0.01$, $\eta_p^2 = 0.11$; passionate: $F[1,196] = 41.72$, $p < 0.001$,

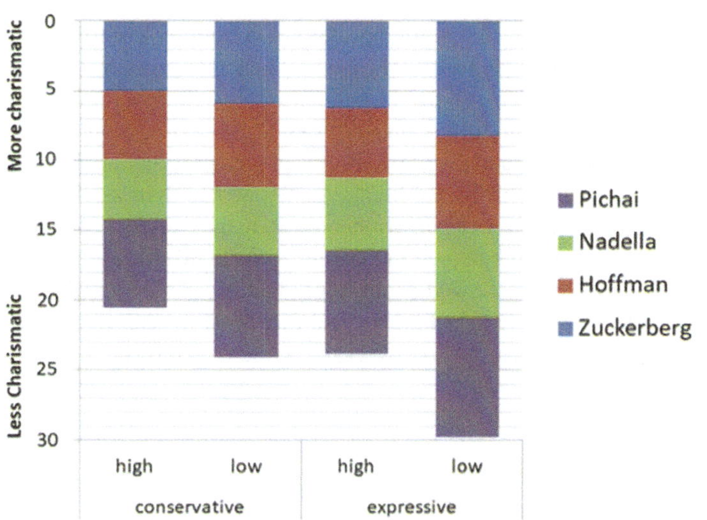

Fig. 11.5 Results of the Prosody conditions *high* and *low* in each Attire condition on the male-speaker assessments

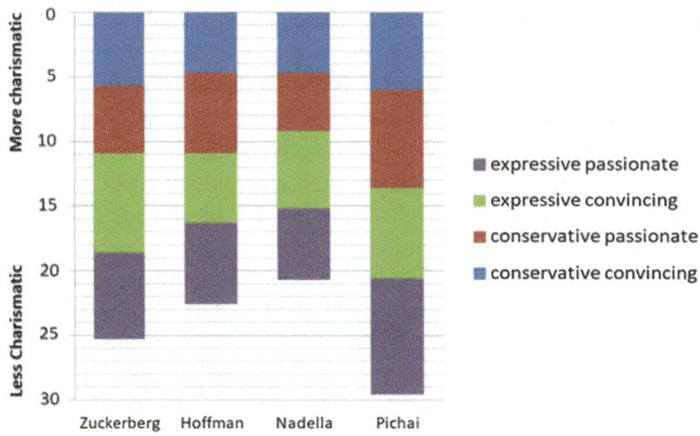

Fig. 11.6 Total assessment of the male speakers on the two charisma scales convincing and passionate

$\eta_p^2 = 0.18$). For a *conservative* attire, the charisma-supporting or reducing effect of Prosody is smaller than for an *expressive* attire. This means that, for a participant's rating of a speaker's charisma, the factor Prosody counts more if the speaker wears an expressive attire. In other words, those who wear a expressive attire (as a man) have to focus more on producing a charismatic speech prosody than those who wear a conservative attire. In fact, the two Attire-Prosody combinations *conservative/low* and *expressive/high* came out as statistically equivalent ($p > 0.05$) in separate t-tests for all 4 male speakers.

Figure 11.5 also shows that some speakers consistently contributed more than others to the cumulative mean values of each bar. That is, some speakers were consistently rated worse than others. Figure 11.6 illustrates this finding more clearly. Across the Attire and Prosody conditions and the two scales convincing and passionate, Zuckerberg and Pichai yielded the highest sums of mean ratings and hence the overall worst charisma ratings, with Pichai being slightly worse than Zuckerberg. Nadella performed best. Reid Hoffman's performance was, in the overall assessment of the 53 participants, somewhere in between Pichai and Nadella. Multiple t-test comparisons between the four speakers showed accordingly that all speakers differed from each other at $p < 0.001$, except for Zuckerberg and Pichai on the convincingness scale ($p > 0.05$) and Nadella and Hoffman on the same scale ($p > 0.05$).

11.3.2 Female Speakers

The results of the four female speakers are different. Unlike for the male speakers, the main effects of Attire and Prosody are not significant. Figure 11.7a, i.e., the counterpart of the male speakers' Fig. 11.5, shows very clearly that the cumulative

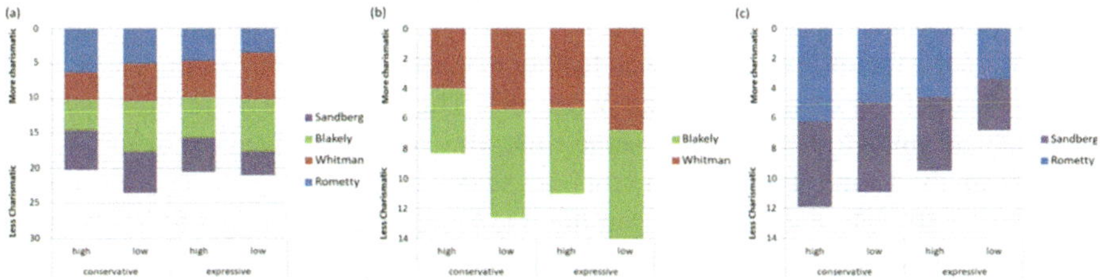

Fig. 11.7 Results of the Prosody conditions *high* and *low* in each Attire condition **a** for all four female speakers, and separately for **b** the Blakely–Whitman speaker pair and **c** the Sandberg–Rometty speaker pair

charisma ratings of the 53 participants are roughly the same for all independent variable conditions. The reason for this becomes obvious in Figs. 11.7b–c and 11.8 (the counterpart of the male speakers' Fig. 11.4): The female speaker sample contains two pairs of speakers whose Attire and Prosody conditions were rated in a diametrically opposed fashion by the 53 participants. This manifests itself in the GLMs in a significant main effect of Speaker (convincing: $F[3,196] = 221.01$, $p < 0.001$, $\eta_p^2 = 0.77$; passionate: $F[3,196] = 100.50$, $p < 0.001$, $\eta_p^2 = 0.61$) and in significant interactions of Speaker and Attire (convincing: $F[3,196] = 199.78$, $p < 0.001$, $\eta_p^2 = 0.75$; passionate: $F[3,196] = 169.39$, $p < 0.001$, $\eta_p^2 = 0.72$) and of Speaker and Prosody (convincing: $F[3,196] = 148.63$, $p < 0.001$, $\eta_p^2 = 0.70$; passionate: $F[3,196] = 276.08$, $p < 0.001$, $\eta_p^2 = 0.81$), each with high Eta-squared effect sizes. The three-way interaction is significant as well (convincing: $F[3,196] = 21.49$, $p < 0.001$, η_p^2

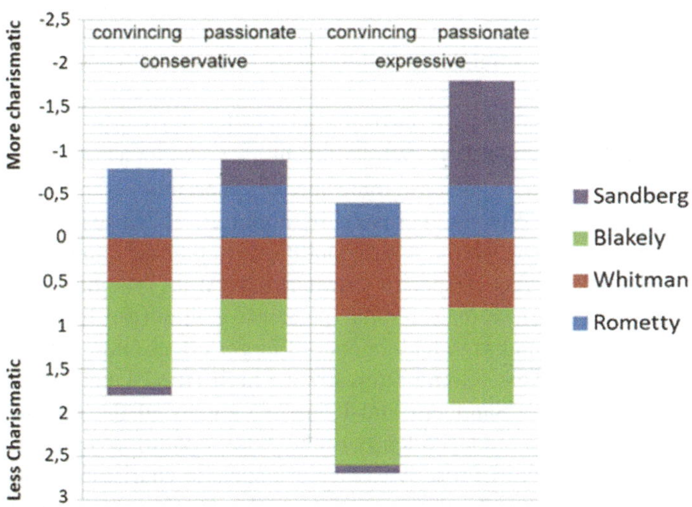

Fig. 11.8 Results of the Attire conditions *conservative* and *expressive* on the female speaker assessments

$= 0.25$; passionate: $F[3,196] = 27.12$, $p < 0.001$, $\eta_p^2 = 0.29$). Multiple-comparisons tests within the factor Speaker showed further that the participants' ratings of Blakely and Whitman differ on neither of the two scales. The same negative result was found for Sandberg and Rometty. At the same time, the latter two speakers differ significantly from the former two speakers on both scales at $p < 0.001$. These test statistics support that Blakely and Whitman on the one hand and Sandberg and Rometty on the other really formed two different pairs of speakers.

In order to look at the two speaker pairs in more detail, we ran separate additional GLMs for the Blakely–Whitman pair and for the Sandberg–Rometty pair.

The results of the two female speakers Blakely and Whitman largely agree with those of the male speakers. That is, the *conservative* Attire condition (dark-colored pantsuits) supports the two speakers' perceived charisma relative to the *expressive* Attire condition (red or pink pantsuits). The corresponding main effects are significant (convincing: $F[1,102] = 59.23$, $p < 0.001$, $\eta_p^2 = 0.45$; passionate: $F[1,102] = 39.86$, $p < 0.001$, $\eta_p^2 = 0.23$). Likewise, the Prosody condition *high*—characterized by increases in speaking rate and pitch level—supports the charisma perception of the two speakers relative to the Prosody condition *low*. The corresponding main effects are significant as well (convincing: $F[1,102] = 61.71$, $p < 0.001$, $\eta_p^2 = 0.51$; passionate: $F[1,102] = 363.44$, $p < 0.001$, $\eta_p^2 = 0.66$).

In contrast, for the two female speakers Sandberg and Rometty, the effects of Attire are exactly inverse and hence also run counter to those of the four male speakers (convincing: $F[1,102] = 121.26$, $p < 0.001$, $\eta_p^2 = 0.80$; passionate: $F[1,102] = 411.41$, $p < 0.001$, $\eta_p^2 = 0.94$). The same applies to Prosody (convincing: $F[1,102] = 50.58$, $p < 0.001$, $\eta_p^2 = 0.37$; passionate: $F[1,102] = 123.77$, $p < 0.001$, $\eta_p^2 = 0.82$). Unlike for Blakely and Whitman and the four male speakers, it is the Prosody condition *low* rather than *high* in which Sandberg and Rometty sound more charismatic in the ears of the 53 participants. Moreover, it is the *expressive* rather than the *conservative* attire condition that makes Sandberg and Rometty look more charismatic in the eyes of the 53 participants.

What all four female speakers have in common is that the overall effect of Attire is smaller than for the male speakers. While the choice between a conservative and an expressive attire was able to increase male speaker charisma by about 50%, female speaker charisma could only be increased by about 20%. A t-test based on absolute difference values between the Attire conditions in the male and female speaker samples shows that this gender-specific effect size of Attire is significant ($p < 0.01$). For the effect of Prosody, it were the female speakers for whom the difference between the two conditions *low* and *high* had an overall larger effect on perceived charisma than for the male speakers. Going from low to high (for Blakely and Whitman) or from high to low (for Sandberg and Rometty) enhanced the female speakers' charisma level by up 50%, independently of the Attire condition. In contrast, for the male speakers, the ability of Prosody to increase speaker charisma was between 10 and 20% and depended on the Attire condition. A t-test based on absolute difference values between the Prosody conditions in the male and female speaker samples shows that this gender-specific effect size of Prosody is also significant ($p < 0.001$).

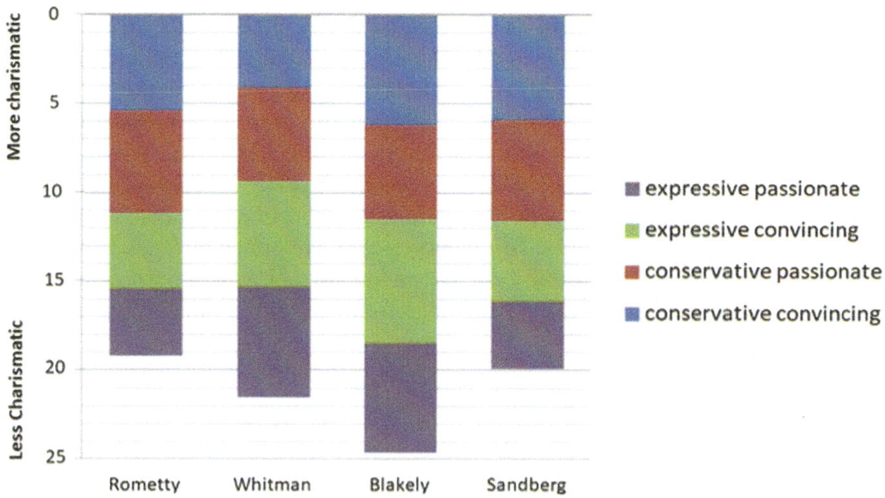

Fig. 11.9 Total assessment of the female speakers' charisma on the two scales convincing and passionate

Regarding the reason why the female speaker sample included two differently rated pairs of speakers, we discovered a parallel between the rating of participants and the speakers' perceived physical attractiveness. These attractiveness ratings (on a scale of 0–10) were made by participants in the feedback questionnaire after the experiment. An analysis of these judgments revealed that Rometty and Sandberg obtained physical-attractiveness values that were, according to within-subjects t-tests, statistically equivalent ($p > 0.05$), but clearly and significantly lower than those obtained by Blakely and Whitman (ø 4.1 vs. 6.6, $t[105] = -12.76$, $p < 0.01$), whose physical-attractiveness judgments were again statistically equivalent ($p > 0.05$). Further questionnaire analyses and even additional acoustic-prosodic measurements and analyses of the keynote-speech excerpts (for the charisma-relevant parameters specified in Niebuhr et al. (2017) showed that attractiveness was the only factor whose statistical results pattern runs exactly parallel to that of the two differently rated female speaker pairs. The total charisma scores shown in Fig. 11.9 yielded a similar, but not exactly parallel results picture as there is a significant difference between Blakely and Whitman ($p < 0.05$) on the one hand, but no significant differences different Whitman and Sandberg and Rometty on the other hand.

11.4 Discussion

The present study investigated the interaction effects of variation in attire and prosody on the perception of male and female speaker charisma. A total of 53 participants took part in the experiment and rated, in individually randomized orders, audio-visual stimuli of eight senior business leaders, four males and four females, on three charisma-related scales that were successfully tried and tested in many previous studies. In the debriefing questionnaire, the participants described the experiment as pleasant and easy, and judged the charisma ratings on the two scales of convincingness and passion as intuitive and applicable (the problematic scale "charming" is discussed in Sect. 11.4.4). Therefore, we view the significant effects of our results as (internally and externally) valid and reliable. This view is also corroborated by the fact that Mark Zuckerberg turned out to be a fairly uncharismatic speaker, which is consistent with previous studies (Niebuhr et al., 2016b). The following discussion is based on this validity and reliability.

11.4.1 Assumptions

We tested three assumptions with our experiment. The first one was whether or not the experiment replicates the known gender-specific effects of pitch level and speaking rate on perceived speaker charisma. This assumption is partially supported by the results of the experiment. The male speakers were rated more charismatic if they spoke with increased pitch and speaking-rate levels (compared to the original prosodic setting of the corresponding speaker). Changes toward lower pitch and speaking-rate levels significantly negatively impacted the charisma of male speakers. For two of the female speakers, Sandberg and Rometty, this influence of prosody on the perceived charisma was exactly inverse. That is, it was the lower pitched, slower way of speaking that was more charismatic, not the higher pitched, faster way of speaking. This gender-specific difference meets the first assumption and is consistent with the results of Berger et al. (2017) and Bachsleitner and Popp (2018). For the other two female speakers, Blakely and Whitman, however, the results were diametrically opposed (i.e., in line with those the male speakers again). Thus, they run counter to what we expected from our assumption (1) for female speakers. In Sect. 11.4.2, we offer an explanation for why the bipartition of our female speakers' results have occurred and why the deviating results for Blakely and Whitman are probably only in apparent contradiction to assumption (1) and the findings of Berger et al. (2017), Bachsleitner and Popp (2018).

Our second assumption was that the experiment would find a gender-specific effect of attire on perceived speaker charisma. This assumption is clearly supported by the findings. Male speaker charisma was enhanced by the conservative style of a dark-colored suit rather than by the expressive style of t-shirt, jeans, and similar casual clothes. The attire effect on female speaker charisma differed from that of the male speakers and was overall more complex. For two female speakers, the assumption was met that an expressive red, as opposed to a conservative dark color, had a charisma-supporting effect. For the other two speakers, it was the other way around.

The third assumption was that the experiment would find the gender-specific effects of attire and prosody to be additive in the perception of speaker charisma. Additive means that an unfavorable attire condition and an unfavorable prosody condition together reduce the perceived speaker charisma more than each unfavorable condition alone. In the opposite direction, a favorable attire condition and a favorable prosody condition together should enhance perceived speaker charisma more than each favorable condition alone. Combinations of favorable and unfavorable attire and prosody conditions should neutralize each other or result in minimally positive or negative charisma effects only. Exactly this overall pattern was found in the experiment for all our eight male and female speakers. For example, it is clearly visible for the male speakers in Fig. 11.5 that *conservative/low* was less charismatic than *conservative/high*, and that *expressive/low* was less charismatic than *expressive/high*. The two extreme pairs of conditions, i.e., the maximally favorable *conservative/high* combination and the maximum unfavorable *expressive/low* combination, yielded the largest overall difference in perceived speaker charisma. The two cross-over combinations *conservative/low* and *expressive/high* neutralized each other statistically. The third assumption was thus clearly met by the data.

11.4.2 The Bipartition of the Female Speaker Group

As was reported in the results section, the bipartition of the female speaker group in terms of charisma ratings runs parallel to the attractiveness ratings of the female speakers. The speaker pair Sandberg/Rometty was perceived most charismatic in the *expressive/low* stimuli and received at the same time relatively low physical-attractiveness ratings (ø 4.1, between-speaker difference <0.5, n.s.). The speaker pair Blakely/Whitman received significantly higher physical-attractiveness ratings (ø 6.6, between-speaker difference <0.5, n.s.) and was perceived most charismatic in the *conservative/high* stimuli. No other differences in speaker judgments, metadata, or personal characteristics (like hair color, age, size, or estimated leadership experience), and no uncontrolled acoustic-prosodic parameter differences matched equally well with the bipartition of the female speaker group as the attractiveness rating. Although it is "a myth that you have to be attractive to be charismatic" (Fox Cabane, 2012: 102), charisma and physical attractiveness are still to some degree related perceptual concepts (Grabo, Spisak, & van Vugt, 2017). Furthermore, it is known that charisma can also be exaggerated and, thus, reversed by an overdose of acoustic or visual triggers. For this reason, Niebuhr et al. (2017) determined so-called "effectiveness windows" that charisma-relevant parameters should neither fall below nor exceed. Against this background we suggest the following explanation for why the bipartition of the female speaker group occurred.

If physically more attractive female speakers already start from an inherently higher perceived charisma level than physically less attractive female speakers, then adding further charisma-enhancing stimuli like a red attire and a slow, low-pitched prosody can result in an overdose and hence in a reversed effect of attire and prosody on perceived charisma. This could have happened for the speaker pair Blakely and Whitman. In contrast, if physically less attractive female speakers start from an inherently lower perceived charisma level, then they can still benefit from adding further charisma-enhancing stimuli like a red attire and a slow, low-pitched prosody to the overall charisma they convey. This could be true of the speaker pair Sandberg and Rometty.

The advantage of this explanation is that it would be consistent with both the assumed charisma-enhancing effect of a red attire and the previously found gender-specific prosodic effects of pitch level and speaking rate in the studies of Berger et al. (2017), Bachsleitner and Popp (2018). Moreover, the provided explanation would also mean that the Attire and Prosody conditions did actually have the same effect on *all* females speakers. It would only be due to the interaction with attractiveness that this uniform effect surfaces differently for the two speaker pairs Blakely/Whitman and Sandberg/Rometty. On this basis, assumption (1) would be fully supported by the present results. It is further in accord with the provided explanation that no attractiveness differences showed up for the four male speakers (all received average ratings between 5.5 and 6.5 on the 10-point scale). Thus, Attire and Prosody were able to influence charisma ratings in a uniform way for the male speakers. In fact, it seems that men are generally rated less critically in terms of physical attractiveness than women, especially in the context of business, leadership, and perceived charisma (Friedman, Riggio, & Casella, 1988). Note in this context that rater gender did not play a significant role in the physical attractiveness ratings of our speakers. Female raters behaved in the same way as male raters.

An alternative but related explanation refers to the experiment of Pearce and Brommel (1972). They found that non-lexical charisma triggers only have a positive effect on attributes of perceived speaker charisma if the audience assesses the speaker as credible and competent. If the same charisma signals are conveyed by a less credible and competent speaker, then they have no effect or even a negative effect on the speaker's charisma. In the light of these findings, the bipartition of the female speaker group in the present experiment could also mean that the 53 participants (i.e., both males and females) assessed the physically more attractive female speakers Blakely and Whitman to be less credible and competent than the less attractive speakers Sandberg and Rometty.

Subsequent studies must continue to investigate which of the two explanations (or maybe a third one) underlies the bipartition of the female speaker group in the present experiment. However, regardless of the explanation, the present findings already have an important practical implication: Female speakers need to pay more attention than men to how many and strong audio-visual charisma triggers they convey, and it is likely that physical attractiveness is an important factor to take into account in this context. More physically attractive women should perhaps rather try to downgrade their remaining charisma triggers, for example, by using a conservative dark-colored outfit and clearly also a less charismatic prosody, whereas for physically less attractive women the opposite can be recommended, i.e., using a more expressively colored outfit and definitely a more charismatic prosody. Why we stress prosody in this connection is stated in 11.4.3, together with further practical implications.

11.4.3 Further Practical Implications

Our results show that the charisma rating of male speakers can be increased or decreased by about 50% through the attire choice alone. The effect of prosody on the charisma rating was smaller and depended on the attire (at least for the two parameters pitch level and speaking rate manipulated here). For women, the effect of prosody was larger than the effect of attire. Like for men, there was an interaction with the choice of attire. However, as we discussed in detail in Sect. 11.4.2, this interaction did not affect the size of the prosody effect, but its direction. The size of the prosody effect was independent of the choice of attire.

Two practical implications can be derived from these findings. First, women benefit more from using the right prosody, while men benefit more from choosing the right attire. Second, in a charisma-supporting conservative attire style (dark suit), men may well afford smaller weaknesses in prosodic charisma performance. In an expressive, casual attire style, on the other hand, men have to take care to deliver a very charismatic prosodic performance if they still want to make a strong charismatic impression. So, anyone who (as a man) has confidence in his excellent delivery can basically also perform in an expressive, casual style of clothing in front of his audience (although a conservative attire would still be better). For those who are insecure and unskilled in their speech performance, a conservative dress style should be a must.

11.4.4 The Scale "Charming"

The inconclusive results of the scale charming and the application problems reported by the participants in the debriefing questionnaire came unexpected. The scale charming was selected, as Rosenberg and Hirschberg (2009) showed that this attribute is even higher correlated with charisma than convincing and passionate and can also be applied more consistently to charisma than convincing and passionate. However, the key difference between our study and that of Rosenberg and Hirschberg is that we presented not just audio stimuli, but multi-modal audio-visual stimuli. It is obvious that charming—unlike convincing and passionate—has both an auditory and a visual rating dimension (to a limited degree, this is also true of passionate, but all passion-related signals of body language were carefully controlled and kept homogeneous in the photographs). In accord with the participants' comments in the debriefing questionnaire, we assume that it was this modality-based ambiguity of the term charming that caused the inconclusive results of the corresponding scale. For example, it turned out that some participants interpreted charming in the sense of a purely visual physical attractiveness and then used it automatically in the sense of sex appeal/attractiveness rather than in the intended sense of speaker charisma.

In summary, the correlated, consistent use of the scales convincing and passionate on the one hand shows that, with the multi-dimensional scaling method, we have a valid and sensitive instrument for the evaluation of speaker charisma. Pilot studies show that charisma is a too complex concept to be directly rated by participants in a consistent way, see Sect. 11.2.4. By breaking down the concept into scales that are highly correlated with each other and with charisma, we can make the rating task easier and more consistent—and still measure the same "thing". However, on the other hand, the inconclusive, inconsistent use of the scale charming also reveals and stresses the current limitations of this instrument. We have not fully understood as yet which facets of charisma are covered by each scale and how complementary and exhaustive this coverage is. Moreover, we have not enough knowledge today to put together a set of scales that are specifically tailored to measuring perceived speaker charisma for different types and modalities of stimuli. Also note in this context that the male speakers were generally rated worse on the passionate scale than on the convincing scale in the present study. For women it was the other way around. That is, independently of the set of scales and the stimuli, special care should be taken when comparing absolute scale levels between experimental conditions.

11.4.5 Generalization

As for all other experiments, our results apply primarily to the conditions under which they were obtained. The simpler and more controlled these conditions are, the lower is the potential generalization of the findings. As we said in the beginning, we selected photos and speech materials from the "field" and, moreover, used multiple speakers per gender to maximize generalizability within the experimental setup. Therefore, we think that our findings are sufficiently generalizable to have a practical use and to give male and female speakers guidance in public-speaking and presentation scenarios. We show with regard to perceived speaker charisma that prosody has an effect, that attire has an effect, that the effect of attire can also be negative (like that of prosody), and that the effects compensate, cancel out, or enhance each other and, in the latter case, can probably also cause overdoses. These facts will be valid in the real world regardless of the current experimental setting.

But, of course, there are many other auditory and visual sources of perceived speaker charisma that play a role, but are not considered or varied here. That is, we expect the strength of the present effects to be shaped by a number of other variables, which themselves may have favorable or unfavorable charisma effects. On the part of the recipients (i.e., the raters), these are, for example, variables from which norms and stereotypes emerge, such as educational attainment, age, cultural background (Power & Galvin, 1997), and the zeitgeist (50 or 100 years ago, a different way of speaking may have been considered more charismatic, cf. Madill, 2015 and the term "vocal zeitgeist" in McCabe & Altman, 2017; also business fashion changes constantly, especially for women, see Sect. 11.1.1). On the part of the speakers, relevant further variables are those that determine competency and prestige attributions, such as race, age, gender, attractiveness, occupation, and social status. Additionally, on the part of both recipient and speaker, there are the linguistic (including dialectal) background and the communication medium, which in our opinion represent secondary variables. These variables do not interact directly with speaker charisma, but indirectly through an influence on primary variables such as competence, stereotypes, etc.

With the exception of some indications on attractiveness, our study cannot make any new conclusions about these additional variables. However, as our male speakers were all rated consistently—despite showing a greater racial diversity than our female speakers—it appears that the factor race plays a subordinate role in speaker charisma, at least among educated raters (like students) and for speakers with a high status and prestige (like business leaders). Age and gender have an effect on charisma. In tendency, those speakers are considered more charismatic who have a similar age as the audience; and men tend to be inherently more charismatic than women (Jokisch et al., 2018; Brooks et al., 2014), in both women's and men's ratings (recall that we, too, have found no gender-specific rating differences). Our own data from a different study (Abidi & Gumpert, 2018) further suggest that the factor second language (L2) does not have to have a negative effect on charisma. Direction and strength of the L2 effect seem to depend less on the comprehensibility of the foreign accent or the command of the foreign language than on the prestige of foreign and native language. Regarding the communication channel, Gallardo and Weiß (2017) found a positive correlation between the signal-compression rate in (mobile) phone communication and listener ratings of charisma-related features. Despite initial emergent answers, there is still a plethora of open questions for all of the factors mentioned above. These open questions must be answered step by step, successively involving more factors. On this basis, we offer a brief outlook.

11.5 Conclusion and Outlook

The present experiment further supports the results of earlier research by identifying attire and prosody as relevant factors in the perception of speaker charisma. In addition, given the considerable effects of the two factors, our findings also support the conclusion of earlier studies that non-lexical factors such as attire and prosody are particularly influential for the perception of speaker charisma; probably more important than the words of a speaker. The paper started with a question: Dress to impress? The answer must clearly be "yes", especially in the case of men. Unlike women, it seems that men are less able to compensate for a charismatically unfavorable attire with prosodic means. Women, in turn, should probably be more careful in combining attire and prosody with other factors such as their own physical attractiveness. Regardless of the gender-specific interactions of attire and prosody, the effects of the two factors in the perception of charismatic speakers are largely additive, both in the positive and in the negative direction.

Based on these new findings, the task of follow-up studies must be to further refine and differentiate the very roughly varied attire and prosody conditions of the present study, and to homogenize the attire variable, for which we had to mix-up style and color in order to be able to use authentic field data of real senior leaders. Using (staged) lab data or field data of less popular speakers (entrepreneurs) could be ways to achieve a greater control of the independent variable conditions. Follow-up studies could also work with A/V videos instead of combinations of photographs and speech, especially if more and richer body-language factors are to be addressed. Two findings are conceivable with using A/V videos. Either the richer body language of videos distracts the raters from the factor attire so much that the latter becomes less relevant than with the photos in this study; or, through the attribution of speaker competence (Pearce & Brommel, 1972), attire functions as a limiting factor, so that any charisma-supporting effects of a richer body language cannot unfold without a favorable attire. In this context, it is also essential to check the charisma attributes used in multi-dimensional rating tasks for their multi-modal suitability. In fact, we believe that the exploration and development of methods for the assessment of speaker charisma or similar socio-communicative concepts is a field of research in its own right. Methods need a solid empirical foundation and have to meet certain standards in terms of their internal validity, exhaustiveness, contextual vulnerability, and sensitivity. Regarding the contributions in this volume as well as the recent developments in human–machine interaction and the growing intercultural and digital communication, it is obvious that the experimental investigation of charisma and similar socio-communicative concepts becomes a topic of growing relevance and urgency.

References

Abidi, M., & Gumpert, K. (2018). *Cross-cultural comparison of speeches and pitches.* Seminar thesis, Friedrich-Alexander-Universität Erlangen-Nürnberg, School of Business and Economics, Department of Technology Management, Germany.

Agno, J., & McEwen, B. (2011). Decoding the executive woman's dress code. Kestly. http://kestlydevelopment.com/hosted/ExecutiveWomansDressCodeRevised.pdf.

Andersson, N. (2009). *Stereotypes of English in hollywood movies. A case study of the use of different varieties of english in Star Wars, The lord of the rings and transformers.* BA thesis, University of Stockholm, Sweden.

Antonakis, J., Fenley, M., & Liechti, S. (2011). Can charisma be taught? Tests of two interventions. *The Academy of Management Learning and Education, 10,* 374–96.

Antonakis, J., Liechti, S., & Fenley, M. (2012). Learning charisma. *Harvard Business Review.* https://hbr.org/2012/06/learning-charisma-2

Antonakis, J., Bastardoz, N., & Jacquart, P. (2016). Charisma: An ill-defined and ill-measured gift. *Annual Review of Organizational Psychology and Organizational Behavior, 3,* 293–319.

Auty, S., & Elliott, R. (1998). Fashion involvement, self-monitoring and the meaning of brands. *Journal of Product & Brand Management, 7,* 109–123.

Awamleh, R., & Gardner, W. L. (1999). Perceptions of leader charisma and effectiveness: The effects of vision content, delivery, and organizational performance. *The Leadership Quarterly, 10,* 345–373.

Bachsleitner, N., & Popp, U. (2018). *Gender-related impact of the speech rate on the perception of charisma.* Seminar thesis, Friedrich-Alexander-Universität Erlangen-Nürnberg, School of Business and Economics, Department of Technology Management, Germany.

Bailey, R. W. (2003). Ideologies, attitudes, and perceptions. In D. R. Preston (Ed.), *Needed research in American Dialects* (pp. 123–150). Durham: Duke University Press.

Bayard, D., Weatherall, A., Gallois, C., & Pittam, J. (2001). Pax Americana: Accent attitudinal evaluations in New Zealand, Australia, and America. *Journal of Sociolinguistics, 5,* 22–49.

Beckman, M. E., Hirschberg, J., & Shattuck-Hufnagel, S. (2005). The original ToBI system and the evolution of the ToBI framework. In S.-A. Jun (Ed.), *Prosodic typology—the phonology of intonation and phrasing* (pp. 9–54). Oxford: Oxford University Press.

Behling, D. U., & Williams, E. A. (1991). Influence of dress on perception of intelligence and expectations of scholastic achievement. *Clothing and Textiles Research Journal, 9,* 1–7.

Berger, S., Niebuhr, O., & Peters, B. (2017, March). Winning over an audience–A perception-based analysis of prosodic features of charismatic speech. In Proceedings of 43rd Annual Conference of the German Acoustical Society, Kiel, Germany (pp. 1454–1457).

Biadsy, F., Rosenberg, A., Carlson, R., Hirschberg, J., Strangert, E. (2008). A cross-cultural comparison of American, Palestinian, and Swedish perception of charismatic speech. In *Proceedings of 4th International Conference of Speech Prosody, Campinas, Brazil* (pp. 579–582).

Boersma, P. (2001). Praat, a system for doing phonetics by computer. *Glot International, 5,* 341–345.

Bosker, H. R. (2007). The role of temporal amplitude modulations in the political arena: Hillary Clinton vs. Donald Trump. In *Proceedings of 18th Interspeech Conference, Stockholm, Sweden* (pp. 1–5).

Brilman, M. (2015). *A multimodal predictive model of successful debaters—Or how i learned to sway votes.* MA thesis, University of Twente, The Netherlands.

Brooks, A. W., Huang, L., Kearney, S. W., & Murray, F. E. (2014). Investors prefer entrepreneurial ventures pitched by attractive men. *Proceedings of National Academy of Sciences of the United States of America (PNAS), 111,* 4427–4431.

Chen, L., Feng, G., Joe, J., Leong, C. W., Kitchen, C., & Lee, C. M. (2014). Towards automated assessment of public speaking skills using multimodal cues. In *Proceedings of 16th International Conference on Multimodal Interaction* (pp. 200–203).

Davis, B. C., Hmieleski, K. M., Webb, J. W., & Coombs, J. E. (2017). Funders' positive affective reactions to entrepreneurs' crowd-funding pitches: The influence of perceived product creativity and entrepreneurial passion. *Journal of Business Venturing, 32,* 90–106.

D'Errico, F., Signorello, R., Demolin, D., & Poggi, I. (2013). The perception of charisma from voice: A cross-cultural study. In *Proceedings of Affective Computing and Intelligent Interaction* (pp. 552–557).

Ernestus, M., & Smith, R. (2018). Qualitative and quantitative aspects of phonetic variation in Dutch eigenlijk. In F. Cangemi, M. Clayards, O. Niebuhr, B. Schuppler, & M. Zellers (Eds.), *Rethinking reduction: Interdisciplinary perspectives on conditions, mechanisms, and domains for phonetic variation* (pp. 129–163). Berlin/Boston: De Gruyter Mouton.

Fox Cabane, O. (2012). *The charisma myth: How anyone can master the art and science of personal magnetism.* New York: Penguin.

Franz, T., & Norton, D. S. (2001). Investigating business casual dress policies: Questionnaire development and exploratory research. *Applied HRM Research, 6,* 79–94.

Friedman, H. S., Riggio, R. E., & Casella, D. F. (1988). Nonverbal skill, personal charisma, and initial attraction. *Personality and Social Psychology Bulletin, 14,* 203–211.

Furnham, A., & Petrova, E. (2010). *Body language in business: Decoding the signals.* Grand Street: Palgrave Macmillan.

Furnham, A., Chan, P. S., & Wilson, E. (2014). What to wear? The influence of attire on the perceived professionalism of dentists and lawyers. *Journal of Applied Psychology, 43,* 1838–1885.

Gallardo, L. F., & Weiß, B. (2017). Towards speaker characterization: Identifying and predicting dimensions of person attribution. In *Proceedings of 18th Interspeech Conference, Stockholm, Sweden* (pp. 904–908).

Gélinas-Chebat, C., Chebat, J. C., & Vaninsky, A. (1996). Voice and advertising: Effects of intonation and intensity of voice on source credibility, attitudes and the intend to buy. *Perceptual and Motor Skills, 83,* 243–262.

Grabo, A., Spisak, B., & van Vugt, M. (2017). Charisma as signal: An evolutionary perspective on charismatic leadership. *The Leadership Quarterly, 28,* 473–485.

Grd, P. (2013). Introduction to age estimation using face images. *Research Papers Faculty of Material Science and Technology Slovak University Bratislava* (vol. 21, pp. 24–30)
Hiroyuki, T., & Rathcke, T. (2016). Then, what is charisma? The role of audio-visual prosody in L1 and L2 political speeches. In *Proceedings of Phonetik & Phonologie im deutschsprachigen Raum, Munich, Germany* (pp. 1–3).
Holladay, S. J., & Coombs, W. T. (1994). Speaking of visions and visions being spoken an exploration of the effects of content and delivery on perceptions of leader charisma. *Management Communication Quarterly, 8,* 165–189.
Hoover, M. (2013). *Dressing to impress: The secrets of proper attire.* Manuscript, the Florida State University, Career Center.
Howlett, N., Pine, K. J., Orakçıoğlu, I., & Fletcher, B. (2013). The influence of clothing on first impressions: Rapid and positive responses to bespoke features in male attire. *Journal of Fashion, Marketing and Management, 17,* 38–48.
Jokisch, O., Iaroshenko, V., Maruschke, M., & Ding, H. (2018). Influence of age, gender and sample duration on the charisma assessment of German speakers. In *Proceedings of 29th Konferenz für Elektronische Sprachsignalverarbeitung, Ulm, Germany* (pp. 1–8).
Jongman, A., Qin, Z., Zhang, J., & Sereno, J. A. (2017). Just noticeable differences for pitch direction, height, and slope for Mandarin and English listeners. The Journal of the Acoustical Society of America, *142*, EL163–EL169.
Karabell, S. (2016). Dressing like a leader: Style tips for women in the spotlight. *Forbes Magazin.* https://www.forbes.com/sites/shelliekarabell/2016/01/16/dressing-like-a-leader-style-tips-for-women-in-the-spotlight/.
Kohler, K. J. (2000). The future of phonetics. *Journal of the International Phonetic Association, 30,* 1–24.
Ladegaard, H. J. (1998). National stereotypes and language attitudes: The perception of British, American and Australian language and culture in Denmark. *Language Communication, 18,* 251–274.
Lee, M. (2014). Transformational leadership: is it time for a recall? *International Journal Of Management and Applied Research, 1*(1), 17–29.
McCabe, D., & Altman, K. W. (2017). Prosody: An overview and applications to voice therapy. *Global Journal of Otolaryngology, 7,* 1–8.
Madill, C. (2015). Keep an eye on vocal fry—It's all about power, status and gender. The conversation. Retrieved October 31, 2018, from http://theconversation.com/keep-an-eye-on-vocal-fry-its-all-about-power-status-and-gender-45883.
Molloy, J. (1977). *The women's dress for success book.* Chicago: Follet.
Molloy, J. T. (1996). *New women's dress for success.* New York: Warner.
Morris, T. L., Gorham, J., Cohen, S. H., & Huffman, D. (1996). Fashion in the classroom: Effects of attire on student perceptions of instructors in college classes. *Communication Education, 45*(2), 135–148.
Moulines, E., & Charpentier, F. (1990). Pitch-synchronous waveform processing techniques for text-to-speech synthesis using diphones. *Speech Communication, 9,* 453–467.
Niebuhr, O., Brem, A., Novák-Tót, E., & Voße, J. (2016). Prosodic constructions of charisma in business speeches—A contrastive acoustic analysis of Steve Jobs and Mark Zuckerberg. In *Proceedings of 8th International Conference of Speech Prosody, Boston, USA* (pp. 1–3).
Niebuhr, O., Voße, J., & Brem, A. (2016). What makes a charismatic speaker? A computer-based acoustic prosodic analysis of Steve Jobs tone of voice. *Computers and Human Behavior, 64,* 366–382.
Niebuhr, O., Tegtmeier, S., & Brem, A. (2017). Advancing research and practice in entrepreneurship through speech analysis—From descriptive rhetorical terms to phonetically informed acoustic charisma metrics. *Journal of Speech Sciences, 6,* 3–26.
Niebuhr, O., Thumm, J., & Michalsky, J. (2018a). Shapes and timing in charismatic speech—Evidence from sounds and melodies. In *Proceedings of 9th International Conference of Speech Prosody, Poznan, Poland* (pp. 582–586).

Niebuhr, O., Skarnitzl, R., & Tylečková, L. (2018b). The acoustic fingerprint of a charismatic voice—Initial evidence from correlations between long-term spectral features and listener ratings. In *Proceedings of 18th International Conference of Speech Prosody, Poznán, Poland* (pp. 359–363).

Novák-Tót, E., Niebuhr, O., & Chen, A. (2017). A gender bias in the acoustic-melodic features of charismatic speech? In *Proceedings of Annual Conference of the International Speech Communication Association* (vol. 18, pp. 2248–2252).

Pearce, W. B., & Brommel, B. J. (1972). Vocalic communication in persuasion. *Quarterly Journal of Speech, 58*(3), 298–306.

Pentland, A. (2008). *Honest signals—How they shape our world*. Cambridge: MIT Press.

Pépiot, E. (2013). Voice, speech and gender: male-female acoustic differences and cross-language variation in English and French speakers. In *Proceedings of 15th Rencontres Jeunes Chercheurs de l'ED 268, Paris, France* (pp. 1–13).

Power, M. R., & Galvin, C. (1997). The culture of speeches: Public speaking across cultures. *Culture Mandala: The Bulletin of the Centre for East-West Cultural and Economic Studies, 2,* 2.

Quené, H. (2004). What is the just noticeable difference for tempo in speech? In H. Quené & V. van Heuven (Eds.), *On Speech and Language: Studies for Sieb G. Nooteboom* (pp. 149–158). Utrecht: Netherlands Graduate School of Linguistics. LOT Occasional Series 2.

Radeloff, D. J. (1990). Role of color in perception of attractiveness. *Perceptual and Motor Skills, 71,* 151–160.

Reeves, B., & Nass, C. I. (1996). *The media equation: How people treat computers, television, and new media like real people and places*. New York: Cambridge University Press.

Rosenberg, A., & Hirschberg, J. (2009). Charisma perception from text and speech. *Speech Communication, 51,* 640–655.

Scherer, S., Layher, G., Kane, J., Neumann, H., Campbell, N. (2012). An audiovisual political speech analysis incorporating eye-tracking and perception data. In *Proceedings of 8th International Conference on Language Resources and Evaluation* (pp. 1114–1120).

Schötz, S. (2006). *Perception, Analysis, and Synthesis of Speaker Age*. Ph.D. thesis, Lund University, Sweden.

Shim, H. S., Park, S., Chatterjee, M., Scherer, S., Sagae, K., Morency, L. -P. (2015). Acoustic and paraverbal indicators of persuasiveness in social multimedia. In *Proceedings of IEEE International Conference on Acoustics, Speech and Signal Processing* (pp. 1–8).

Signorello, R., D'Errico, F., Poggi, I., Demolin, D. (2012). How charisma is perceived from speech: A multidimensional approach. Privacy, security, risk and trust (PASSAT). In *International Conference on Social Computing (SocialCom), Amsterdam, The Netherlands* (pp. 435–440).

Silverberg, D. (2017). Do the colours you wear at work matter? *BBC Business*. https://www.bbc.co.uk/news/business-41003867

Ten Bosch, L., Oostdijk, N., & de Ruiter, J. P. (2004). Turn-taking in social dialogues: Temporal, formal and functional aspects. In *Proceedings SPECOM, St. Petersburg*.

Terhardt, E. (1974). Pitch, consonance, and harmony. *Journal of the Acoustic Society of America, 55,* 1061–1069.

Touati, P. (1993). Prosodic aspects of political rhetoric. In *Proceedings of ESCA Workshop on Prosody, Lund, Sweden,* 168–171.

Towler, A. J. (2003). Effects of charismatic influence training on attitudes, behavior, and performance. *Personnel Psychology, 56,* 363–381.

van Santen, J. P. H. (1994). Assignment of segmental duration in text-to-speech synthesis. *Computer Speech & Language, 8,* 95–128.

Weber, M. (1947). *The theory of social and economic organization*. New York: The Free Press of Glencoe.

Weirich, M., & Simpson, A. P. (2014). Differences in acoustic vowel space and the perception of speech tempo. *Journal of Phonetics, 43,* 1–10.

Xu, Y., & Sun, X. (2002). Maximum speed of pitch change and how it may relate to speech. *Journal of the Acoustical Society of America, 111,* 1399–1413.

Chapter 12
Birds of a Feather Flock Together But Opposites Attract! On the Interaction of F0 Entrainment, Perceived Attractiveness, and Conversational Quality in Dating Conversations

Jan Michalsky and Heike Schoormann

Abstract Dating conversations are especially influenced by the interlocutors' perceived attractiveness. As visual attractiveness determines the course and nature of the interaction, the perceived overall quality of the conversation may also be influenced by the perceived attractiveness and simultaneously also affect the further development of the conversation. Accordingly, perceived attractiveness and conversational quality constantly interact in dating conversations. Studies focusing on the effects of both impressions on a speaker's vocal behavior in terms of prosodic entrainment, i.e., the adaptation of a speaker's prosodic features relative to his/her interlocutor, suggest that higher visual attractiveness leads to a greater divergence in f0 in mixed-sex pairs, while greater conversational quality results in larger degrees of f0 entrainment. In this paper, we further investigate the effects of both perceived attractiveness and conversational quality on prosodic entrainment of f0 in dating conversations with a special focus on their interaction. We conducted a dating experiment with 20 young heterosexual singles who engaged in 100 short spontaneous mixed-sex dating conversations. The results suggest that f0 entrainment correlates with both perceived attractiveness and conversational quality. Prosodic entrainment decreased with higher ratings of perceived attractiveness and increased with higher ratings of perceived conversational quality. Additionally, the results indicate that f0 entrainment not only depends on the impressions of attractiveness and conversational quality but also affects them. Furthermore, seemingly conflicting effects may be resolved by emphasizing one effect over the other, e.g., quality over attractiveness. This emphasis seems to depend on speaker sex and may also change during the course of the conversation. The details of this complex interaction, their interdependence, the importance of speaker sex, as well as possible implications are discussed.

J. Michalsky (✉) · H. Schoormann
University of Oldenburg, Ammerländer Heerstraße 114-118, 26129 Oldenburg, Germany
e-mail: j.michalsky@uni-oldenburg.de

H. Schoormann
e-mail: heike.schoormann@uni-oldenburg.de

B. Weiss et al. (eds.), *Voice Attractiveness*, Prosody, Phonology and Phonetics,
https://doi.org/10.1007/978-981-15-6627-1_12

Keywords Attractiveness · Conversational quality · Likability · Dating · Entrainment · Accommodation · Adaptation · F0

12.1 Introduction

12.1.1 Prosodic Entrainment and Its Role in Interaction

Most if not all prosodic features bear a high functional load on several communicative levels. Pitch, intensity, or speaking rate, for example, can convey linguistic functions such as focus (cf. Ladd 2008), paralinguistic meanings such as a speaker's emotions or attitudes (cf. Scherer, Ladd, & Silverman, 1984; Ladd, Silverman, Tolkmitt, Bergmann, & Schere, 1985), while simultaneously providing extra-linguistic information such as the sex or age of a speaker (cf. Linville, 1996) within the same stretch of speech. A change in prosodic features, such as increasing the speaking rate, reducing intensity, or raising the f0 mean, can reflect the social relationship of two speakers, e.g., in terms of social status (cf. Gregory, 1996) or dominance (cf. Puts, Gaulin, & Verdolini, 2006), while signaling attitudes and emotions that in turn affect and influence the interpersonal relationship. However, a phenomenon that has been linked to signaling and influencing interpersonal relationships has not been observed in the way prosodic features vary by themselves, i.e., in absolute terms, but in the way they change relative to the correspondent prosodic features of the interlocutor.

Entrainment, also often referred to as accommodation, convergence, or adaptation among others, describes this observed interdependence, i.e., speakers adjusting their linguistic features to those of the interlocutor particularly by becoming more similar (cf. Levitan, 2014). Entrainment can occur on all linguistic levels and may lead to an adaptation of the lexical choice (Brennan & Clark, 1996) and the syntactic structure (Reitter & Moore, 2007) but it can also influence prosodic features by matching speaking rate (Schweitzer, Lewandowski & Duran, 2017), intensity (cf. Levitan, 2014), or aspects of the fundamental frequency (cf. Levitan, 2014). Edlund, Heldner, and Hirschberg (2009) as well as Levitan (2014; see also Sect. 11.2.3) distinguish three types of prosodic entrainment which need to be differentiated. Proximity describes two interlocutors becoming similar with respect to a prosodic feature during a conversation, convergence describes two interlocutors becoming increasingly more similar during the course of a conversation, and synchrony describes a relative adaptation to the dynamics of an interlocutor's prosodic feature without necessarily becoming more similar in absolute terms.

There are two explanatory approaches to the occurrence of entrainment in human communication. Although they are often considered to be competing and mutually exclusive, we suggest that both approaches complement each other. According to the *communication model* (Natale, 1975) as well as the *perception behavior link* (Chartrand & Bargh, 1999), entrainment can be regarded as a device to enhance intelligibility by matching speaking styles and thus facilitating the identification of phonological categories by reducing phonetic variability. Accordingly, entrainment is a more or less automatic human behavior. This approach is supported by the fact that we also find entrainment in non-social interaction with synthetic voices used by machine applications (cf. Gessinger et al., 2018). The *communication accommodation theory* (Giles, Coupland, & Coupland, 1991) among others, however, assumes an iconic relationship between entrainment and social distance with smaller linguistic differences signaling closeness on a social level. This approach thus suggests that entrainment is not a mere automatism in interaction but it is dependent on the social relationship.

The focus of this paper lies on the role of f0 in signaling the relationship between interlocutors and a speaker's perceived attitude toward an interlocutor, respectively. Specifically, we study the connection between f0 entrainment and social distance in the situational setting of dating conversations. One factor that has a stronger influence in the current setting compared to other communicative situations is the perceived visual attractiveness of the interlocutor as dating conversations involves mating intention. Although especially important in mating contexts, the perceived attractiveness affects most if not every kind of conversation from everyday small talk to business communication (cf. Cialdini, 2009; Brooks, Huang, Kearney, & Murray, 2014). As of yet, it is largely unknown how perceived attractiveness interacts with prosodic entrainment and the perceived pleasantness of a conversation, henceforth referred to as conversational quality. The connection between the perceived visual attractiveness of the interlocutor, the perceived conversational quality, and a speaker's change in fundamental frequency constitutes the objective of the study at hand.

12.1.2 Prosodic Entrainment and Perceived Conversational Quality

Assuming a link between prosodic entrainment and social distance, the question arises how social distance was measured in previous studies. Rather than measured directly, social distance was approached as a construct derived from a wide variety of social features associated with closeness such as mutual liking (Levitan et al., 2012), support (Street, 1984; Levitan et al., 2012), giving encouragement (Nenkova, Gravano, & Hirschberg, 2008), or higher degrees of collaboration and cooperation (Lubold & Pon-Barry, 2014). We can assume that social closeness is reflected in the perceived quality of the conversation and will thus regard conversational quality as a predictor for social distance in the framework of this study. As we assess conversational quality through the subjective evaluation of the interlocutors, any further mention of conversational quality refers to the perceived conversational quality.

First evidence for the connection between entrainment and conversational quality stems from it signaling a closer connection between interlocutors and resulting in higher degrees of communicative success. Entrainment as an indicator for task success has been described for several different tasks. Thomason, Nguyen, and Litman (2013) report that student engineering groups that showed higher degrees of entrainment also showed better task results. Similar observations also hold for map task experiments (Reitter & Moore, 2007) as well as student tutoring programs (Friedberg, Litman, & Paletz, 2012). According to the theory of alignment (Pickering & Garrod, 2006), entrainment is a crucial contributor to communicative success in general. Lubold and Pon-Barry (2014) suggest that entrainment is connected to collaboration and rapport in learning tasks which also positively affects communicative success and thereby task success. Similarly, Taylor (2009) and Beňuš (2014) propose that task success greatly depends on the establishment of a common situational model, a process which is facilitated by the coordination of behavior. Accordingly, becoming closer with respect to verbal and non-verbal behavior might facilitate the construction of a common situational model.

How does task success relate to conversational quality? In other words, what is the goal of a non-task-oriented conversation? Although this is a rather difficult question to answer extensively, we can regard the establishment of a social bond as a major goal of verbal communication (cf. Dunbar, 2020). This is even more apparent in dating conversations where the establishment of a social bond serves as the basis for a romantic relationship that can be regarded as an explicit rather than an implicit goal (Hewstone, Stroebe, & Jonas, 2012: 870ff).[1] We can assume that the quality of a conversation greatly affects a conversations' ability to establish and/or improve the social relationship of two interlocutors. Accordingly, conversational quality can be regarded as the non-task-oriented equivalent of collaboration, affecting communicative success by affecting the social relationship.

Although previous studies on entrainment have for the most part been linked to either a speaker's perception of his/her interlocutor or the previously mentioned task success, there are some studies on meanings more closely related to conversational quality. Gonzales, Hancock, and Pennebaker (2009) found entrainment to be correlated to overall dialogue quality. Ireland et al. (2011) report that entrainment predicts the probability of initiating romantic relationships as well as the stability of existing relationships. In marriage counseling dialogues, Lee et al. (2010) found higher degrees of entrainment when couples were talking about positive rather than negative topics. Furthermore, entrainment was reported to result in smoother conversation with respect to turn latencies and fewer interruptions which can be regarded as attributes of high quality in conversations (Nenkova et al., 2008). Lastly, Michalsky et al. (2018) also found conversational quality and entrainment to be connected in dating conversations with smaller differences in f0 occurring in conversations that were perceived as more pleasant.

In conclusion, although conversational quality has rarely been assessed explicitly within the respective studies, we expect conversations that are perceived as better or more pleasant to show a higher degree of prosodic entrainment. This expectation applies to conversations in general and specifically to dating conversations.

[1]However, this is only true if we restrict our investigation to dating conversations which aim at finding a potential partner, which of course is not true for every kind of dating conversation.

12.1.3 Prosodic Entrainment and Perceived Attractiveness

The topic of vocal attractiveness received a lot of attention not only from a phonetic or even linguistic perspective but also from a sociological and psychological perspective. Furthermore, the immediate connection between attraction and aspects of evolutionary biology has generated assumptions that lead to specific linguistic hypotheses. Although this paper focuses on how speakers react prosodically to perceived attractiveness, i.e., the attracted voice, the underlying assumption is that we react to attractiveness by trying to sound more attractive (cf. Hughes, Farley, & Rhodes, 2010; Fraccaro et al., 2011). Accordingly, speakers would try to imitate features of attractive voices when perceiving their interlocutor as more attractive. To this end, a short overview on the prosodic features of attractive voices will be provided.

What prosodic features contribute to the impression of vocal attractiveness is a complex topic and cannot be solely and maybe not even primarily attributed to voice pitch. However, fundamental frequency as the acoustic correlate of voice pitch is the commonly studied feature of vocal attractiveness. The main reason for this can be found in the frequency code (Ohala, 1983, 1984) which assumes an evolutionary connection between pitch and attractiveness. In animal mating behavior, female individuals show the general tendency to select bigger and stronger male individuals to ensure protection as well as survival of their offspring. Accordingly, size is a biological factor in natural selection. While many species developed strategies to project size visually, others employ strategies to signal largeness through vocal features. Since due to physiological reasons larger individuals generally have a lower fundamental frequency, certain species such as wolves use lower pitch to suggest largeness. As largeness plays a role in selecting a partner for female individuals rather than males, it is associated with masculinity while smallness and high pitch are associated with femininity.

In general, studies confirmed these findings for human communication. Female listeners were found to evaluate male voices as significantly more attractive when they were realized with a lower f0 mean (Collins, 2000; Feinberg, Debruine, Jones, & Perrett, 2005; Hodges-Simeon, Gaulin, & Puts, 2010; Jones Feinberg, Debruine, Little, & Vukovic, 2010; Xu, Lee, Wu, Liu, & Birkholz, 2013) while male listeners judged female voices with a higher f0 mean as more attractive (Collins & Missing, 2003; Feinberg et al., 2008; Jones et al., 2010; Xu et al., 2013). However, the results for male listeners and thus female voices were not consistent. Oguchi and Kikuchi (1997) as well as Leaderbrand et al. (2008) suggest that female voices are perceived as more attractive when realized with a lower f0 mean. One explanation for this contradiction is provided by Karpf (2006) who proposed two different types of female attractiveness. Following Karpf's (2006) distinction, lower pitch is associated with the concept of sexiness and seductiveness while high pitch is associated with femininity. However, both are perceived as attractive female voices in general. Another explanation may be found in the communicative setting and thus the communicative intent. There are several goals such as intimacy goals, identity goals, or status goals sought in a relationship (cf. Zimmer-Gembeck, Hughes, Kelly, & Connoly, 2011) that can affect whether individuals are looking for anything from short-term flings to long-term relationships as well as different qualities sought in a partner associated with different goals which may lead to different concepts of attractiveness. However, this assumption has never been incorporated into experimental studies on vocal attractiveness.

In addition to the general features of male and female vocal attractiveness, the following findings are of relevance to the study at hand. Firstly, Vukovic et al. (2010) report that the perception of pitch as a cue to attractiveness not only depends on the speaker's absolute pitch but also on the listener's own average pitch. Furthermore, Borokowska and Pawlowski (2011) found a threshold at which an increase or decrease in mean fundamental frequency, respectively, does not increase perceived attractiveness any further. Lastly, Fraccaro et al. (2011) point toward the importance of naturalness and context when investigating perceived attractiveness as this feature seems to be especially susceptible to artificiality.

Most studies on vocal attractiveness commonly avoid defining the concept of attractiveness altogether. As evident from the inconsistent findings for female voices, listeners may employ a variety of different concepts of attractiveness when judging vocal attractiveness. However, we propose that investigating the prosodic effects of perceived visual attractiveness allow us to dispense with this problem and the need for defining the concept. Although listeners may still have a variety of reasons to perceive another person as more or less attractive, the result of the perceived attractiveness should always be attractive which can be connected to a physiological reaction and should therefore be more or less consistent across individuals (cf. Fraccaro et al., 2011). Although speakers may still employ different vocal strategies to express attraction, those differences are most likely not caused by differences in the concept of attractiveness that caused said attraction. Accordingly, the concept of attractiveness should be largely independent of the effects found for perceived attractiveness.

The effects of perceived attractiveness of an interlocutor on a speaker's f0 seem to confirm the assumption that speakers react to perceived attractiveness by mimicking the features of attractive voices and thereby trying to sound more attractive themselves. Male speakers who interacted with more attractive female interlocutors were found to lower their f0 mean (Hughes et al., 2010). For female speakers, however, we again find contradicting results. Female speakers were found to lower their f0 mean (Hughes et al., 2011) when talking to a more attractive male interlocutor as well as to raise their f0 mean under the same conditions (Fraccaro et al., 2011). According to Fraccaro et al. (2011) this may be explained through different experimental settings with varying degrees of contextual naturalness. In addition, the differences could again be related to the two different concepts of female attractiveness suggested by Karpf (2006). However, this assumption not only implies that male listeners have two different concepts of attractiveness associated with female voices but also that female speakers readily employ these two different concepts when signaling attraction.

How the prosodic effects of perceived visual attractiveness relate to prosodic entrainment has not been studied prior to Michalsky and Schoormann (2017) but there are some conclusions to be drawn from the research described above. Studies suggest that male speakers lower their f0 while female speakers, at least in some cases, raise their f0 when interacting with a more attractive interlocutor. Since male speakers on average have a lower f0 mean than female speakers for physiological reasons, both effects result in the speakers increasing the distance in f0 and thus showing what is called *prosodic disentrainment*. Michalsky and Schoormann (2017, 2018) suggest that this connection of prosodic disentrainment and perceived attractiveness can indeed be found in spontaneous dating conversations. A recent study by Beňuš et al. (2018) suggests that disentrainment can lead to the impression of dominance, which, according to the frequency code (Ohala, 1983, 1984), can be associated with masculinity. Yet, these results obtained from human–machine interaction would only support the hypothesis for the female listeners and not for the male listeners. A study by Schweitzer et al. (2017) suggest that there might also be effects of entrainment related to attractiveness. However, their findings are restricted to speaking rate and not f0 and furthermore focused on the concept of social attractiveness in same-sex dialogues.

In conclusion, we expect perceived attractiveness to result in prosodic disentrainment, directly contradicting the effects we expect for conversational quality.

12.1.4 The Dilemma: Good Conversations with Attractive Interlocutors

Regarding the effects of conversational quality and perceived attractiveness on prosodic entrainment we arrive at the preliminary expectation that higher conversational quality would result in social closeness and thus smaller differences in prosodic features, i.e., prosodic entrainment, while perceived attractiveness results in larger prosodic differences and hence prosodic disentrainment. This contradiction poses a challenge since conversational quality and perceived attractiveness not only operate on the same prosodic feature (f0 mean) while pointing in opposite directions but also because we expect both social parameters to highly influence dating conversations and thus to frequently co-occur and even interact. As such, we need to ask what happens with a speaker's f0 in conversations with high conversational quality and high perceived attractiveness, i.e., in conversations in which we would expect prosodic entrainment as well as prosodic disentrainment?

One hypothesis is that one effect overrules the other, i.e., signaling either conversational quality or perceived attractiveness is more important in dating conversations and thus only one of the contradicting effects is observable in this conversational setting.

A second hypothesis would be that the effects of perceived attractiveness are sensitive to the naturalness and context of the interaction. Higher perceived attractiveness may result in disentrainment only when investigated specifically in a mating context with scripted messages as done by Fraccaro et al. (2011) while possibly enhancing the effects of conversational quality by strengthening the social bond and thereby leading to more entrainment in spontaneous dating scenarios. However, this would contradict our previous finding on perceived attractiveness in dating conversations (Michalsky & Schoormann 2017, 2018).

Thirdly, the effects of perceived attractiveness and conversational quality may cancel each other out. This assumption would, however, entail a very uneconomic use of social signals. Accordingly, where f0 fails to signal conversational quality and perceived attractiveness simultaneously, other prosodic parameters may assume this function. Unfortunately, the scope of this paper is limited to f0.

Lastly, although effects of perceived attractiveness and conversational quality may co-occur within the same conversation, they need not occur simultaneously. One possibility is that perceived attractiveness is based on a first impression and the signaling of attraction hence decisive for the initiation of a conversation. Accordingly, the effects may be restricted to the first part of a conversation. Conversational quality on the other hand, develops over time and peaks during the course of the conversation. The effects of conversational quality may, therefore, be the strongest in the later parts of the conversation when the effects of perceived attractiveness have declined. Another distribution might regard different topics or even different intentions during the conversation. There may be phases where interlocutors are predominantly flirting, showing stronger effects of perceived attractiveness, and others where interlocutors are bonding, showing stronger effects of conversational quality. While we investigate only the former question by looking at different time points of the conversation, the latter remains for future research.

The study at hand was designed to improve on three shortcomings encountered in the previous research. Firstly, most studies investigate either perceived attractiveness or conversational quality. We suggest that if not explicitly asked to separate the two, speakers are inclined to let the two notions influence each other. Accordingly, we expect the judgement on perceived attractiveness to be influenced by the overall conversational quality and in return the impression of conversational quality to be compromised by the attractiveness of the interlocutor. Although this interdependence can never be totally excluded, explicitly instructing participants to judge both impressions on different scales is a first approach to telling them apart by raising awareness of the potential conflict.

Secondly, perception ratings are often taken from external observers rather than from the subjects participating in the study. Since the perception of attractiveness as well as conversational quality can and will greatly vary between participants actually partaking in the respective conversations and external observers, all judgements in this study are taken directly from the interlocutors.

Lastly, there are two possible perspectives regarding the connection of prosodic entrainment and social variables with respect to causality that are frequently separated and rarely both investigated within the same studies. On the one hand, the social relationship of two interlocutors can manifest itself in prosodic entrainment which thus serves as an indicator for the social relationship. On the other hand, prosodic entrainment may in return affect the social relationship and even facilitate the establishment of social bonds. Accordingly, we can either ask how the relationship affects prosodic entrainment but also how prosodic entrainment affects a social relationship. In this study, we incorporate both views to shed some light on the question of correlation and causality, although a definite answer to that question is categorically impossible.

This study is based on the same corpus as some of our previous work on the topic (cf. Michalsky, 2017; Michalsky & Schoormann, 2016, 2017, 2018; Michalsky et al., 2018, 2018). We would like to inform the reader about the possibility of conflicting information. Our previous research on the topic constitutes work in progress on a growing corpus with changing normalization methods and shifting focus regarding the f0 parameters in question. Since the results presented in this paper constitute the final state of the analysis, the information presented in this paper explicitly replaces older information.

Perceived visual attractiveness

1. Does the perception of visual attractiveness in an opposite-sex interlocutor systematically correlate with a speaker's f0 entrainment in accordance with previous findings?
2. Do changes in a speaker's f0 entrainment correlate with an interlocutor's perception of the speaker in terms of visual attractiveness?
3. Are these two effects connected in a systematic way?

Perceived conversational quality

1. Does the perceived conversational quality systematically correlate with a speaker's f0 entrainment in accordance with previous findings?
2. Do changes in a speaker's f0 entrainment correlate with an interlocutor's perception of the conversational quality?
3. Are these two effects connected in a systematic way?

Perceived visual attractiveness and conversational quality

1. Do the effects of perceived attractiveness and conversational quality interact in their influence on f0 entrainment?
2. Does f0 entrainment show contradicting or complementing effects on the perception of attractiveness and conversational quality?

12.2 Method

12.2.1 Subjects

The study was conducted with 20 participants, 10 female, and 10 male, all paid volunteers and at the time of the experiment students at the University of Oldenburg. All subjects were aged between 19 and 28 and monolingual speakers of High German who spent the majority of their lives in Lower Saxony. Furthermore, all subjects reported to be heterosexual as well as single during the whole course of the study. With the exception of two speakers, whose conversation was excluded from the experiment, all subjects were previously unacquainted. All subjects were informed about the nature of the experiment as a dating situation.

12.2.2 Procedure

All participants were informed about the dating setting of the experiment prior to their recordings. Female and male participants waited in separated rooms and were led to the recording rooms via separate staircases to avoid any interaction prior to their actual conversations. Each participant was paired with every other participant of the opposite sex resulting in a total of 100 opposite-sex conversations, all recorded in two parallel recording sessions in two quiet separate university office rooms. All recordings were done within two weeks during spring break. The use of the phonetic laboratory was explicitly avoided to ensure a more natural setting based on the importance of naturalness in evaluating attractiveness by Fraccaro et al. (2011). The participants were encouraged to engage in spontaneous conversations of 15–20 min without any restrictions or guidelines regarding the choice of conversational topics. However, example topics were provided in case conversations were stalling and subjects needed inspiration.

Immediately before each conversation all participants judged their respective interlocutor on a 10-point Likert scale with respect to their perceived visual attractiveness and general likability. The participants were separated by a screen that allowed them to see each other's faces but concealed the questionnaire so that the evaluations were not revealed to the interlocutor. The screen was removed at the beginning of the conversation. Directly after each conversation, the participants received another questionnaire and repeated the covert evaluation of perceived visual attractiveness and general likability. Furthermore, a third scale was added to this second questionnaire to evaluate how pleasant the subjects perceived the conversation as a whole to assess conversational quality.

Recordings were made in stereo using head-mounted microphones (DPA 4065 FR) to ensure an optimal balance between recording quality and naturalness. We used a portable digital recorder (Tascam HD P2) at a sampling rate of 48 kHz and 24-bit resolution.

12.2.3 Types and Measurements of Entrainment

According to Edlund et al. (2009) and Levitan (2014) we can distinguish three different types of entrainment: *proximity*, *convergence*, and *synchrony*.

Proximity covers what is usually referred to as entrainment, accommodation, or adaptation and describes two speakers being more similar with respect to a linguistic feature when talking to each other than when not talking to each other. Accordingly, proximity needs some sort of reference value by either operating on a local level and comparing the differences of prosodic features at adjacent turns with non-adjacent turns (Levitan, 2014) or globally by comparing the differences during a conversation with differences to other speakers or conversations. In this study we combine the two by comparing the general differences at adjacent turns in correlation with perceived attractiveness as well as conversational quality across conversations.

Convergence describes increasing proximity over time during the course of a single conversation. Accordingly, we again measure the difference in a linguistic feature at adjacent turns but with respect to their changes during the conversation. Convergence can either be assessed locally by tracking changes from turn to turn or globally, e.g. by comparing the first and second half of a conversation.

Synchrony constitutes a categorically different type of entrainment that is either not considered at all or assumed as the primary type of entrainment. Synchrony describes the relative adaptation of a speaker's linguistic features to the respective feature of his/her interlocutor by adjusting their values relative to each other without necessarily becoming more similar. For example, a speaker may react to a raised f0 mean of his/her interlocutor by raising his/her own f0 mean by the same amount, thus imitating his/her interlocutor's prosodic behavior without a decrease in the differences between the two as it is the case for proximity or convergence. To measure synchrony, we check for correlations between the prosodic feature of the turn-taking speaker and the turn-passing speaker in adjacent turns of a speaker change inducing turn break. A positive correlation generally points toward synchrony while a negative correlation is often linked to an effect of increased or decreased proximity.

12.2.4 Acoustic Analysis

For the acoustic analysis we used Praat (Boersma & Weenink, 2016). Since the recordings were made in stereo, we separated the audio tracks for each speaker. The audio tracks were manually annotated for interpausal units (IPU, cf. Levitan, 2014). We analyzed all IPUs adjacent to a turn break inducing speaker change. IPUs were defined mechanically by stretches of speech preceded or followed by a pause with a pause defined as an interruption of speech by silence or non-speech noise of at least 500 ms. Accordingly, we made no difference between pauses at phrase boundaries and hesitation pauses in favor of interlabeler reliability. The corpus consists of 14.687 IPUs from 98 conversations. One conversation had to be excluded due to prior acquaintance of the participants another one was lost to a recording error. We extracted the f0 mean from the interpausal units as we suggest that the f0 mean captures the register better than the median (cf. Michalsky & Schoormann, 2016; Michalsky et al., 2018). Furthermore, range features at phrase final boundaries are heavily distorted by pragmatic functions and therefore unreliable in capturing entrainment in this specific data set (cf. Michalsky, 2014, 2015). For synchrony, we measured the f0 mean of the IPU of the turn-passing and the turn-taking speaker and converted it to semitones to a reference value of 50 Hz. We z-transformed the data by speaker by subtracting the IPUs' f0 mean values from the average f0 mean value of all IPUs of each speaker across all conversations and dividing it by the standard deviation of the same set. For proximity and convergence, we calculated the absolute difference between the f0 mean of IPUs adjacent to turn breaks in semitones. Furthermore, we tagged the IPUs occurring in the first five minutes as well as the last five minutes of the conversations.

12.2.5 Statistical Analysis

For the statistical analysis, we conducted linear mixed effects models using *R* (R Core Team 2017) with the *lme4*-package (Bates, Maechler, Bolker & Walker, 2015) and the *lmerTest*-package (Kuznetsova, Brockhoff, & Christensen, 2016). Model fit was determined by maximum likelihood ratio tests. P-values were calculated using the Satterthwaite approximation. We calculated different models for the effects of prosodic entrainment on the investigated social variables and the effects of these social variables on prosodic entrainment.

For the effects of perceived ATTRACTIVENESS and CONVERSATIONAL QUALITY on prosodic entrainment we used different dependent variables with respect to the type of prosodic entrainment. For *synchrony*, we calculated Pearson correlation coefficients (*f0 correlation*) between the f0 mean of the turn-passing speaker and the turn-taking speaker, which resembles the degree of synchrony (cf. Edlund et al., 2009; Levitan, 2014), and used it as the dependent variable. As fixed factors, we used perceived visual attractiveness (ATTRACTIVENESS), perceived conversational quality (QUALITY), speaker sex (SEX) and all interactions. In both the proximity model and the convergence model, we used the difference between the f0 means of the IPUs adjacent to turn breaks (*f0 difference*) as dependent variables. For the proximity model, fixed factors were identical to the synchrony model. For the convergence model, we added the TIME (in seconds) of the respective turn break as a fixed factor.

For the effects of prosodic entrainment on perceived *attractiveness* and *conversational quality*, we calculated two different models for each of the three types of entrainment with either perceived attractiveness (*attractiveness*) or perceived conversational quality (*conversational quality*) as dependent variables with the respective counterpart serving as fixed factor (ATTRACTIVENESS or QUALITY). In the synchrony model, we used the correlation coefficients (F0 CORRELATION) described above as a fixed factor as well as SEX and all interactions. In the proximity model we used the difference between the IPUs adjacent to turn breaks (F0 DIFFERENCE) as well as SEX and their interaction as fixed factors. For the convergence model, we again expanded the proximity model by the fixed factor TIME and the additional possible interactions.

Since we suggest that the effects of perceived attractiveness and conversational quality may affect different parts of the conversation to different degrees, we conducted post hoc tests for every model described above separated by conversational part. The variable conversational part splits the data set into turns occurring within the first five minutes of each conversation and turns occurring within the last five minutes of each conversation to see whether the effects are restricted to certain conversational parts. Note that this leads to a substantial reduction of the data set and may result in statistically insignificant effects due to insufficient data points. However, this was only done for proximity as effects for convergence were already absent from the entire conversation and the Pearson correlation coefficients calculated for synchrony were not robust enough for splitting the data set into thirds.

12.3 Results

12.3.1 Effects of Perceived Attractiveness and Conversational Quality on Prosodic Entrainment

12.3.1.1 Proximity

Table 12.1 presents the results for the effects of perceived ATTRACTIVENESS and conversational QUALITY on proximity. We find significant interactions between the two factors as well as a three-way-interaction with SPEAKER SEX illustrated in Fig. 12.1. Accordingly, we conducted post hoc tests separated by SPEAKER SEX to investigate the nature of this three-way-interaction. Tables 12.2 and 12.3 present the post hoc results for the female and the male speakers, respectively. Table 12.2 shows that the male speakers show significant effects for both perceived ATTRACTIVENESS and QUALITY without interactions although marginal effects are suggested by Fig. 12.1. Male speakers decrease their *f0 differences* between turns with increasing conversational QUALITY and increase these differences with increasing visual ATTRACTIVENESS of the interlocutor. For the female speakers we find a significant interaction between ATTRACTIVENESS and conversational QUALITY (s. Table 12.3). Female speakers also decrease their *f0 differences* with increasing CONVERSATIONAL QUALITY and increase *f0 differences* with visual ATTRACTIVENESS. However, as shown in Fig. 12.1, the effects for ATTRACTIVENESS become smaller with increasing conversational QUALITY. This means that female speakers do react less to the perceived ATTRACTIVENESS of their interlocutor when the conversation is perceived as highly positive. In conversations with below average QUALITY, however, ATTRACTIVENESS significantly correlates with the degree of *proximity*.

Table 12.1 Significant main effects and interactions of perceived ATTRACTIVENESS and CONVERSATIONAL QUALITY on *proximity*

Fixed factors	*b*	*SE*	*df*	*t*	*p*
ATTRACTIVENESS	0.74	0.09	14560.00	8.32	<0.001
SEX	3.11	0.99	129.90	3.16	<0.01
ATTRACTIVENESS * QUALITY	−0.07	0.01	14570.00	−5.27	<0.001
ATTRACTIVENESS * SEX	−0.84	0.15	14570.00	−5.72	<0.001
QUALITY * SEX	−0.34	0.12	14570.00	−2.79	<0.01
ATTRACTIVENESS * QUALITY * SEX	0.10	0.02	14570.00	4.83	<0.001

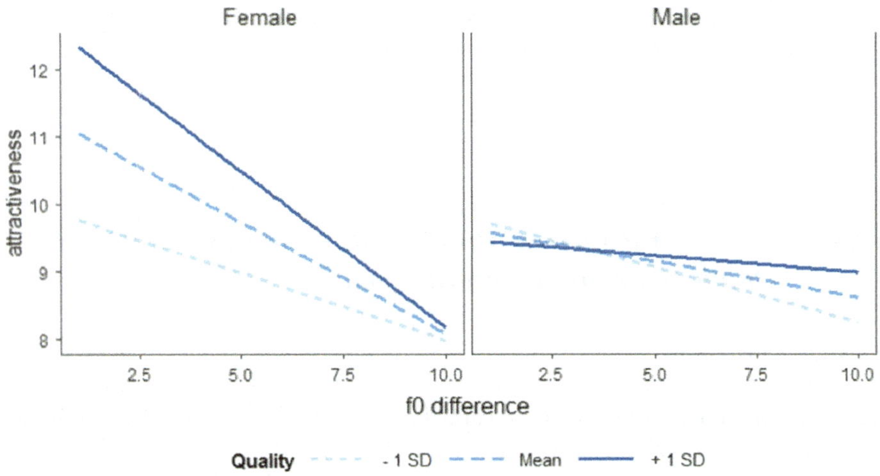

Fig. 12.1 Interaction of the effects of perceived ATTRACTIVENESS, QUALITY, and SEX on *f0 difference*

Table 12.2 Post hoc significant main effects and interactions of perceived ATTRACTIVENESS and CONVERSATIONAL QUALITY on *proximity* for the male speakers

Fixed factors	*b*	*SE*	*df*	*t*	*p*
ATTRACTIVENESS	0.11	0.04	7375.06	2.86	<0.01
QUALITY	−0.10	0.04	7372.77	−2.65	<0.01

Table 12.3 Post hoc significant main effects and interactions of perceived ATTRACTIVENESS and CONVERSATIONAL QUALITY on *proximity* for the female speakers

Fixed factors	*b*	*SE*	*df*	*t*	*p*
ATTRACTIVENESS	0.74	0.09	7100.57	8.10	<0.001
ATTRACTIVENESS * QUALITY	−0.07	0.01	7179.94	−5.17	<0.001

Table 12.4 Post hoc significant main effects and interactions of perceived ATTRACTIVENESS and CONVERSATIONAL QUALITY on *proximity* for the male speakers for the first five minutes of a conversation

Fixed factors	*b*	*SE*	*df*	*t*	*p*
QUALITY	−0.41	0.16	2661.52	−2.49	<0.05
QUALITY * ATTRACTIVENESS	0.05	0.03	2661.27	1.97	<0.05

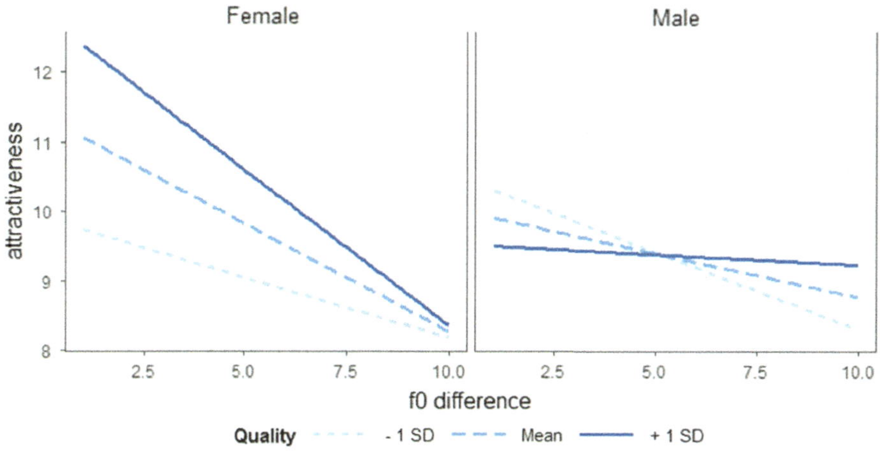

Fig. 12.2 Interaction of the effects of perceived ATTRACTIVENESS, QUALITY, and SEX on *f0 difference* in the first five minutes

The post hoc investigation of the conversational parts reveals that female speakers behave consistently throughout the entire conversation and show the same effects as reported above for the first as well as the last five minutes. For the male speakers, however, the effects change over the course of the conversation. While the effects for the last five minutes are identical to the effects we found for the whole conversation, we find a deviation in the first five minutes. As shown in Table 12.4, male speakers show a significant interaction between ATTRACTIVENESS and conversational QUALITY in the first five minutes. Although this interaction also seems to be present in the whole conversation when comparing Figs. 12.1 and 12.2, it only reaches significance for the first five minutes. This interaction is different to the one found for the female speakers. Figure 12.2 shows that the effects of conversational QUALITY become smaller with increasing perceived ATTRACTIVENESS. Accordingly, while for the female speakers conversational QUALITY overruled ATTRACTIVENESS throughout the whole conversation, for the male speakers, ATTRACTIVENESS overrules conversational QUALITY. In other words, male speakers do entrain less in pleasant conversations with attractive women. However, this effect is restricted to the first five minutes and is found for neither the last 5 min nor the conversational as a whole.

Table 12.5 Significant main effects for *convergence*

Fixed factors	*b*	*SE*	*df*	*t*	*p*
TIME	−0.0005	0.0001	14550.0000	−4.2530	<0.001

12.3.1.2 Convergence

Table 12.5 shows that there is a significant effect for TIME on the *f0 differences* at turn breaks. Accordingly, we find a general effect for *convergence* with speakers becoming more similar to each other over time. However, this effect shows no interaction with either perceived ATTRACTIVENESS or conversational QUALITY. Hence, while we find effects of conversational QUALITY and perceived ATTRACTIVENESS on entrainment, the observed general convergence is not enhanced by the social variables investigated.

12.3.1.3 Synchrony

Table 12.6 reports the effects of perceived ATTRACTIVENESS and conversational QUALITY on *synchrony*. In contrast to *proximity*, we find no significant effects for SPEAKER SEX or any interaction with SEX. Accordingly, we find main effects for ATTRACTIVENESS and conversational QUALITY as well as their interaction for both sexes. Figure 12.3 illustrates the interaction between ATTRACTIVENESS and conversational QUALITY. We find that increasing ATTRACTIVENESS is correlated with greater synchrony if conversational QUALITY is low but correlates with less synchrony if conversational QUALITY is high. The same is true for the opposite perspective. Increasing conversational QUALITY is correlated with stronger *synchrony* if the perceived ATTRACTIVENESS is low but is correlated with lower synchrony if the perceived ATTRACTIVENESS is high.

Table 12.6 Significant main effects and interactions of perceived ATTRACTIVENESS and CONVERSATIONAL QUALITY on *synchrony*

Fixed factors	*b*	*SE*	*df*	*t*	*p*
ATTRACTIVENESS	0.05	0.02	174.50	2.90	<0.01
QUALITY	0.05	0.01	190.18	3.44	<0.001
ATTRACTIVENESS * QUALITY	−0.01	0.00	187.27	−3.04	<0.01

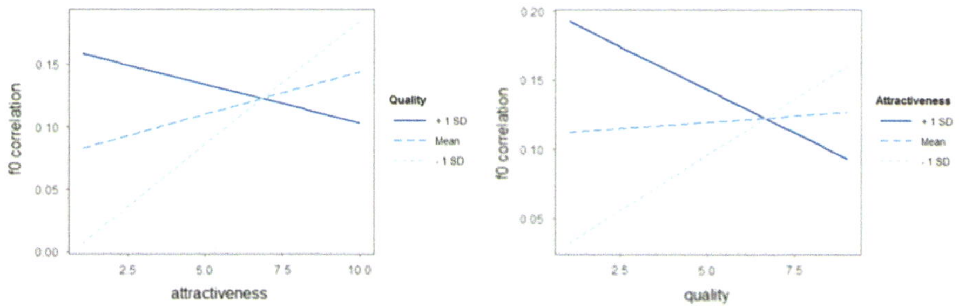

Fig. 12.3 Interaction of the effects of perceived ATTRACTIVENESS and QUALITY on *f0 correlation*

12.3.2 Effects of Prosodic Entrainment on Perceived Attractiveness and Conversational Quality

12.3.2.1 Proximity

Table 12.7 presents the effects of PROXIMITY on perceived *attractiveness*. We find significant effects for the three-way-interaction between F0 DIFFERENCE, QUALITY, and SEX. Figure 12.4 illustrates this three-way-interaction while Tables 12.8 and 12.9 report the post hoc results separated by SEX. For both sexes, we find a significant interaction between F0 DIFFERENCE and QUALITY. In general, both female and male speakers show a tendency to judge speakers as more *attractive* if they show greater F0 DIFFERENCES and hence a greater degree of disentrainment. However, female speakers show strong effects of F0 DIFFERENCE for *attractiveness* if QUALITY is low or average but close to no effects if QUALITY is high. Male speakers on the other hand show noticeable effects for F0 DIFFERENCE if QUALITY is high and less pronounced effects if QUALITY is average or low.

Table 12.10 presents the effects of PROXIMITY on perceived *quality*. Comparable to *attractiveness*, we find significant effects for the three-way-interaction between F0 DIFFERENCE, ATTRACTIVENESS, and SEX. Figure 12.5 illustrates this three-

Table 12.7 Significant main effects and interactions of PROXIMITY on perceived *attractiveness*

Fixed factors	*b*	*SE*	*df*	*t*	*p*
F0 DIFFERENCE	0.10	0.01	14570.00	9.39	<0.001
QUALITY	0.58	0.02	14580.00	34.78	<0.001
SEX	1.03	0.32	38.96	3.19	<0.01
F0 DIFFERENCE * QUALITY	−0.01	0.00	14570.00	−6.98	<0.001
F0 DIFFERENCE * SEX	−0.12	0.02	14570.00	−7.06	<0.001
F0 DIFFERENCE * QUALITY * SEX	0.02	0.00	14570.00	6.58	<0.001

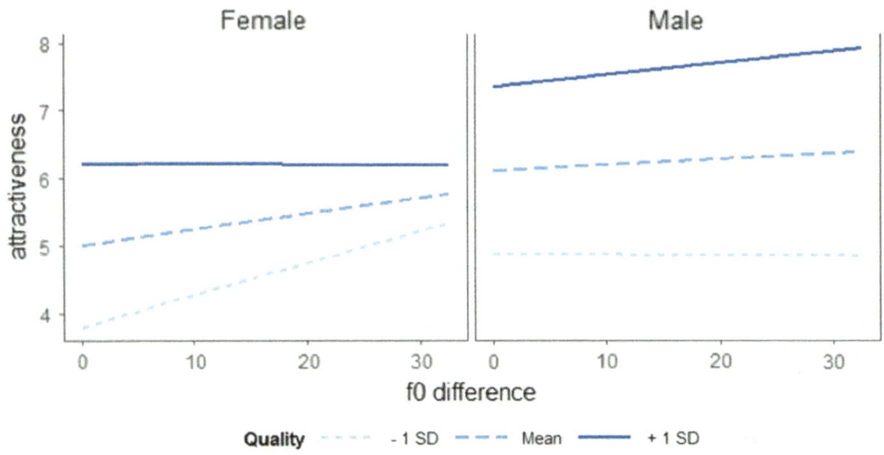

Fig. 12.4 Interaction plot for the effects of F0 DIFFERENCE, QUALITY, and SEX on perceived *attractiveness*

Table 12.8 Post hoc significant main effects and interactions of PROXIMITY on perceived *attractiveness* for the female speakers

Fixed factors	*b*	*SE*	*df*	*t*	*p*
F0 DIFFERENCE	0.10	0.01	7272.00	9.63	<0.001
QUALITY	0.58	0.02	7274.00	35.64	<0.001
F0 DIFFERENCE * QUALITY	−0.01	0.00	7271.00	−7.15	<0.001

Table 12.9 Post hoc significant main effects and interactions of PROXIMITY on perceived *attractiveness* for the male speakers

Fixed factors	*b*	*SE*	*df*	*t*	*p*
QUALITY	0.59	0.02	7306.00	32.14	<0.001
F0 DIFFERENCE * QUALITY	0.00	0.00	7302.00	2.43	<0.05

Table 12.10 Significant main effects and interactions of PROXIMITY on perceived *conversational quality*

Fixed factors	*b*	*SE*	*df*	*t*	*p*
ATTRACTIVENESS	0.93	0.02	14580.00	40.55	<0.001
SEX	1.14	0.38	38.24	3.01	<0.01
F0 DIFFERENCE * ATTRACTIVENESS	−0.01	0.00	14570.00	−4.48	<.001
ATTRACTIVENESS * SEX	−0.27	0.03	14580.00	−8.20	<0.001
F0 DIFFERENCE * ATTRACTIVENESS * SEX	0.01	0.00	14570.00	3.19	<0.01

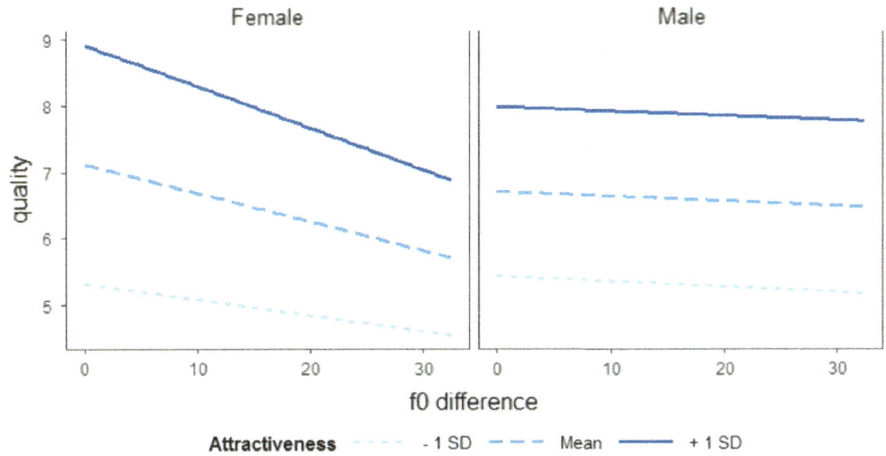

Fig. 12.5 Interaction plot for the effects of F0 DIFFERENCE, ATTRACTIVENESS, and SEX on perceived *quality*

Table 12.11 Post hoc significant main effects and interactions of PROXIMITY on perceived *conversational quality* for the female speakers

Fixed factors	*b*	*SE*	*df*	*t*	*p*
ATTRACTIVENESS	0.93	0.03	7279.00	36.81	<0.001
F0 DIFFERENCE * ATTRACTIVENESS	−0.01	0.00	7273.00	−4.07	<0.001

Table 12.12 Post hoc significant main effects and interactions of PROXIMITY on perceived *conversational quality* for the male speakers

Fixed factors	*b*	*SE*	*df*	*t*	*p*
ATTRACTIVENESS	0.66	0.01	7304.00	72.57	<0.001

way-interaction while Tables 12.11 and 12.12 report the post hoc results separated by SEX. Again, we see a common general tendency for both sexes but in contrast to perceived *attractiveness*, both sexes judge the conversation as better if the interlocutor shows smaller F0 DIFFERENCE and hence greater degrees of entrainment. Again, these effects interact with the other social variable, in this case perceived ATTRACTIVENESS. For the female speakers, Fig. 12.5 shows that the effects of F0 DIFFERENCE on *quality* increase with perceived ATTRACTIVENESS, which is statistically significant in the post hoc test reported in Table 12.11. Male speakers show the same tendency but as shown in Fig. 12.5, the effects are much smaller and do not reach statistical significance in the post hoc test (s. Table 12.12).

The post hoc investigation of the conversational parts shows that female speakers show the same effects for *attractiveness* as for the conversations as a whole within both the first and the last five minutes of the conversation. For the male speakers,

Table 12.13 Post hoc significant main effects and interactions of PROXIMITY and ATTRACTIVENESS on perceived *conversational quality* for the male speakers for the last five minutes of a conversation

Fixed factors	*b*	*SE*	*df*	*t*	*p*
F0 DIFFERENCE	−0.02	0.01	2615.00	−2.81	<0.01
ATTRACTIVENESS	0.65	0.02	2615.00	43.00	<0.001

however, we only find significant effects for the conversation as a whole but not for the conversational parts. With respect to Fig. 12.4, we suspect that the effects for the conversation as a whole are already too small and are hence lost when splitting the data set.

Comparable to perceived *attractiveness*, the effects for *quality* are robust for the female speakers within the conversational parts. Female speakers show the same positive effects of entrainment on *quality* for both the start and the end of the conversation. The male speakers, however, show deviating effects from the conversation as a whole, which also point in the opposite direction. While we find effects on *attractiveness* for the whole conversation but not for the parts, we find the opposite for *quality*. While entrainment does not significantly correlate with *quality* in the conversation as whole, we find a significant effect of F0 DIFFERENCE on *quality* in the part subsets (s. Table 12.13). Furthermore, these effects only occur in the last five minutes of the conversation but not in the first five minutes. Lastly, F0 DIFFERENCE does not interact with perceived ATTRACTIVENESS in contrast to any other entrainment effects reported in this chapter.

12.3.3 Convergence and Synchrony

Comparable to the effects of perceived attractiveness and conversational quality on *convergence* (s. chapter 12.3.1.2) we find no significant effects for convergence on either variable. However, in contrast to the effects found for perceived attractiveness and conversational quality on *synchrony* (s. chapter 12.3.1.3) we also find no significant effects for synchrony on either variable.

12.4 Discussion

The results show that there is a strong connection between prosodic entrainment and both perceived visual attractiveness and conversational quality. We find that prosodic entrainment reflects the social relationship by showing effects for the perceived visual attractiveness of an interlocutor as well as effects for the perceived quality of the conversation. Furthermore, the degree of prosodic entrainment correlates with how pleasant a speaker perceives a conversation as well as how visually attractive s/he perceives his/her interlocutor. However, while there are core effects that suggest a direct interpretation and are in accordance with previous studies as well as the expectations given in chapter one, there are several findings that pose a challenge for future research. Especially the synchrony effects, the reciprocity of the connection between entrainment and social variables, as well as the interaction of perceived attractiveness and conversational quality leave many open questions as discussed in the following.

12.4.1 Effects of Perceived Attractiveness and Conversational Quality on Prosodic Entrainment

Perceived attractiveness and conversational quality both significantly correlate with the degree to which a speaker entrains to his/her interlocutor. However, both variables correlate with entrainment differently and with notable differences depending on speaker sex. In general, both female and male speakers show greater degrees of entrainment in terms of proximity if they perceive the conversation as better. This is compatible with our expectations from the link between prosodic entrainment and conversational quality as well as social distance in general (cf. Nenkova et al., 2008; Gonzales et al., 2009; Levitan et al., 2012). In contrast, both sexes show greater degrees of disentrainment in conversations with more visually attractive interlocutors. This is also in accordance with our expectations from previous research on the effects of visual attractiveness on prosody in general (cf. Hughes et al., 2010; Fraccaro et al., 2011). However, this also means that the effects are indeed diametrically opposing each other.

For the female speakers, this results in a significant interaction between attractiveness and conversational quality with respect to entrainment. The effects of attractiveness decrease with higher degrees of conversational quality and are even absent in conversations that are perceived as very pleasant. Accordingly, female speakers emphasize conversational quality over attractiveness in terms of entrainment. This is consistent across the entire conversation. The opposite is true for the male speakers. Here we also find a significant interaction but male speakers show decreasing effects of conversational quality as attractiveness increases. Accordingly, male speakers emphasize visual attractiveness over conversational quality. However, this is only true for the first five minutes of the conversation and neither for the last five minutes nor the conversation as a whole. Hence, male speakers emphasize visual attractiveness when first engaging in a conversation but show more balanced prosodic effects for both variables as the conversation emerges.

The picture is less clear for the other types of prosodic entrainment. Perceived visual attractiveness and conversational quality both correlate with synchrony. However, the effects are difficult to interpret. Both variables show a positive correlation with the degree of synchrony if the respective other variable is low, negatively if the other variable is high, and marginally or not at all if the other one is average. We suggest that synchrony as measured in this study reflects the complex relationship between attractiveness and conversational quality and cannot be interpreted in its own right. Furthermore, we find a general convergence effect, i.e., a general trend for speakers to become more similar over time. However, this trend is independent from either social variable.

12.4.2 Effects of Prosodic Entrainment on Perceived Attractiveness and Conversational Quality

The effects of prosodic entrainment on perceived attractiveness and conversational quality show a nearly reciprocal relationship with the effects reported above. Both sexes judge conversations as better where the interlocutor shows a greater degree of prosodic entrainment in the form of proximity. Although the literature on the effects of entrainment on perception is scarce, these effects are in line with our expectations (cf. Nenkova et al., 2008; Gonzales et al., 2009; Levitan et al., 2012). Furthermore, both male and female speakers perceive interlocutors who show greater degrees of disentrainment as more visually attractive. This is also in line with our expectations since disentrainment generally leads male speakers to lower their voices and female speakers to raise their voices, which was found to increase perceived attractiveness (cf. Collins, 2000; Collins and Missing, 2003; Feinberg et al., 2005, 2008; Hodges-Simeon et al., 2010; Jones et al. 2010; Xu et al., 2013). However, it is not the mere distinction between low and high which is connected to attractiveness but specifically the distance caused by disentrainment. An interlocutor's pitch is thus evaluated within his/her own natural register and not in absolute terms as comparable across speakers. Again, the effects of perceived attractiveness and conversational quality are contradicting.

For the female speakers, the effects of perceived attractiveness and conversational quality interact significantly. The effects of entrainment on conversational quality become stronger with an increased perceived visual attractiveness of the interlocutor. Simultaneously, the effects of attractiveness become weaker the better the conversation. Both interactions are consistent across the entire conversation. Accordingly, we find the same dominance of conversational quality over visual attractiveness reported above. Again, the picture is vastly different for the male speakers. While the effects of entrainment on perceived attractiveness are statistically independent from conversational quality, the effects of entrainment on conversational quality become weaker the more attractive the interlocutor. Accordingly, the male speakers again show a dominance of attractiveness over conversational quality. Furthermore, we find another effect compatible with the results reported above. While the dominance of visual attractiveness over conversational quality on entrainment disappears after the first 5 min of the conversation, the effects of entrainment on conversational quality only appear after the first 5 min. Hence, although the male speakers show a general dominance of visual attractiveness over conversational quality, these effects shift over time with attractiveness being emphasized when first engaging in a conversation while the perception of conversational quality manifests its effects in the later parts of the conversation.

In contrast to the effects of the social variables on entrainment, we find no effects of synchrony on either visual attractiveness or conversational quality. Again, we suggest that this may be related to the complex interaction between the two social factors. Furthermore, the data set may be too small for the reliable application of Pearson correlation coefficients as measurements for synchrony (cf. Edlund et al., 2009; Levitan, 2014), since we found effects of synchrony using other although cruder measurements. The size of the data set could also explain the absence of convergence effects.

12.4.3 The Dilemma: Good Conversations with Attractive Interlocutors

This chapter has shown that high degrees of conversational quality and visual attractiveness within the same conversation do indeed lead to contradicting effects as expected from the introduction. However, with respect to our initial expectations, we do not find one factor completely canceling out the other. Instead, the results suggest a weighting of the two variables. While female speakers emphasize conversational quality, male speakers generally emphasize perceived attractiveness. Accordingly, if both perceived attractiveness and conversational quality are high, female speakers tend to show stronger entrainment while male speakers show stronger disentrainment. This observation is complemented by the finding that male speakers also show a shift in weighting. While female speakers consistently emphasized conversational quality over attractiveness, male speakers show a tendency to emphasize perceived attractiveness when first engaging in a conversation and then shifting the focus to conversational quality as the conversation progresses. Accordingly, with respect to our initial expectation we find both one factor overruling the other as well as differences in distribution across a conversation. However, we did not find an association of different types of entrainment with different social variables.

A factor not considered within this study concerns differences in the weighting of these social variables not only in their distribution by time but also by topic. Accordingly, there may be conversational topics that are thematically closer to mating and hence show a higher demand for signaling attractiveness versus topics closer related to forming stronger bonds and hence related to signaling conversational quality comparable to the effects found for positive versus negative topics by Lee et al. (2010). Such a topic related shift in signaling social variables would also mirror and thus add to the interpretation of the effects observed for the male speakers as a higher density of mating related topics in the first half of the conversation compared to the more bonding related topics in the last half seems likely.

Lastly, the weighting of perceived attractiveness may also be related to personality. Emphasizing perceived conversational quality over perceived visual attractiveness could be related to factors such as emotional empathy or agreeableness. Consequently, the differences we find for speaker sex may actually not be related to speaker sex itself but to gender-related personality attributes.

12.4.4 Additional Thoughts and Further Implications

Before discussing some further implications of the results, we would like to address two observations regarding the experiment itself to clarify the possible generalizability of our results. Prior to the experiment, we assessed personal data from all participants including their intent to participate in the study. As reported above, all participants were informed that the experiment was designed as a dating study. However, only three participants stated that they were actually interested in dating and eventually finding a partner. Furthermore, all of these three participants were male. The remaining participants all declared to be merely interested in having good conversations and meeting new people. Accordingly, the majority of the participants did not intent to date prior to the conversations or at least did not admit it. Inspecting the conversations with respect to content leads to a mixed result. While most conversations confirm the assessment by a lack of flirting, several conversations suggest a strong intent for dating. One pair even requested to exchange contact information although both participants did not declare to be interested in finding a partner. Accordingly, engaging in a dating conversation is not necessarily something that happens intentionally. Furthermore, participants may just not be willing to admit their intent when participating in a scientific study. The fact, that we do find strong effects for perceived attractiveness may support this conclusion. However, as pointed out above, perceived attractiveness may play a strong role even in non-dating conversations which shifts the focus of the generalizability of this study.

The second observation regards the naturalness of the conversations. Although initially most participants were irritated by the setting and often commented on the recording situation, this issue quickly dissipated in most conversations. Overall, we perceive the majority of the conversations as resembling natural interactions. The participants engaged freely in spontaneous dialogues, choosing a wide variety of different topics and transitioning fluently between them. There are also several cases of participants talking about the researchers, other university staff, or even sharing personal information which they were explicitly instructed not to reveal, suggesting that participants quickly forgot about being recorded.

With respect to the ongoing debate about the function of entrainment, our study supports both categories of assumptions. The effects of perceived conversational quality strongly support the *communication accommodation theory* (Giles et al., 1991) and related models which link social closeness to greater entrainment. Greater conversational quality can be related to a stronger social bond between the interlocutors and hence a greater degree of social closeness. However, we also find an effect of categorical convergence, which is not affected by either conversational quality or attractiveness. Accordingly, speakers become categorically closer with respect to f0 over time. These effects support the assumption of entrainment as an automatism, for example, to enhance intelligibility by matching speaking styles as suggested by the *communication model* (Natale 1975) or the *perception behavior link* (Chartrand & Bargh, 1999) among others. Lastly, the effects of perceived attractiveness allow for two possible interpretations. On the one hand, perceived attractiveness may primarily affect f0 lowering or raising with disentrainment just being a logical consequence and not a feature in itself. On the other hand, the effect of social distance, which is linked to disentrainment (cf. Giles et al., 1991), may actually be the primary effect. Accordingly, there may be sociological/psychological reasons why a higher degree of social distance is linked to greater perceived attractiveness.

We suggest that our findings should be generalizable to non-dating conversations to some degree. As described above, the participants mostly stated that they did not intend to flirt or date. Accordingly, we can characterize the conversations as a hybrid of natural conversations in a dating setting leading to real dating conversations in some cases. Hence, we expect the effects of perceived attractiveness and conversational quality to be slightly less pronounced in real non-dating mixed-sex conversations and more pronounced in real intended dating conversations but present in both.

Another follow-up question concerns the generalizability to same-sex dating conversations. The particular question regards the two possible interpretations of the findings on perceived attractiveness as primarily leading to a raising or lowering in f0 or to an effect of disentrainment with respect to the interlocutor. Accordingly, for same-sex dating conversations we would either expect both female speakers to raise and both male speakers to lower their f0 or both speakers to move away from the interlocutor's f0. In the latter case, we would expect the speaker with the higher register to raise his/her f0 and the other speaker to lower his/her f0. The fact that female speakers consistently raised their f0, although both lowered and raised f0 is perceived as attractive by male listeners (cf. Karpf, 2006), supports the assumption that indeed disentrainment and not primarily f0 movement is linked to perceived attractiveness.

With respect to other prosodic cues, the effects observed for f0 are not easily generalizable. The effects found for f0 entrainment and conversational quality are in line with studies on other prosodic parameters. For example, Schweitzer et al. (2017) observe a link between social attractiveness and speaking rate. However, there are no studies on the effects of visual attractiveness or any observations of disentrainment regarding anything but f0. If the disentrainment in f0 is a secondary effect of raising or lowering f0, then those effects are linked to the natural sex differences expected from the frequency code (Ohala, 1983, 1984) and should not transfer to anything other than f0. However, if disentrainment and hence signaling social distance is the primary cue, we could expect other prosodic features to show similar effects. Accordingly, taking other prosodic features into consideration could also further our understanding concerning what to expect in same-sex conversations for reasons explained above.

12.5 Conclusion

This paper shows that the perceived quality of a conversation and the perceived visual attractiveness of an interlocutor are linked to f0 entrainment. This relationship is largely reciprocal with f0 entrainment both apparently affecting and reflecting the social variables. Regarding the different types of entrainment (cf. Edlund et al., 2009; Levitan, 2014), the effects are mainly restricted to f0 proximity with no systematic effects for synchrony or convergence. As expected from the literature, we find contradicting effects with conversational quality being linked to more entrainment and attractiveness being linked to more disentrainment. Additionally, both variables depend on as well as affect each other and the respective effects on and of entrainment. This contradiction is primarily resolved by emphasizing one over the other with female speakers emphasizing conversational quality over attractiveness and male speakers doing the opposite. However, male speakers also show a shift from emphasizing attractiveness to conversational quality over the course of the conversation. Future research needs to investigate how the connection of f0 entrainment and perceived attractiveness and conversational quality relates to conversational topics as well as personality profiles, as well as take other prosodic features such as speaking rate, intensity variation, or voice quality into consideration. Furthermore, the role of synchrony leaves several open questions for further investigation.

References

Bates, D., Maechler, M., Bolker, B., & Walker, S. (2015). Fitting linear mixed-effects models using lme4. *Journal of Statistical Software*, *67*(1), 1–48.

Belz, M., Mooshammer, C., Fuchs, S., Jannedy, S., Rasskazova, O., & Żygis, M. (Eds.). (2018). *Proceedings of the Conference on Phonetics & Phonology in German-Speaking Countries*. Berlin: Humboldt Universität.

Beňuš, S. (2014). Social aspects of entrainment in spoken interaction. *Cognitive Computation*, *6*(4), 802–813.

Beňuš, S., Trnka, M., Kuric, E., Matrák, L., Gravano, A., Hirschberg, J., & Levitan, R. (2018). Prosodic entrainment and trust in human-computer interaction. In *Proceedings of Speech Prosody 9, Poznań, Poland* (pp. 220–224).

Boersma, P., & Weenink, D. (2016). Praat: Doing phonetics by computer. Retrieved from http://www.fon.hum.uva.nl/praat/

Borkowska, B., & Pawlowski, B. (2011). Female voice frequency in the context of dominance and attractiveness perception. *Animal Behaviour*, *82*(1), 55–59.

Brennen, S. E., & Clark, H. H. (1996). Conceptual pacts and lexical choice in conversation. *Journal of Experimental Psychology: Learning, Memory and Cognition*, *22*(6), 1482–1493.

Brooks, A. B., Huang, L., Kearney, S. W., & Murray, F. E. (2014). Investors prefer entrepreneurial ventures pitched by attractive men. *Proceedings of the National Academy of Sciences of the United States of America*, *111*, 4427–4431.

Chartrand, T. L., & Bargh, J. A. (1999). The chameleon effect: The perception-behavior link and social interaction. *Journal of Personality and Social Psychology*, *76*(6), 893–910.

Cialdini, R. B. (2009). *Influence: Science and Practice* (5th ed.). Boston: Allyn & Bacon.

R Core Team. (2017). R: A language and environment for statistical computing. Vienna, Austria: R Foundation for Statistical Computing. Retrieved from https://www.R-project.org/.

Collins, S. A. (2000). Men's voices and women's choices. *Animal Behaviour*, *60*, 773–780.

Collins, S. A., & Missing, C. (2003). Vocal and visual attractiveness are related in women. *Animal Behaviour, 65*, 997–1004.
Dunbar, R. I. M. Coevolution of neocortex size, group size and language in humans. *Behavioural Brain Science, 16*, 681–735.
Edlund, J., Heldner, M., & Hirschberg, J. (2009). Pause and gap length in face-to-face interaction. In *Proceedings of INTERSPEECH 2009*.
Feinberg, D. R., Debruine, L. M., Jones, B. C., & Perrett, D. I. (2005). Manipulations of fundamental and formant frequencies influence the attractiveness of human male voices. *Animal Behaviour, 69*, 561–568.
Feinberg, D. R., Debruine, L. M., Jones, B. C., & Perrett, D. I. (2008). The role of femininity and averageness of voice pitch in aesthetic judgements of women's voices. *Perception, 37*, 615–623.
Fraccaro, P. J., Jones, B. C., Vukovic, J., Smith, F. G., Watkins, C. D., Feinberg, D. R., et al. (2011). Experimental evidence that women speak in higher voice pitch to men they find attractive. *Journal of Evolutionary Psychology, 9*(1), 57–67.
Friedberg, H., Litman, D., & Paletz, S. (2012). Lexical entrainment and success in student engineering groups. In *Spoken Language Technology Workshop (SLT) 2012* (pp. 404–409). IEEE.
Gessinger, I., Schweitzer, A., Andreeva, B., Raveh, E., Möbius, B., & Steiner, I. (2018). Convergence of pitch accents in a shadowing task. In *Proceedings of Speech Prosody, Poznań, Poland* (vol. 9, pp. 225–229).
Giles, H., Coupland, N., & Coupland, J. (1991). Accomodation theory: Communication, context, and consequence. *Contexts of Accomodation. Developments in Applied Sociolinguistics, 1*.
Gonzales, A. L., Hancock, J. T., & Pennebaker, J. W. (2009). Language style matching as a predictor of social dynamics in small groups. *Communication Research*.
Gregory, S. W. (1996). A nonverbal signal in voices of interview partners effectively predicts communication accommodation and social status perceptions. *Journal of Personality and Social Psychology, 70*, 1231–1240.
Hewstone, M., Stroebe, W., & Jonas, K. (2012). *An Introduction to Social Psychology* (5th ed.). Hoboken, New Jersey: BPS Blackwell.
Hodges-Simeon, C. R., Gaulin, S. J. C., & Puts, D. A. (2010). Different vocal parameters predict perceptions of dominance and attractiveness. *Human Nature, 21*, 406–427.
Hughes, S. M., Farley, S. D., & Rhodes, B. C. (2010). Vocal and physiological changes in response to the physical attractiveness of conversational partners. *Journal of Nonverbal Behavior, 34*, 1–13.
Ireland, M. E., Slatcher, R. B., Eastwick, P. W., Scissors, L. E., Finkel, E. J., & Pennebaker, J. W. (2011). Language style matching predicts relationship initiation and stability. *Psychological Science, 22*, 39–44.
Jones, B. C., Feinberg, D. R., Debruine, L. M., Little, A. C., & Vukovic, J. (2010). A domain-specific opposite-sex bias in human preferences for manipulated voice pitch. *Animal Behaviour, 79*(57–62)
Karpf, A. (2006). *The Human Voice*. New York, NY: Bloomsbury Publishing.
Kuznetsova, A., Brockhoff, P. B., & Christensen, R. H. B. (2016). lmerTest: Tests in linear mixed effects models. R package version 2.0-30. Retrieved from https://CRAN.R-project.org/package=lmerTest
Ladd, R. D., Silverman, K., Tolkmitt, F., Bergmann, G., & Scherer, K. (1985). Evidence for the independent function of intonation contour type, voice quality, and f0 range in signaling speaker affect. *Journal of the Acoustical Society of America, 78*, 435–444.
Leaderbrand, K., Dekam, J., Morey, A., & Tuma, L. (2008). The effects of voice pitch on perceptions of attractiveness: Do you sound hot or not. *Winona State University Psychology Student Journal*.
Lee, C. C., Black, M. P., Katsamanis, A., Lammert, A. C., Baucom, B. R., Christensen, A., et al. (2010). Quantification of prosodic entrainment in affective spontaneous spoken interactions of married couples. *Proceedings of Interspeech*, 793–796.
Levitan, R. (2014). Acoustic-prosodic entrainment in human-human and human-computer dialogue. Columbia University. Ph.D. thesis.

Levitan, R., Gravano, A., Willson, L., Beňuš, S., Hirschberg, J., & Nenkova, A. (2012). Acoustic-prosodic entrainment and social behavior. In *Proceedings of the 2012 Conference of the North American Chapter of the Association for Computational Linguistics: Human Language Technologies* (pp. 11–19).

Linville, S. E. (1996). The sound of senescence. *Journal of Voice, 10*(2), 190–200.

Lubold, N., & Pon-Barry, H. (2014). Acoustic-Prosodic Entrainment and Rapport in Collaborative Learning Dialogues. Proceedings of the. (2014). *ACM Workshop on Multimodal Learning Analytics Workshop and Grand Challenge, November 12–12, 2014*, Turkey, Istanbul.

Michalsky, J. (2017). Pitch synchrony as an effect of perceived attractiveness and likability. In: *Proceedings of DAGA 2017*.

Michalsky, J., & Schoormann, H. (2016). Effects of perceived attractiveness and likability on global aspects of fundamental frequency. In *Proceedings of P&P12* (120–124).

Michalsky, J., & Schoormann, H. (2017). Pitch convergence as an effect of perceived attractiveness and likability. In *Proceedings of INTERSPEECH, 2017* (pp. 2253–2256).

Michalsky, J., Schoormann, H., & Niebuhr, O. (2018a). Conversational quality is affected by and reflected in prosodic entrainment. In: *Proceedings of Speech Prosody, Poznań, Poland* (vol. 9).

Michalsky, J., Schoormann, H., & Niebuhr, O. (2018b). Turn transitions as salient places for social signals—Local prosodic entrainment as a cue to perceived attractiveness and likability. In M. Belz, C. Mooshammer, S. Fuchs, S. Jannedy, O. Rasskazova, & M. Żgis (Eds.), *Proceedings of the Conference on Phonetics & Phonology in German-Speaking Countries* (pp. 169–172). Berlin: Humboldt Universität.

Natale, M. (1975). Convergence of mean vocal instenity in dyadic communication as a function of social desirability. *Journal of Personality and Social Psychology, 32*(5), 790–804.

Nenkova, A., Gravano, A., & Hirschberg, J. (2008). High frequency word entrainment in spoken dialogue. In *Proceedings of the 46th Annual Meeting of the Association of Computational Linguistics on Human Language Technologies: Short Papers* (pp. 169–172).

Oguchi, T., & Kikuchi, H. (1997). Voice and interpersonal attraction. *Japanese Psychological Research, 39*, 56–61.

Ohala, J. (1983). Cross-language use of pitch. An ethological view. *Phonetica, 40*, 1–18.

Ohala, J. (1984). An ethological perspective on common cross-language utilization of f0 in voice. *Phonetica, 41*, 1–16.

Pickering, M. J., & Garrod, S. (2006). Alignment as the basis for successful communication. *Research on Language and Computation, 4*, 203–228.

Puts, D. A., Gaulin, S. J. C., & Verdolini, J. (2006). Dominance and the evolution of sexual dimorphism in human voice pitch. *Evolution and Human Behavior, 27*, 283–296.

Reitter, D., & Moore, J. D. (2007). Predicting success in dialogue. *Annual Meeting - Association for Computational Linguistics, 45*, 808.

Scherer, K., Ladd, R. D., & Silverman, K. (1984). Vocal cues to speaker affect: testing two models. *Journal of the Acoustical Society of America, 76*, 1346–1356.

Schweitzer, A., Lewandowski, N., & Duran, D. (2017). Social attractiveness in dialogs. In *Proceedings of INTERSPEECH 2017* (pp. 2243–2247).

Street, R. L. (1984). Speech convergence and speech evaluation in fact-finding interviews. *Human Communication Research, 11*(2), 139–169.

Taylor, J. G. (2009). Cognitive computation. *Cognitive Computation, 1*, 4–16.

Thomason, J., Nguyen, H. V., & Litman, D. (2013). Prosodic entrainment and tutoring dialogue success. *Artificial Intelligence in Education*, 750–753.

Vukovic, J., Jones, B. C., Debruine, L. M., Feinberg, D. R., Smith, F. G., Little, A. C., et al. (2010). Women's own voice pitch predicts their preferences for masculinity in men's voices. *Behavioral Ecology, 21*(4), 767–772.

Xu, Y., Lee, A., Wu, W.-L., Liu, X., & Birkholz, P. (2013). Human vocal attractiveness as signaled by body size projection. *PLoS ONE, 8*(4),

Zimmer-Gembeck, M. J., Hughes, N., Kelly, M., & Connoly, J. (2011). Intimacy, identity and status: Measuring dating goals in late adolescence and emerging adulthood. *Motivation and Emotion, 36*(3), 311–322.

Part IV
Databases

Chapter 13
Acoustic Correlates of Likable Speakers in the NSC Database

Benjamin Weiss, Jürgen Trouvain, and Felix Burkhardt

Abstract Speech stimuli from scenario-based conversations were analyzed regarding acoustic correlates of likability. Utterances from the pizza ordering scenario of the NSC corpus were selected, and the confederate's turns were excluded. These stimuli were recorded in high quality and were subjected to third-party listeners' ratings. Six promising acoustic parameters from related work are tested applying methods of correlation, regression, and regression trees. These parameters are average fundamental frequency, articulation rate, standard deviation of both and of intensity, as well as spectral center of gravity. The amount of variance explained remains below 50%. Results confirm variability of the fundamental frequency as dominating correlate of likable voices in male and female speakers. It is concluded that the promising acoustic parameters are not robust to stimulus duration and scenario. Therefore, it is argued to explore the applicability of locally defined and linguistically motivated parameters.

Keywords Voice · Acoustic parameters · Likability · Rating test · Database · Analysis · Modelling

13.1 Introduction: Likability of Speakers

The aim of this chapter is twofold: First, acoustic correlates of likability ratings for the common stimulus length of a single utterance are presented as brief literature survey with a focus on re-occurring results. The second aim is to check whether such

B. Weiss (✉)
Technische Universität Berlin, Ernst-Reuter-Platz 7, 10405 Berlin, Germany
e-mail: benjamin.weiss@tu-berlin.de

J. Trouvain
Saarland University, Campus C7.2, 66123 Saarbrücken, Germany
e-mail: trouvain@coli.uni-saarland.de

F. Burkhardt
audEERING GmbH, Friedrichstraße 68, 10117 Berlin, Germany
e-mail: fburkhardt@audeering.com

B. Weiss et al. (eds.), *Voice Attractiveness*, Prosody, Phonology and Phonetics,
https://doi.org/10.1007/978-981-15-6627-1_13

confirmed correlates can be applied for longer stimuli provided with a database for studying attributions and preference of speakers. The domain hereby is limited to evaluations of unacquainted speakers in order to maximize the impact of the first impression obtained from voice and speaking style of speakers.

The question of whether a person likes another person or not represents one of the most crucial social attitudes in humans, as it lays the basis for own social behavior (Chap. 1). In the most extremes of cases, liking somebody or not can determine not only the kind of social interaction, but whether there is avoidance or approach in the first place. In the research topic of first impressions, studies investigated the potential impact of surface signals, like clothing and facial expressions, but also voice and speaking style, on the formation of likability (Ambady & Skowronski, 2008). While people might not be inclined to immediately judge whether they truly like a person from a few seconds of interaction of recorded voice samples, listeners can express their gradual preference of the voice of a speaker, and thus his or her likability.

Such explicit ratings already show significant consistency between raters. For example, a standard measure of consistency between multiple raters is the intraclass correlation (ICC) with values between 0 and 1. It ranges for likability ratings from ICC = 0.76 (Burkhardt, Schuller, Weiss, & Weninger, 2011) to ICC = 0.93 (Weiss & Burkhardt, 2010). This strong consistency documents that not only visual, but also acoustic information has a systematic relationship with a first impression. As the first impression is persistent over time and has predictive power (Ambady, Bernieri, & Richeson, 2000, 2006; Peterson, Cannito, & Brown, 1995; Hecht & LaFrance, 1995), it also potentially affects relationship building. As such acoustic or visual data in first encounters are sparse and superficial, attributions and even stereotypes play an important role in the formation of likability judgments. Salient attributions of regional or social background of speakers or disfluencies are relevant for likability (Scherer & Giles, (Scherer and Giles, 1979); Giles, 1980, Weiss & Burkhardt, 2012; McCroskey & Mehrley, 1969).

When studying acoustic correlates of likable voices, the effect of such attributions should therefore be minimized by providing homogeneous groups of speakers in terms of social and regional background, age, speech pathology, physical attractiveness, or gender (Murry, Singh, & Sargent, 1977; Murry & Singh, 1980; Linville, 2001; Brückl, 2011; Kreiman & Gerratt, 1996). Ideally, homogeneous groups of raters/listeners should be selected as well, or a diverse group that is balanced for these influencing factors can be recruited (Deal & Oyer, 1991). Of course, this statement holds not for the case of explicitly aiming to test for the effects of those attributions. Physical attractiveness that is inferred from voice and speaking style is such an example. Attractiveness is, like aesthetics for many other domains, a well-known and important factor for preference and liking, regardless of the sexual preference. The aim of this chapter is to present work that identifies which acoustic characteristics affect likability of unacquainted speakers, apart from the aforementioned attributions of age, gender, regional, and social background or speech-related pathologies.

13.2 A Review on Acoustic Correlates

In search for acoustic correlates of likability ratings, the dominant perspective is a global one, spanning the entire stimulus: The chosen stimuli are acoustically analyzed as a whole, and parameter values are aggregated, for example, to obtain the average fundamental frequency (F_0) for a complete utterance. As a consequence, vowel formants are typically only analyzed if the stimulus consists only of a single vowel (e.g., Bruckert et al., 2006).

Early research has brought a body of results on the so-called suprasegmentals, which are F_0, intensity, and duration (Lehiste, 1970). For these suprasegmentals, analysis and re-synthesis studies have been conducted to identify acoustic parameters and test their impact on listeners' ratings. As outlined in Chap. 1, likability can be used as a synonym for social attractiveness. It can be related to unacquainted speakers to the attribution of warmth or benevolence—although there are social situations, in which competence might play a bigger role. Therefore, some of the results mentioned here stem not directly from ratings of likability but were elicited by questionnaires with related items or scales that also contribute to social attractiveness. Examples are friendliness, sympathy, or pleasantness (Weiss & Möller, 2011). If available in the studies, results for competence are also mentioned.

At least for male speakers and sometimes for females as well, the **fundamental frequency** (F_0) correlates negatively with ratings of benevolence, trust, likability, or pleasantness (Brown, Strong, & Rencher 1974; Apple, Streeter, & Krauss, 1979; Bruckert et al., 2006; Gravano et al., 2011; Weirich, 2010; Chattopadhyay, Dahl, Ritchie, & Shahin, 2003; Weiss & Burkhardt, 2012; Weiss, 2013). Noteworthy, however, are contradicting results of a positive correlation reported for German male speakers (Scherer, 1979). A similar opposing effect was found for the brief greeting "hello" in Scottish English (McAleer, 2014). Both results can be interpreted in a different communicative context, in which a raised average pitch is more appropriate, maybe to signal arousal.

The observed general tendency of a lower F_0 being evaluated more positively concurs with a positive association of perceived "darkness" for male speakers and the attribution of "being relaxed" for voices of the two genders considered (Weiss et al., 2018b). Variability or range of F_0 shows a positive effect on likability-related concepts of benevolence or warmth (Brown et al., 1973, 1974; Ray, 1986) but also on competence (Ray, 1986). There is also evidence for a positive effect of a rising F_0 contour (Bruckert et al., 2006; Weiss & Burkhardt, 2012, McAleer, 2014). However, this seems to be a more problematic acoustic correlate due to its dependency on the linguistic material.

Intensity as the second aspect reveals a negative correlation with benevolence but a positive one with competence (Ray, 1986), although this relationship might be more complex (Scherer, 1979). Effects caused by speech manipulation also have shown to add up or cancel each other out, dependent on the sign of correlation, which means that they can be considered as being independent of each other (Ray, 1986).

With respect to **articulation rate**, an ideal-point relation was found for likability-related concepts, especially for benevolence. This relation can be visualized as an inverted U-shaped line. This separates its effect from the more linear positive correlation with competence (Brown et al., 1974, 1975; Smith, Brown, Strong, & Rencher, 1975; Apple et al., 1979; Street, Brady, & Putnam, 1983). For articulation rate, an additional effect could be found. Apparently, the raters' own intrinsic rates affect the evaluation of speakers' rates (Street et al., 1983; Feldstein, Dohm, & Crown, 2001). One interpretation is that listeners perceive articulation rate according to their own reference as high or low. A second interpretation would be that there is an effect of similarity preference, which may interfere with a linear relation between rate and likability. In all cases, the result shows a positive correlation with moderate or slightly increased rates, confirming the rough inverted U shape of relationship, and thus a saturation for very fast conditions (Street & Brady, 1982). This is why Table 13.1 gives results on both, positive correlations and an ideal-point relation with moderate or similar rates.

All these suprasegmentals are relatively easy to manipulate, e.g., Trouvain et al. (2006) could convincingly model personality dimensions such as sincerity, competence, and excitement with speech synthesis. These suprasegmentals are also easy to measure automatically (maybe apart from articulation rate). This may be the reason that they have been studied extensively. More recently, spectral measures have been moved into focus with the aim to study voice quality, but also other, yet understudied, anatomical and articulatory sources of spectral aspects of speech. For example, shimmer, i.e., the local variability in amplitude, correlates positively with likability (Gravano et al., 2011), while measures of energy distribution, such as spectral tilt or center of gravity, show positive evaluation with less energy in higher frequencies (Weiss et al., 2017; Weiss, 2015). One reason could be a co-variation with the average F_0, i.e., the perception of "dark" or "relaxed" voices (Weiss et al., 2016; Weiss, 2018b). However, a summary of many studies on this topic reveals non-significant results for spectral parameters (cf. Table 13.1).

There is some kind of tendency to be found in this summary. First of all, there are studies showing no effects, which are mostly analysis studies and thus might represent a non-sufficient variability in parameter values to show an effect. But also, F_0 mean, F_0 variability, and articulation rate seem to form a kind of majority vote to have a systematic effect, despite some contradicting results. For other parameters, such as variability in intensity or articulation rate, but also for spectral measures, such a systematic pattern is not obvious.

A particular issue is the status of the stimuli used in listening-and-rating experiments that are typically applied in this line of research. For example, the very short stimulus "hello" was rated and acoustically analyzed (McAleer, 2014). In this study, step-wise regression models of likability for male voices include average F_0 and, negatively, the harmonic-to-noise ratio (HNR). For female voices, a similar model is made up of HNR (negative sign), a rising F_0 contour, and the F_0 range. The positive, and thus contradicting, result for pitch might have not appeared in the case of presenting the full utterances the "hello" was cut out from. While other studies used even shorter stimuli, i.e., vowels that have been excluded in this chapter, likability

Table 13.1 Summary of results from literature on acoustic correlates of likability and similar concepts. Positive, negative, and non-significant relations are depicted by +, −, and o, respectively. Gender of the speakers is indicated by *m*, *f*, respectively, as is language, and if the stimuli were re-synthesized. Reports on non-significant results may be incomplete

Reference	Gender	Language	Re-synthesized	F_0 mean	F_0 variation	F_0 raise	Intensity mean	Intensity variation	Arti-culation rate mean	Ideal articula-tion rate	Arti-culation rate variation	Harmonic-to-noise ratio	Shimmer	Jitter	Voiced–unvoiced ratio	Formant frequen-cies	Formant disper-sion	Spectral tilt	Spectral center of gravity	Spectral skew-ness
Brown et al. (1973)	m	en	R		+					+										
Brown et al. (1974)	m	en	R	−	+					+										
Bruckert et al. (2006)	m	fr*		−		+										o	o			
Duran (2017)	mf	de		o	+							o	o	o						
Feldstein et al. (2001)	mf	en								+										
Fernández Gallardo and Weiss (2016)	m	de		o	o		o	o	−			o						+	−	
	f	de		o	o		o	+	o			o						o	o	
Gravano et al. (2011)	mf	en		−	−	o	+		o			o	−	o	o					
McAleer (2014)	m	en		+	o	o						−	o	o			o	o		
	f	en		o	+	+						−	o	o			o	o		
Ray (1986)	m	en	R		+		−		−											
Smith et al. (1975)	m	en	R							+										
Street and Brady (1982)	m	en	R							+										
Street et al. (1983)	m	en							+	+										
Weiss (2015)	m	de		o	o				o										o	o
	f	de		o	o				+										o	+
Weiss & Burkhardt (2010)	m	de		−					+										−	o
	f	de		o					+										−	+
Weiss et al. (2010)	m	de											−							
	f	de											o							
Weiss (2013)	m	de		−	o	o		−	o		o									
	f	de		−	o	o		o	+		−									
Weiss and Burkhardt (2012)	mf	de		−	o	+		o	+			o		+				o	o	
Weiss et al. (2017)	f	de	R															+		

*vowels only, female raters only

is a concept that emerges in social situations. It should therefore be studied not only concerning the voice quality but also the speaking style. In order to use more realistic data and to identify or verify correlates that emerge only with longer stimuli, such as variability of F_0, the Nautilus Speaker Characteristics (NSC) database was recorded and used. It is described in the next section.

13.3 Material

The aim of this new analysis is to extend the insight into acoustic correlates of likable voices by avoiding several limitations of earlier research. First of all, the *number of speakers analyzed has often been very small*, about 20–30, for example. Secondly, the *social situation has been unclear*. Examples are the aforementioned utterance "hello" or reading aloud single sentences. And thirdly, the *stimuli have been very short*. Therefore, the Nautilus[1] database was created. It features 300 German speakers (aged 18–35, of which are 126 males) and was recorded with the aim to study speaker characteristics (Fernández Gallardo & Weiss, 2018). During recruitment, the speakers were subjectively checked for neither exhibiting a strong regional or social accent, nor displaying signs of a voice-related sickness or speech disorder. Although all speakers display Standard German, some speakers do exhibit some regional features and suprasegmental non-modal voice qualities. Hearing issues were not reported during collecting speakers' details, which is important for properly conducting the interactive scenarios and understanding the instructions of the experimenter. The database and documentation of the Nautilus Speaker Characteristics (NSC) have been compiled by Fernández Gallardo (2018). NSC includes recordings from simulated telephone conversations, read passages, and read sentences in high signal quality.[2] From this database, telephone scenarios were chosen as appropriate material, as it contained a typical and well-defined social situation of unacquainted dyads that can be judged by third-party listeners. The scenario used for analysis here is ordering something to eat from a pizza service with a phone call. It stems from a list of pre-defined scenarios used for evaluating audio network transmission quality (Rec & P.805, 2007). The invited and recorded speakers all took over the role of the caller, while a student confederate played the pizza service. The caller obtained the following task information: a fake surname, address, and phone number. The instruction was to order a single pizza for two people, preferably a vegetarian option. During the conversation, the caller is asked to note down the exact final toppings, price, and duration until delivery. Such a conversation typically took about 60 s to complete.

[1]Nautilus is the recording booth name used in the laboratory.

[2]The ISLRN of this corpus is `157-037-166-491-1`. Is has been made available at the CLARIN repository: hdl.handle.net/11022/1009-0000-0007-C05F-6 under the CLARIN ACA+BY+NC+NORED license (freely available for scientific research).

In preparation of the stimuli for the listening-and-rating test on likability, all parts of the confederate in the pizza scenario were removed from the recordings. The resulting stimuli have an average duration of 23 s (SD = 3.3 s). Based on a questionnaire with 34 items that was developed to assess voice-based personality attributions (Weiss & Möller, 2011; Fernández Gallardo & Weiss, 2017b), a final version was created with only minimal changes (Fernández Gallardo & Weiss, 2018).[3] For the evaluation, each stimulus was rated by 15.1 listeners on average (sd = 1.17, due to splitting the students into groups). Altogether, 114 students, in the frame of a lecture's exercise, took part in this test (44 females, 70 males, aged on average 24.5 years with an SD of 3.4). 93 of these were native German speakers, and the remaining participants were fluent in German. Each listener rated male and female stimuli in separate blocks with sliders on continuous scales. On average, each rater listened to about 16.9 males (sd = 0.49) and 23.2 females (sd = 2.33, due to splitting the data into sets). A single session took about 50 min.

The questionnaire itself includes items to cover major concepts of personality attributions. It is based on existing instruments for the personality circumflex (Wiggins, Trapnell, & Phillips, 1988) for the first impression of warmth and agency, the OCEAN personality taxonomy (Rammstedt & John, 2007), the three-dimensional model of emotional states with valence, activity, and potency (Osgood, Suci, & Tannenbaum, 1957), and estimation of physical attractiveness that is affecting personality attributions and frequent attributions observed empirically for unacquainted voices (Weiss et al., 2018b). The questionnaire also includes the item of likability. A screenshot shows all scales with sliders on one page Fig. 13.1.

13.4 Analysis

Data analysis is presented in four sections. First, the comprehensive questionnaire responses are reduced in dimensionality to obtain values for the concept of likability. The subsequent correlation analysis aims at testing promising acoustic parameters from Sect. 13.2 on the new stimuli. Two simple modeling approaches are presented with different aims, mainly to find out how much variance the acoustic correlates of likability can explain. In order to inspect potentially non-linear relationships, a regression tree is applied.

---

[3] likable/non-likable, insecure/secure, unattractive/attractive, sympathetic/unsympathetic, decided/indecisive, obtrusive/unobtrusive, close/distant, interested/bored, unemotional/emotional, irritated/not irritated, passive/active, unpleasant/pleasant, characterful/characterless, reserved/sociable, nervous/relaxed, distant/affectionate, conformable/dominant, affected/unaffected, cold/hearty, young/old, factual/not factual, excited/calm, competent/incompetent, beautiful/ugly, unfriendly/friendly, feminine/masculine, offensive/submissive, committed/indifferent, boring/interesting, compliant/cynical, genuine/artificial, stupid/intelligent, adult/childish, bold/modest.

Fig. 13.1 Screenshot of the first page of the rating interface. After pressing "Start" a new playback and continue button appears, while the scales remain

13.4.1 Factor Analysis

A factor analysis of the personality questionnaire was conducted to identify the most relevant basic dimension that explains the ratings. With this method, co-variabilities are represented by a smaller number of underlying factors each representing multiple questionnaire items for subsequent analysis. As human social evaluation concepts can be expected to be correlated to some degree, a non-orthogonal method was applied. The result of the factor analysis reveals five factors. These are named after inspecting the items that contribute to each one as warmth, attractiveness, confidence, compliance, and maturity (Fernández Gallardo & Weiss, 2017a, 2018). The first two show a strong correlation with each other ($r = 0.77$). Not only because of this correlation, but also due to the single questionnaire item "likability" correlating with these two dimensions (with warmth: $r = 0.87$, with attractiveness: $r = 0.83$), these two dimensions are apparently related to the attitude toward speakers. Considering the small number and inconsistent groups of raters, the first principal component of warmth and attractiveness is used to represent the concept of likability more robust than the single item "likability". This principal component is used as target for identifying acoustic correlates and represents likability on values from -3 to $+3$.

13.4.2 Correlation Analysis

For the first analysis, we tested the most important and promising acoustic parameters that can be derived from Table 13.1.[4] The chosen candidates are F_0 mean, F_0 SD, intensity SD, articulation rate mean, articulation rate SD, and Center of Gravity (CoG). Although variability in intensity and rate are not very promising candidates according to Table 13.1, they were chosen nevertheless. This was done to test whether the claim of Ketrow (1990) can be supported that variability in suprasegmentals generally is signaling benevolence and positively affects likability. Except articulation rate, all acoustic parameters were measured with Praat (Boersma, 2001). Average articulation rate and its SD were estimated by an acoustic model (Weiss et al., 2018a) that was trained on the perceptually motivated "perceived local speaking rate" (PLSR) (Pfitzinger, 1999). The reason for applying this method is that stimulus duration would not be appropriate because of the varying linguistic material of the spontaneous utterances and that other established methods (De Jong & Wempe, 2009) sometimes have issues with the detection of silence and of unstressed syllables.

The results of linear bivariate correlations are presented in Table 13.2, separately for females and males. This separation reflects different value ranges of acoustic parameters but also the potentially different references and relations in likability formation. Gender of the raters was not analyzed due to the small number of listeners for each stimulus. False discovery rate approach is used to adjust for multiple testing (Benjamini & Hochberg, 1995). It is not as conservative as Bonferroni correction. The false discover rate sorts all p-values from lowest ($i = 1$) to highest ($i = max(i)$) and adjusts the *alpha*-level by $i/max(i) \cdot \alpha$. There are only two significant results for male and female speakers, respectively (see Sect. 13.5), indicated by bold p-values. Using the divergence from a global mean in articulation rate in order to represent an ideal-point relation is not significant in either gender. Before discussing these results, simple modeling of the data is conducted.

[4]While articulation rate is not an acoustic parameter in a narrow sense, the estimates used here are a prediction result based on spectral data, and it is also called acoustic parameter for convenience.

Table 13.2 Pearson's correlation between selected acoustic parameters and likability

Parameter	Female speakers		Male speakers	
	Pearson's r	p-value	Pearson's r	p-value
F_0 mean	0.25	**0.0008**	0.16	0.0688
F_0 SD	0.44	**<0.0001**	0.52	**<0.0001**
Intensity SD	−0.04	0.5975	−0.07	0.4427
Artic. Rate mean	0.05	0.4937	0.30	**0.0006**
Artic. Rate SD	0.02	0.7845	0.15	0.0913
CoG	0.05	0.5521	0.18	0.0445

Table 13.3 Linear models for Likability: Females ($p < 0.0001$, $R^2 = 0.206$); males ($p < 0.0001$, $R^2 = 0.345$). Parameters not included into a model are represented by "—" . (significance levels of < .05*, < .01*, *and* < .001*** are applied)

Parameters	Females	Males
Intercept	**−0.56****	**−0.15**
F_0 mean	**0.38**	**−0.67***
F_0 SD	**0.49*****	**0.83*****
Artic. rate mean	–	**0.22****
Artic. rate SD	–	–
Intensity SD	–	–
CoG	–	–

13.4.3 *Linear Regression Analysis*

As a second step, describing likability ratings with these selected acoustic parameters can shed a light on the amount of variance explained. Due to the relatively large number of stimuli, acoustic modeling can furthermore help to identify additional candidates of acoustic correlates that have non-linear relationships or meaningful interaction effects with other parameters, as attempted in the next subsection. As linear baseline, linear regression with step-wise inclusion of parameters was performed.[5] Overall, the resulting models are significant but explain only about 1/5 of the variance for female and about 1/3 for male speakers (Table 13.3). While, for males, articulation rate mean is included in addition to the two pitch-related parameters, F_0 mean does contribute significantly to the model with F_0 SD, most likely due to a cross-correlation between them ($r = 0.34^{***}$). The resulting estimates are depicted in Fig. 13.2.

[5]Based on AIC, and single inclusion and exclusion of variables; only main effects.

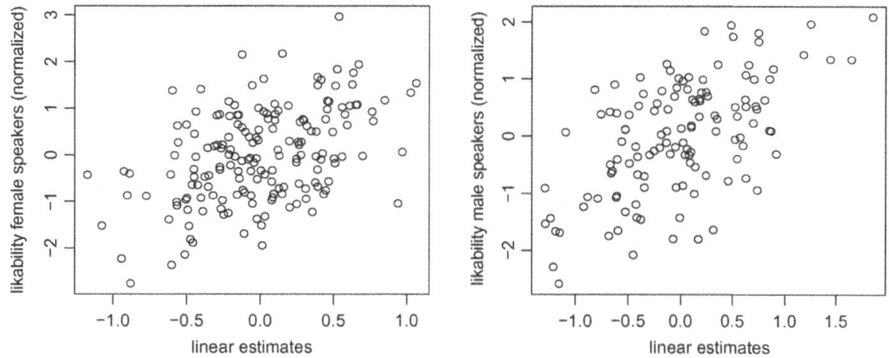

Fig. 13.2 Results of the two linear models for likability of female (left) and male (right) speakers. Average likability values versus model estimates from acoustics

13.4.4 *Non-linear Modeling*

A second simple approach to modeling is using regression trees to better take non-linearities into account. The main aim is not a better fit, but a better view for identifying acoustic correlates of likability in voices. Such trees, pruned on cross-correlation errors to avoid overfitting, improve the amount of variance a little, but without succeeding 50% of variance. Figure 13.3 shows the two regression trees, with the target likability value and the percentage of data points given in the boxes, while the joints are labeled with the conditional value of the acoustic parameter used for splitting the data.

13.5 Discussion

The positive correlation of likability with F_0 SD for both speaker sets and articulation rate for males is in line with results of other studies presented in the literature survey in Sect. 13.2. However, other promising parameters are not significantly correlated for the stimuli from our pizza order scenario, in particular, CoG and, for males, F_0 mean. Actually, average F_0 is even correlated negatively for female speakers, which is in contrast to the state of the art. One reason for these results may be the small number and inconsistently selected raters (Fernández Gallardo & Weiss, 2018). However, a more probable cause can be suspected in the much longer stimulus durations and the different genres compared to other experiments. In particular, the strong effect of variability in the fundamental frequency (F_0 SD) may mask smaller effects, for example, CoG as timbre-related parameter. The two modeling approaches indicated that F_0 mean is relevant for male speakers.

The unexpected positive correlation with average F_0 for females is still surprising, as other work with shorter German stimuli repeatedly showed an opposite effect. This

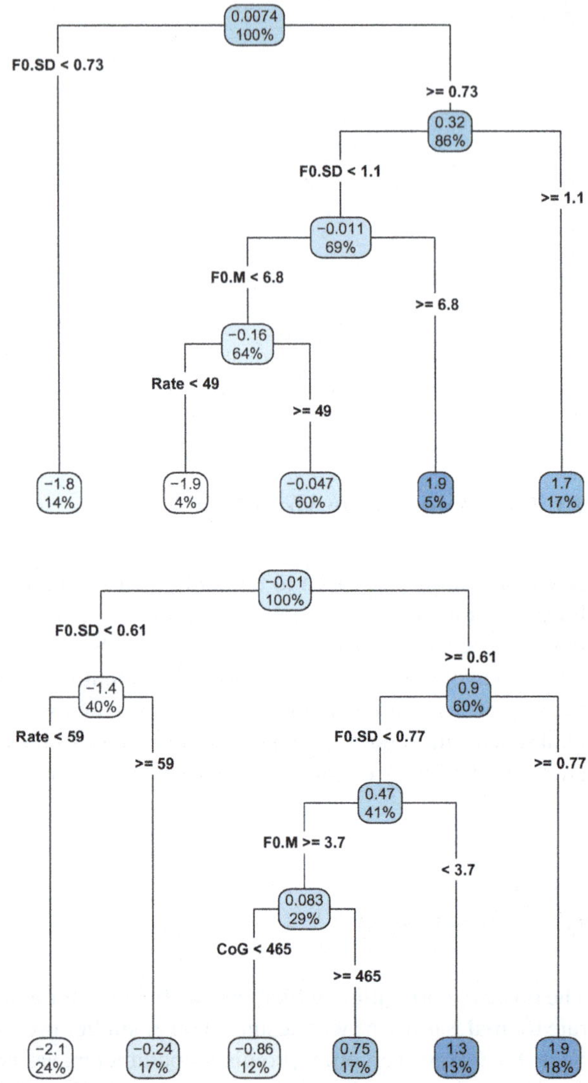

Fig. 13.3 Results of the two regression trees for likability of female (top; $R^2 = 0.312$) and male (bottom; $R^2 = 0.437$) speakers. **F0.M** refers to F_0 mean in ERB (as normalization attempt from Hz); values of **Rate** are in the units of PLSR (Perceived Local Speaking Rate, usually between 50 and 150) instead of syllables/s

even holds in combination with a positive impact of higher F_0 SD, which seems to exclude the possibility of articulatory grounding of raising F_0 mean by increasing F_0 variability. This contradicting result may indicate a situational difference between reading single utterances in a rather factual tone, where low articulatory tension or even larger vocal folds in males may be rewarded, whereas in a truly conversational situation a social signal of interest (a raised voice due to higher tension), benevolence, or even a biological signal of female attractiveness might be positively perceived. This kind of speculation has of course to be tested, for example, with re-synthesis experiments for both kinds of situations. At least, for the non-significant correlation in males, the two models solve this issue by including F_0 both times with a negative relationship.

The attempt to describe likability ratings with simple models reveals only low performing results. Not even half of the variance is explained, despite applying approaches that use all data for training. More interesting are the systematics incorporated in the models. First of all, parameters, which are non-significant in the correlation analysis, are included in order to explain more variance, i.e., male F_0 mean in males for the linear and the tree model, and CoG in the regression tree. For females, rate is included in the regression tree. For one split in the female data, the positive impact of higher F_0 mean is confirmed. For male speakers, a lower F_0 has a positive impact, just as expected from literature. Additionally, still very simple regression trees perform better than the linear baseline, indicating non-linearities observed elsewhere (Weiss & Burkhardt, 2012).

13.6 Conclusion

Despite some agreement in English and German studies, the attempt to confirm a set of potential acoustic correlates of likable voices was not overall successful. The analysis of the Nautilus data confirms only F_0 variation and articulation rate as relevant parameters. Especially, F_0 variation seems to be a very salient parameter in this conversational data. The role of F_0, or pitch level in general, has to be re-examined. Currently, a stereotype of low-pitched male voices and high-pitched female ones seem to be too simple for German. In light of other studies, there seems to be a pool of potential correlates that not necessarily show a relation in each analysis. However, generalization seems not to be possible from the given results.

This reveals a more general issue with the material. Most data referred to as related work are single short sentences, which are sometimes difficult to discern as read or not, but for which simple aggregated values are intuitive parameter choices. With longer durations, as in the Nautilus database, not only more material, including several sentences and utterances, are available, but also a specific social situation is evident. Apart from obvious differences due to this kind of styles, aggregating simple acoustic parameter values over time could result in unreliable correlates. As almost all parameters are globally defined, they seem to be fragile for changes in material. In order to better compare acoustics between for example brief greetings ("hello") (McAleer, 2014) to longer utterances or even short conversations, the value of locally defined or dynamic parameters has to be tested.

In order to define more robust parameters and even more automatic measurement, segment-based and articulatorily defined candidates should be defined to better represent perceptually salient aspects that are relevant for likability, especially when studying timbre. One example is the so-called speakers' or actors' formant to assess a potentially positive effect of trained voices. It manifests as a peak in the acoustic spectrum: 3–4 kHz for males (Nawka, Anders, Cebulla, & Zurakowski, 1997), and

4–5 kHz for females (Tayal, Stone, & Bisrkholz, 2017). This resonance seems to be caused by an epi-laryngeal narrow and pharyngeal wide configuration that is evident in professional speakers, and it is considered as pleasant also for non-trained speakers (Leino, Laukkanen, & Radolf, 2011). The issue is to properly re-synthesize this phenomenon in a valid and salient way. A recent analysis shows a relation for males voices (Weiss, 2015) that is even stronger than the typical average F_0. However, this effect was not confirmed by a first attempt of overall spectral manipulation (Karnop & Weiss, 2016), maybe due to missing representation in other acoustic features that are perceptually relevant for stimulating the acoustic effect of this configuration. Other, more phonetically or phonologically defined parameters such as vowel formant dispersion as a measure of articulatory precision or aspects of intonation, have not yet been studied in depths for likability, simply because they require manual or automatic phonetic analysis for segmental selection.

There are further factors in the research of likability of voices that remain under-explored or simply ignored. Among them is the question of how audible smiling in voices has an influence on whether somebody likes a formerly unknown person. Certain types of smiling are perceived as displays of happiness (Krys, 2016). For instance, in a Brazilian study smiling faces were considered as happier and even as more attractive than a neutral expression (Otta, Abrosio, & Hoshino, 1996). However, there is evidence that in some cultures visually transmitted smiling faces of unknown persons may have a *negative* image on side of the viewers (Krys, 2016). Thus, it could be that similar patterns could occur for audibly transmitted smiling.

As mentioned in the introduction, the level of speech fluency can also have an effect on the perceived attractiveness of voices and thus might affect social attractiveness as well. For instance, Zuta (2007) showed that in retold narratives male voices were considered least attractive by female listeners when comparatively many disfluencies occurred, along with less varied $F0$ and a high degree of nasality. Also, the number and the duration of pauses is a strong marker of fluency but also of the valence of speech (Tisljár-Szabó and Pléh, 2014). Too long pauses seem to have a tendency toward a negative and less likable image of the speaker, also in dialogs.

With regard to intonation contours, the impression of politeness and pleasantness obviously depend on the sentence mode. For instance, Uldall (1960) found that declarative sentences were perceived with a high degree of pleasantness when produced with either a falling or rising pitch at the end; however, questions and commands tend to be felt pleasantly only when they showed a final rise.

Audible smiling, fluency, pauses, sentence accents, and phrase tones can be considered as local phenomena of spoken sentences and longer stretches of speech. In contrast, a regional or a foreign accent is always a global phenomenon. Regarding accents, people sometimes have more or less strong attitudes which can heavily influence the likability of the speakers in a negative and likewise in a positive way.

Lastly, acoustic correlates of sexual preference and physical attractiveness have been mostly neglected in this line of research. While there are some cross-correlations found for likability as social attractiveness and subjective estimates of physical attractiveness from voice or ratings of vocal attractiveness directly (McAleer, 2014), well-founded correlates, such as formant dispersion in males (Fitch & Giedd, 1999; Bruckert et al., 2006) might increase insight into the cause of a likable first impression in speech.

References

Ambady, N., & Skowronski, J. J. (Eds.). (2008). *First Impressions*. New York: Guilford Press.

Ambady, N., Bernieri, F. J., & Richeson, J. A. (2000). Toward a histology of social behavior: Judgmental accuracy from thin slices of the behavioral stream. In M. P. Zanna (Ed.), *Advances in experimental social psychology* (Vol. 32, pp. 201–272). San Diego: Academic Press.

Ambady, N., Krabbenhoft, M. A., & Hogan, D. (2006). The 30-sec sale: Using thin slice judgments to evaluate sales effectiveness. *Journal of Consumer Psychology*, *16*, 4–13.

Apple, W., Streeter, L. A., & Krauss, R. M. (1979). Effects of pitch and speech rate on personal attributions. *Journal of Personality and Social Psychology*, *37*(5), 715–727.

Benjamini, Y., & Hochberg, Y. (1995). Controlling the False Discovery Rate: A Practical and Powerful Approach to Multiple Testing. *Journal of the Royal Statistical Society - Series B (Methodological)*, *57*, 289–300.

Boersma, P. (2001). Praat, a system for doing phonetics by computer. *Glot International*, *5*, 341–345.

Brown, B. L., Strong, W. J., & Rencher, A. C. (1973). Perceptions of personality from speech: Effects of manipulations of acoustical parameter. *Journal of the Acoustical Society of America*, *54*(1), 29–35.

Brown, B. L., Strong, W. J., & Rencher, A. C. (1974). Fifty-four voices from two: The effects of simultaneous manipulations of rate, mean fundamental frequency, and variance of fundamental frequency on ratings of personality from speech. *Journal of the Acoustical Society of America*, *55*(2), 313–318.

Brown, B. L., Strong, W. J., & Rencher, A. C. (1975). Acoustic determinants of perceptions of personality from speech. *Linguistics*, *13*(166), 11–32.

Bruckert, L., Liénard, J.-S., Lacroix, A., Kreutzer, M., & Leboucher, G. (2006). Women use voice parameter to assess men's characteristics. *Proceedings of the Royal Society B: Biological Sciences*, *237*(1582), 83–89.

Brückl, M. (2011). Altersbedingte Veränderungen der Stimme und Sprechweise von Frauen. Dissertation. Mündliche Kommunikation, S. Sendlmeier (ed.), Bd. 7, Berlin: Logos.

Burkhardt, F.; Schuller, B.; Weiss, B. & Weninger, F. (2011). Would you buy a car from me?—On the likability of telephone voices. In *Proceedings of 12th Interspeech* (pp. 1557–1560), Florence.

Chattopadhyay, A., Dahl, D. W., Ritchie, R. J., & Shahin, K. N. (2003). Hearing voices: The impact of announcer speech characteristics on consumer response to broadcast advertising. *Journal of Consumer Psychology*, *13*(3), 198–204.

De Jong, N. H., & Wempe, T. (2009). Praat script to detect syllable nuclei and measure speech rate automatically. *Behavior Research Methods*, *41*, 385–390.

Deal, L. V., & Oyer, H. J. (1991). Ratings of vocal pleasantness and the aging process. *Folia Phoniatrica*, *43*, 44–48.

Duran, D., Lewandowski, N., Bruni, J., & Schweitzer, A. (2017). Akustische Korrelate wahrgenommener Persönlichkeitsmerkmale und Stimmattraktivität. In *Proceedings of Elektronische Sprachsignalverarbeitung* (pp. 91–98). TUD Press.

Feldstein, S., Dohm, F.-A., & Crown, C. L. (2001). Gender and speech rate in the perception of competence and social attractiveness. *Journal of Social Psychology*, *141*, 785–806.

Fernández Gallardo, L. (2018). The Nautilus Speaker Characterization Corpus. ISLRN: 157-037-166-491-1. hdl.handle.net/11022/1009-0000-0007-C05F-6.

Fernández Gallardo, L., & Weiss, B. (2016). Speech likability and personality-based social relations: A round-robin analysis over communication channels. In *Proceedings of 17th Interspeech* (pp. 903–907), San Francisco.

Fernández Gallardo, L., & Weiss, B. (2017a). Perceived interpersonal speaker attributes and their acoustic features. In *13. Tagung Phonetik & Phonologie im deutschsprachigen Raum* (pp. 61-64), Berlin.

Fernández Gallardo, L., & Weiss, B. (2017b). Towards speaker characterization: Identifying and predicting dimensions of person attribution. In *Proceedings of 18th Interspeech* (pp. 904–908), Stockholm.

Fernández Gallardo, L., & Weiss, B. (2018). The Nautilus Speaker characterization corpus: Speech recordings and labels of speaker characteristics and voice descriptions. In *Proceedings of 11th Language Resources and Evaluation Conference (LREC)* (pp. 2837–2842), Miyazaki.

Fitch, W. T., & Giedd, J. (1999). Morphology and development of the human vocal tract: A study using magnetic resonance imaging. *Journal of the Acoustical Society of America*, *106*, 1511–1522.

Giles, H. (1980). Accommodation theory: Some new directions. *York Papers in Linguistics*, *9*, 105–136.

Gravano, A., Levitan, R., Willson, L., Beňuš, Š., Hirschberg, J., & Nenkova, A. (2011). Acoustic and prosodic correlates of social behavior. In *Proceedings of Interspeech* (pp. 97–100).

Hecht, M. A., & LaFrance, M. (1995). How (fast) can I help you? Tone of voice and telephone operator efficiency in interactions. *Journal of Applied Social Psychology*, *25*(23), 2086–2098.

ITU-T Rec. P.805 (2007). *Subjective evaluation of conversational quality*. Geneva: International Telecommunication Union.

Karnop, C., & Weiss, B. (2016). Zum Effekt von Tempo, Tonhöhe und Sprecherformant auf Sympathiebewertungen: Ein Resyntheseexperiment. In *27. Konferenz Elektronische Sprachsignalverarbeitung* (pp. 206–213), Leipzig.

Ketrow, S. M. (1990). Attributes of a telemarketer's voice and persuasiveness: A review and synthesis of the literature. *Journal of Direct Marketing*, *4*, 7–21.

Kreiman, J., & Gerratt, B. R. (1996). The perceptual structure of pathologic voice quality. *Journal of the Acoustical Society of America*, *100*, 1787–1795.

Krys, et al. (2016). Be careful where you smile: Culture shapes judgments of intelligence and honesty of smiling individuals. *Journal of Nonverbal Behavior*, *40*, 101–116.

Lehiste, I. (1970). *Suprasegmentals*. Cambridge, Massachusetts: MIT Press.

Leino, T., Laukkanen, A.-M., & Radolf, V. (2011). Formation of the actor's/speakers' formant: A study applying spectrum analysis and computer modeling. *Journal of Voice*, *25*(2), 150–158.

Linville, S. E. (2001). *Vocal aging*. San Diego: Singular Thomson Learning.

McAleer, P., Todorov, A., & Berlin, P. (2014). How do you say 'Hello'? Personality impressions from brief novel voices. *PLOS ONE, 9*(3).

McCroskey, J. C., & Mehrley, R. S. (1969). The effects of disorganization and nonfluency on attitude change and source credibility. *Speech Monographs*, *36*, 13–21.

Murry, T., & Singh, S. (1980). Multidimensional analysis of male and female voices. *Journal of the Acoustical Society of America*, *68*, 1294–1300.

Murry, T., Singh, S., & Sargent, M. (1977). Multidimensional classification of abnormal voice qualities. *Journal of the Acoustic Society of America*, *61*, 1630–1635.

Nawka, T., Anders, L. C., Cebulla, M., & Zurakowski, D. (1997). The speaker's formant in male voices. *Journal of Voice*, *11*(4), 422–428.

Osgood, C. E., Suci, G., & Tannenbaum, P. (1957). *The measurement of meaning*. Urbana, IL: University of Illinois Press.

Otta, E., Abrosio, F. F. E., & Hoshino, R. L. (1996). Reading a smiling face: Messages conveyed by various forms of smiling. *Perceptual and Motor Skills*, *82*, 1111–1121.

Peterson, R., Cannito, M., & Brown, S. (1995). An exploratory investigation of voice characteristics and selling effectiveness. *Journal of Personal Selling & Sales Management, 15*(1), 1–15.

Pfitzinger, H. R. (1999). Local speech rate perception in German speech. In *Proceedings of ICPhS* (pp. 893–896).

Rammstedt, B., & John, O. P. (2007). Measuring personality in one minute or less: A 10-item short version of the big five inventory in English and German. *Journal of Research in Personality, 41*, 203–212.

Ray, G. B. (1986). Vocally cued personality prototypes: An implicit personality theory approach. *Journal of Communication Monographs, 53*(3), 266–176.

Scherer, K. R. (1979). Personality markers in speech. In K. Scherer & H. Giles (Eds.), *Social markers in speech* (pp. 147–209). Cambridge: Cambridge University Press.

Scherer, K. R., & Giles, H. (1979). *Social markers in speech*. Cambridge: Cambridge University Press.

Smith, B. L., Brown, B. L., Strong, W. J., & Rencher, A. C. (1975). Effects of speech rate on personality perception. *Language and Speech, 18*, 145–152.

Street, R. L, Jr., & Brady, R. M. (1982). Speech rate acceptance ranges as a function of evaluative domain, listener speech rate and communication context. *Communication Monographs, 49*(4), 290–308.

Street, R. L, Jr., Brady, R. M., & Putnam, W. B. (1983). The influence of speech rate stereotypes and rate similarity or listeners' evaluation of speakers. *Journal of Language and Social Psychology, 2*(1), 37–56.

Tayal, S., Stone, S., & Birkholz, P. (2017). Towards the measurement of the actor's formant in female voices. In *Proceedings of Elektronische Sprachsignalverarbeitung* (pp. 286–293). TUD Press.

Tisljár-Szabó, E., & Pléh, C. (2014). Ascribing emotions depending on pause length in native and foreign language speech. *Speech Communication, 56*, 35–48.

Trouvain, J., Schmidt, S., Schröder, M., Schmitz, M., & Barry, W. J. (2006). Modelling personality features by changing prosody in synthetic speech. In *Proceedings of Speech Prosody*.

Uldall, E. (1960). Attitudinal meanings conveyed by intonation contours. *Language and Speech, 3*, 223–234.

Weirich, M. (2010). *Die attraktive Stimme: Vocal stereotypes. Eine phonetische Analyse anhand akustischer und auditiver Parameter*. Saarbrücken: Verlag Dr. Müller.

Weiss, B. (2013). Prosodische Elemente vokaler Sympathie. In *Konferenz Elektronische Sprachsignalverarbeitung* (Vol. 24, pp. 212–217), Bielefeld.

Weiss, B. (2015). Akustische Korrelate von Sympathieurteilen bei Hörern gleichen Geschlechts. In *26. Konferenz Elektronische Sprachsignalverarbeitung* (pp. 165–171), Eichstädt.

Weiss, B. (2016). Voice descriptions by non-experts: Validation of a questionnaire. In *12. Tagung Phonetik & Phonologie im deutschsprachigem Raum* (pp. 228–231), München.

Weiss, B., & Burkhardt, F. (2010). Voice attributes affecting likability perception. In *11th Interspeech* (pp. 1934–1937), Makuhari.

Weiss, B., & Burkhardt, F. (2012). Is 'not bad' good enough? Aspects of unknown voices' likability. In *13th Interspeech* (pp. 1–4), Portland.

Weiss, B., & Möller, S. (2011). Wahrnehmungsdimensionen von Stimme und Sprechweise. In *Konferenz Elektronische Sprachsignalverarbeitung* (Vol. 22, pp. 261–268), Aachen.

Weiss, B., Möller, S., & Polzehl, T. (2010). Zur Wirkung menschlicher Stimme auf diewahrgenommene Sympathie—Einfluss der Stimmanregung. In *Konferenz Elektronische Sprachsignalverarbeitung* (Vol. 21, pp. 56–63), Berlin.

Weiss, B., Hacker, A., Moshona, C., Rudawski, F., & Ruhland, M. (2017). Studying vocal social attractiveness by re-synthesis—Results from two student projects applying acoustic morphing with Tandem-Straight. In *28. Konferenz Elektronische Sprachsignalverarbeitung* (pp. 316–323), Saarbrücken.

Weiss, B., Hillmann, S., & Michael, T. (2018a). Kontinuierliche Schätzung von Sprechgeschwindigkeit mit einem Rekurrenten Neuronalen Netzwerk. In *29. Konferenz Elektronische Sprachsignalverarbeitung* (pp. 186–191). Ulm.

Weiss, B., Estival, D., & Stiefelhagen, U. (2018b). Studying vocal perceptual dimension of non-experts by assigning overall speaker (dis-)similarities. *Acta Acustica united with Acustica*, *104*, 174–184.

Wiggins, J. S., Trapnell, P., & Phillips, N. (1988). Psychometric and geometric characteristics of the revised interpersonal adjective scales (IAS-R). *Multivariate Behavioral Research*, *23*(4), 517–530.

Zuta, V. (2007). Phonetic criteria of attractive male voices. In *Proceedings of 16th ICPhS, Saarbrücken* (pp 1837–1840).

Chapter 14
Ranking and Comparing Speakers Based on Crowdsourced Pairwise Listener Ratings

Timo Baumann

Abstract Speech quality and likability is a multi-faceted phenomenon consisting of a combination of perceptory features that cannot easily be computed nor weighed automatically. Yet, it is often easy to decide which of two voices one likes better, even though it would be hard to describe why, or to name the underlying basic perceptory features. Although likability is inherently subjective and individual preferences differ, generalizations are useful and there is often a broad intersubjective consensus about whether one speaker is more likeable than another. We present a methodology to efficiently create a likability ranking for many speakers from crowdsourced pairwise likability ratings which focuses manual rating effort on pairs of similar quality using an active sampling technique. Using this methodology, we collected pairwise likability ratings for many speakers (>220) from many raters (>160). We analyze listener preferences by correlating the resulting ranking with various acoustic and prosodic features. We also present a neural network that is able to model the complexity of listener preferences and the underlying temporal evolution of features. The recurrent neural network achieves remarkably high performance in estimating the pairwise decisions and an ablation study points toward the criticality of modeling temporal aspects in speech quality assessment.

Keywords Ranking · Speech quality · Likability ratings · Found data
Crowdsourcing · Sequence modelling

14.1 Introduction

Speaker traits (such as age or gender), emotional coloring (such as anger or distress), socio-cultural aspects (such as accent or dialects), conscious or subconscious coloring toward the addressee (such as friendliness or positivity), and other paralinguistic aspects (such as clarity and comprehensibility) are expressed through various prosodic, suprasegmental, segmental, and non-segmental aspects of one's speech

T. Baumann (✉)
Universität Hamburg, Language Technology Group, Hamburg, Germany
e-mail: mail@timobaumann.de

B. Weiss et al. (eds.), *Voice Attractiveness*, Prosody, Phonology and Phonetics,
https://doi.org/10.1007/978-981-15-6627-1_14

and voice, where the combination of features and their temporal evolution are far from trivial. Intermittent deficiencies (e.g., a lisp) or deviations limited to a few features (e.g., nasalisation) can already have strong influences on the perceived quality. Together, they form the 'quality' of speech. It is important to note that no one 'best' combination of all features exists that would constitute 'ideal' speech.

Voice is a highly personal and subjective matter such that a multitude of combinations of these features result in a 'good' voice. This often makes likability comparisons hard and inherently subjective. Despite subjective preferences, *intersubjective* agreement on the preferences can often be found by-and-large, making generalizations useful. Generalizations are also necessary, for example, to cast news speakers, readers, or other speaking roles that need to approximate an intersubjective consensus. Such castings are typically performed by small expert jurys (potentially limiting the universality of decisions) and for small numbers of speaker candidates (for practical reasons).

In our work, we use rankings to analyze the influencing factors of speaker likability for broad speaker populations, or to eventually 'score' a voice sample along a range of speakers. Hence, we are interested in full rankings rather than in who is the best speaker for a task. Our aim is to create rankings for large speaker populations, by large and diverse jurys, and while keeping the effort as low as possible.

To simplify the human effort involved in creating the ranking, we have participants take many pairwise decisions on which of two stimuli is better. We then create a ranking from the pairwise comparisons (see below). The number of possible pairs grows quadratically with the number of the stimuli compared. Thus, while full comparisons for each rater are possible for small speaker groups (10 speakers $\rightarrow$ 45 rating pairs), these are infeasible for large speaker groups (225 speakers $\rightarrow$ 25200 rating pairs), in particular when relying on volunteer raters. Thus, our method must be able to build rankings from incomplete comparisons. Note, however, that many of the ratings will have predictable outcomes if one known-strong and one known-weak speaker are paired. It will be helpful to not waste too much human effort on such pairs; in contrast, human input on speakers of similar (or unknown) quality is most informative.

The main idea is to start from an initial ranking (based on some initial ratings) which is iteratively revised as more evidence becomes available with more ratings. Once the initial ranking is available, rating outcomes can be predicted and human effort can be directed away from comparisons with clear outcomes and toward the most informative pairs; this will be described in detail in Sect. 14.2.

Section 14.3 describes the corpus developed via crowdsourcing and based on the iterative method, both in terms of the stimuli used, as well as the resultant preference ratings. Section 14.4 examines the overall preference ranking derived from all pairwise ratings and finds some explaining factors in terms of high-level properties of the speech stimuli (and their speakers) via linear correlations.

As outlined above, however, prosody is a highly non-linear phenomenon and we hence build a recurrent neural network-based model that successfully identifies listener preferences using non-linear (but opaque) aggregation functions. Via an ablation study we find that the tunes in to phone-specific prosodic aspects given phonetic identity as additional features. Section 14.5 describes model for estimating the preferences of raters and analyzes the importance of features for modeling speaker preference. We conclude that modeling the *temporal aspects* of speech is critical for preference estimation.

14.2 Rankings from Pairwise Comparisons

Rankings have a long history in competitive sports, where individuals or teams play against each other in order to determine who's best. Two common forms, elimination and round-robin tournaments, both require a high degree of control over who plays who, which is not always possible. In addition, they may lead to only partial rankings. In chess, Elo's system (Elo, 1978) was designed to overcome these issues: a player's skill is estimated based on prior match outcomes, and skills are updated after each match. Skill changes correspond to the surprisal of the system by the match outcome. A ranking can be derived by ordering players by their skill. Microsoft TrueSkill™ (Herbrich, 2007) uses a Bayesian estimation of rankings from pairwise comparisons originally developed for ranking players of online games (based on their win/loss performance). TrueSkill models skill as a normal distribution, i.e., it makes the system's uncertainty about skill explicit, which enables smoother updates and more robust results when few match outcomes are available.

Most work in speech quality estimation has used direct scalar ratings of individual stimuli (Burkhardt, Schuller, Weiss, & Weninger, 2011) or required each subject to assign a complete ranking for all stimuli. Gallardo (2016) feeds paired comparisons into a Bradley–Terry–Luce model (Bradley & Terry, 1952) and finds similar results to direct scaling. Both of these methods have been limited to few raters and/or few stimuli. We extend the methodology introduced by Sakaguchi, Post, and Van Durme (2014) who created rankings for machine translation systems from pairwise comparisons using Microsoft TrueSkill™. In our metaphor, we view each rating as a 'match' in which the preferred stimulus wins against the dispreferred stimulus. We then compute the 'skill' of stimuli and their ranking. TrueSkill also provides *match making* capabilities that, given one player, select an opponent that has the most similar skill and where uncertainty of the skill difference is low (technically, TrueSkill estimates the probability of a draw and prefers matches with high draw probability). This is meant to lead to interesting matches with similarly skillful opponents. We use match making to select stimulus pairs for human rating in an iterative fashion which uses the ratings collected so far to steer our *active sampling* approach to select among the possible stimulus pairs to be compared. We actively select stimulus pairs that are expected to be informative for the full ranking based on a preliminary ranking of all ratings performed so far.

In our application, we found the abovementioned strategy for match making to be flawed: as scores tend to get more certain with more data, stimuli are preferred that already participated in many comparisons. As a result, the number of comparisons is not balanced on all stimuli but accumulates on few, well-known anchor points.[1] We use an approach that better balances the number of ratings per stimulus: We (1) pick a first stimulus based on the system's uncertainty about its ranking and (2) compute the match quality for all opponents and pick the opponent based on the predicted match quality with a dampening factor for the number of comparisons that the opponent has played so far. As a result, we (a) favor little-tested stimuli over well-tested ones and (b) select informative games over predictable ones. We randomly select pairs weighted by the criteria mentioned above which enables us to sample multiple 'interesting' pairs at once.

In comparison to Sakaguchi et al. (2014), which ranked 13 translation systems for which complete evaluation data had already been collected, we rank a total of 223 speakers, thus well over an order of magnitude more, in a live setting without external reference ranking.

14.3 Stimuli and Rating Collection via Crowdsourcing

We limit our likability judgements to one specific reading genre: the reading of encyclopaedic entries in Wikipedia. We use recordings from the Spoken Wikipedia.[2] as a broad sample of read *speech in the wild* The Spoken Wikipedia project unites volunteer readers who devote significant amounts of time and effort into producing read versions of Wikipedia articles as an alternate form of access to encyclopaedic content. It can thus be considered a valid source of speech produced by ambitious but not always perfect readers. The data has been prepared as a corpus (Baumann, Köhn, & Hennig, 2018) and the German subset of the corpus, which we use here, contains ∼300 h of speech read by ∼300 speakers.

To avoid rating preferences based on *what* is spoken rather than how, we choose as stimuli the opening that is read for every article in the Spoken Wikipedia, which is (supposed to be) identical for all articles except for the article lemma.[3] We extract that stimulus for every speaker in the German subset of the Spoken Wikipedia Corpus using the alignment information given in Baumann et al. (2018). As some alignment information was missing or clearly wrong, our stimulus pool is reduced to 227 speakers. We then masked the article lemma with noise in a length that matches the average reading speed of the stimulus. The mean/median duration per stimulus is 4.7/4.57 s with 5/95% quantiles at 3.74/6.03 s.

For every rating pair, participants were asked to rate which of the two voices they would prefer for having a Wikipedia article read out to them. We realized a web-based rating experiment on the basis of BeaqleJS (Kraft and Zölzer 2014) which we

[1]This may not be a problem when using TrueSkill for match assignment, as participation in games is limited by the players' availability.

[2]https://en.wikipedia.org/wiki/Wikipedia:WikiProject_Spoken_Wikipedia.

[3]Expected reading: 'Sie hören den Artikel *article lemma* aus Wikipedia, der freien Enzyklopädie.' (You are listening to the article *article lemma* from Wikipedia, the free encyclopedia.).

extended to allow for an open number of pairwise ratings for each participant. The experiment operated with a mini-batch cache of 1000 rating pairs from which clients sampled randomly. The cache was updated manually whenever more than 200 ratings had been submitted by re-creating a new best ranking and selecting stimulus pairs as outlined above. We opted against an active backend with immediate update and selection of the next most relevant rating pair to ensure availability in times of high system usage (e.g., during the minutes after a mailing list advertised our experiment).

We solicited participants to our experiment via the German Wikipedia 'off-topic bulletin board' and various open mailing lists of student organizations (particularly CS students), as well as the Chaos Computer Club in Germany, Austria, and Switzerland in order to reach a wide variety of dialect and age groups. We deliberately did not explicitly invite the Spoken Wikipedia community to participate, as they could have been particularly biased.

Statistics of the participants' self-reported meta data are shown in Table 14.1. As can be seen, Northern Germans, males, and 20–30 years olds are over-represented in our data (presumably computer science students at Universität Hamburg). However, almost all other demographic groups are included as well, at least to some extent. In total, we collected 5440 ratings from 168 participants. Participation was strictily voluntary and without compensation and hence the resulting ratings are unlikely to be prone to vandalistic behavior.

Although participants could perform as many ratings as they liked, they were instructed that 10 ratings are sufficient, 30–50 preferable, and that they should take a break after 100 ratings (and possibly return the next day). We excluded participants who submitted a single rating only. The median ratings per participant were 26 with half the participants between 11 and 43 ratings and 5/95% quantiles at 4 and 101 ratings, respectively.

Participants were asked to always state a preference, even if unsure, and did not explicitly have the option to state that they could not decide. It is more informative for our setup to get contradicting preferences than to explicitly invite the participants to omit a decision. As our method steers toward 'difficult' comparisons, many omitted decisions could otherwise have been expected. Our software, however, did allow to skip ahead without making a decision and sometimes participants did not provide a decision (accidentally or on purpose). These instances were ignored in further processing, as no rating has been recorded.

We also measured the time taken for each rating. The median time per rating is 14.3 s with half the ratings between 11.3 and 21.3 s and 5/95% quantiles at 6.3 and 39.7 s, respectively. 6.3 seconds can still be considered a reasonable lower bound for listening to both stimuli and then taking the decision quickly. In total, participants spent $\sim$26 h on rating stimulus pairs.[4]

[4]We substitute the median for the slowest 2.5% of ratings, as participants were obviously sidetracked who spent more than 55 s for a single rating.

Table 14.1 Breakdown of self-reported meta information of participants and their rating counts

	Total	Participants	Ratings
		168	5440
Gender	Female	41	1665
	Male	109	3221
	Unreported	18	554
Age	<20	18	358
	20–30	78	2593
	30–40	34	1030
	40–60	24	886
	>60	6	418
	Unreported	8	155
Dialectal origin	Northern Germany	83	2656
	Berlin/Brandenburg	8	128
	Northrhine-Westphalia	11	464
	Middle Germany	9	443
	Rhine-/Saarland	3	82
	Baden-Wurttemberg	15	432
	Bavaria	8	405
	Austria	5	179
	Switzerland	0	0
	Unsure/other	26	651

The stimulus ordering was randomized. Participants have a slight tendency for stimulus B over A (2784 versus 2656, n.s.: sign test, $p = .09$), which could be interpreted as a recency effect.

We measure the degree of disagreement by constructing a directed acyclic graph of the preference relation expressed through all ratings (i.e., the stimuli are nodes and one edge is introduced per rating). If ratings were consistent, there would not be any rating circles (a < b, b < c but c < a) and the proportion of feedback arcs can be taken as a measure of consistency. We heuristically compute the minimum feedback arc set of all ratings (Eades, Lin, & Smyth, 1993) and find the proportion to be 29%. In a preliminary experiment using only 10 stimuli and all 45 possible comparisons, only one rater was 'perfect' in not producing any circles. Hence, we know that both within-rater and across-rater inconsistencies occur. In addition, our stimulus selection process is tailored towards choosing pairs that are expected to be hard to rate (and the disagreeing proportion grew over the runtime of the experiment).

14.4 Ranking Analyses

We feed all pairwise ratings into TrueSkill™ to derive rankings. In TrueSkill, more recent ratings are more influential for the final ranking due to the iterative update mechanism.[5] As proposed by Sakaguchi et al. (2014), we use the fact that rankings depend on the rating order to validate our method: we permute the ratings and create many rankings for the same set of ratings (below: N = 300). We then take the median ranking as the final decision. Thus, we are also able to report ranking confidence levels.[6]

Rankings can be compared using correlation coefficients like Kendall's Tau (Langville & Meyer, 2012, Chap. 16). We find that pairwise correlations of the 300 rankings result in $\tau > 0.92$ and that each ranking against the median ranking gives $\tau > 0.95$. Thus, we conclude that TrueSkill leads to consistent rankings (within bounds) and that the median ranking is a meaningful middle ground for all rankings.

The final median ranking with quartile and 5/95% confidence ranks is shown in Fig. 14.1. As can be seen in the figure, there is no one clear ranking of all speakers. While there is a best and worst stimulus shared among all rankings, variability is larger in the middle. Overall, the average rank variability is 6.7 ranks within the 25–75% confidence interval and 16.4 ranks within the 90% confidence interval. Interestingly, some clusters of similarly 'good' stimuli emerge, e.g., as highlighted in the green circled area where 11 stimuli share similar ranks with a high variability that are delimited with high confidence to higher ranks (upper right of circled area) and slightly less to lower ranks.

Finally, we use rankings to predict the outcome of ratings as another way of testing the ranking validity. We assume that a rating will be 'won' by the better ranked stimulus (although similarly ranked stimuli could easily have any outcome). We use 100-fold cross-validation and find that on average, the prediction performance is 68%. Given that 29% of ratings can be expected to be mis-predicted due to the rating inconsistencies, the rankings have a high level of predictive value. As described above, TrueSkill can compute match quality, effectively describing how likely a rating will lead to disagreement among raters. We find that prediction performance highly correlates with that score (Kendall's $\tau = -0.81$, $p < .001$).

We investigate which stimulus pairs have been selected for comparison to find out whether the method proposed in Sect. 14.2 works effectively. The rated pairs are presented in Fig. 14.2. We find that pairs along the diagonal (i.e., with similar ranks) have been tested more densely than pairs further apart. Furthermore, the plot shows that 'better' stimuli (as per the ranking) win more often against inferior stimuli (green/blue division of the plot) and multiple controversial ratings (red) mostly occur along the diagonal. Overall, our 5440 ratings spread over 4000 different pairs, that is,

[5]This is a feature when ranking human players, as their true performance may change over time – but this is not the case in our experiment.

[6]The confidence is about TrueSkill producing a preference ordering given another permutation of ratings. We cannot make any guarantee with respect to some 'gold' ranking, which does not exist for our data.

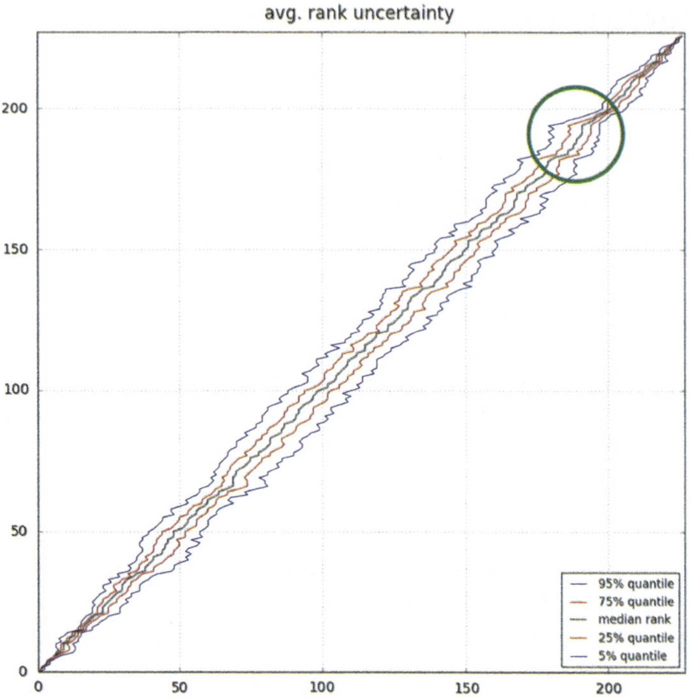

Fig. 14.1 Ranking results (both axes ordered by median ranking) including rank confidence on the x-axis; the circled area is further discussed in the text

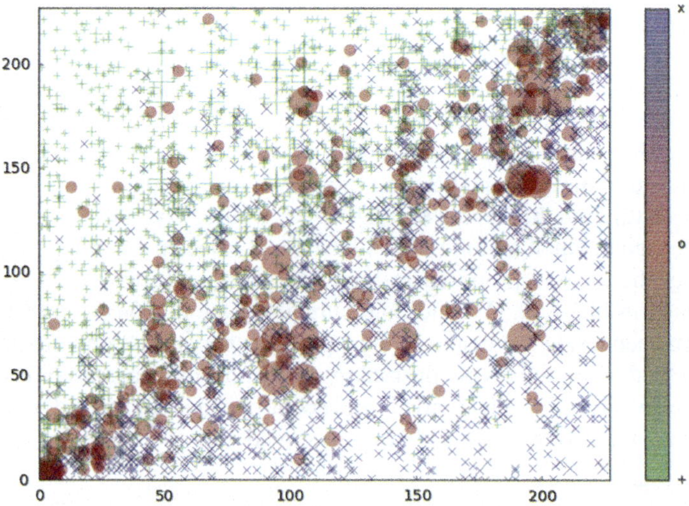

Fig. 14.2 Scatter plot of pairs compared: the plot is more dense along the diagonal, as stimuli are compared more often when they are of comparable rank; axes ordered by median ranking, color-coding indicates the avg. outcome of comparisons

Fig. 14.3 Line graph comparison of median rankings for female (top) and male (bottom) raters (stimuli spoken by females are shown in red)

7,7% of all possible comparisons. 3057 pairs have been tested once, 666 pairs twice, and the remaining pairs up to 9 times (which seem to be artefacts of older versions of pair selection). Overall, the average stimulus has been rated 46 times with the 5/95% quantiles at 39 and 56 ratings. Thus, our rating pair selection strategy successfully balances stimulus selection and opponent assignment.

14.4.1 The Influence of Rater Population on Ranking Outcome

Finally, we analyze the rankings wrt. to gender. We produce one median ranking each for ratings from female and male listeners (randomly subsampling the male ratings to the number of female ratings; see Table 14.1). We find only a moderate correlation ($\tau = 0.44$, $p \ll .001$) between female and male listener rankings, which indicates different preferences between these listener groups. We further analyze the ranking wrt. to speaker gender of the stimuli.[7] The rank assigned to a female speaker is on average 12.7 ranks better for female than for male listeners (half of the stimuli between -32 and $+60$ ranks), indicating that one major difference between female and male listeners is their preference toward female voices.

Figure 14.3 compares the gender-dependent rankings (each line corresponds to a stimulus, female stimuli in red). The less inclined a line, the more similar the rank for female/male listeners. As can be seen, preferences differ both in ranking female speakers as for male speakers. It is interesting to note that Dykema, Diloreto, Price, White, and Schaeffer (2012) find that male speakers respond more truthfully to questions posed by female voices, yet they seem to disprefer them in our data. The results highlight the importance of gender-appropriate voice selection for reading encyclopaedic, and possibly other factual information.

We also divide our data by age (<30 versus >30) and dialect (Northern German versus all other dialects as there is insufficient data to further differentiate among dialects). In both cases, correlation between the groups is stronger (age: $\tau = 0.50$, dialect: $\tau = 0.54$) than in the gender partition. No age or dialect information is available for the speakers, hence we cannot compare within/across-group effects (e.g., we would expect matched dialects of speaker and listener being preferred).

---

[7]Unfortunately, just 20 of 227 stimuli (9%) were spoken by females.

14.4.2 Acoustic Correlates of Ranking Quality

We finally experiment with acoustic factors that could explain the speaker likability expressed by the median ranking shown in Fig. 14.1. First, we compute the perceptual quality of audio stimuli as standardized by ITU-T P.563 (Malfait, Berger, & Kastner, 2006). We find a low (but significant) correlation ($\tau = 0.14$, $p < .002$) of achieved median ranking and estimated MOS for the audio transmission quality.[8] We conclude that carefully arranged recording conditions could coincide with better speech quality, or that listener judgements are influenced by encoding quality – in contrast to Burkhardt et al. (2011) where no such influence was found in a similar task.

We estimate the liveliness of the speaker's prosody as it might be a relevant factor of likability. We compute the pitch range in semi-tones and take the 50% (25–75%) and 90% (5–95%) ranges of the measured pitch. On average, the 50/90% ranges are 4.3/12.8 semi-tones for all speakers. We find very slight but non-significant correlations between either liveliness measure and the ranking. As this could be due to very little data from each short stimulus, we also extract pitch from the full articles. This allows us to estimate each speaker's liveliness *in general*, not just in the opening of the article. Here we find that the inter-quartile (50%) pitch range correlates somewhat ($\tau = 0.10$, $p < .03$) with the ranking.

14.5 Listener Preference Classification

In previous work (Eyben, Wöllmer, & Schuller, (Burkhardt et al., 2011)), speaker likability has been modeled using OpenSmile (Eyben et al., 2010) features based on linear and non-linear aggregation functions (such as means and medians) to aggregate over the duration of the stimulus. Features were used to train classifiers such as SVMs which resulted in moderately high (better than chance) performance in classifying speakers as being above or below median likability (Burkhardt et al., 2011). Like the analyses in Sect. 14.4.2, the abovementioned aggregation functions cannot take into account the context of feature characteristics in the stimulus, and are unlikely to accurately express more fine-grained details relevant for speech quality (such as where and how a pitch accent is realized, beyond mean pitch). In this section, we experiment with neural sequence-learning methods (RNNs) to encode the complex temporal evolution of features of speech quality into a latent feature space and use the difference in these for pairs of speech stimuli to train our classifier.

[8] We must mention that all speech in the Spoken Wikipedia is distributed as OGG/Vorbis, with varying bit rates and under diverse recording conditions.

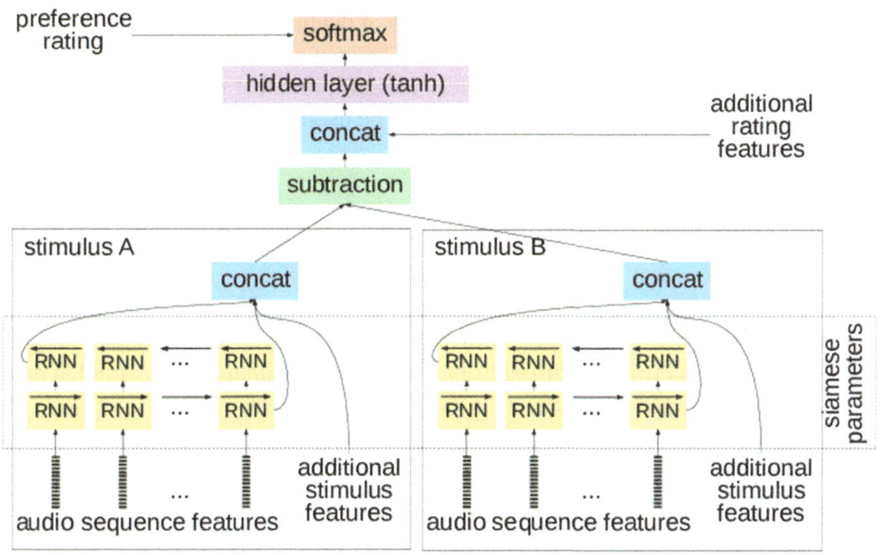

Fig. 14.4 Diagram of the neural architecture for speech likability preference. The task is symmetric (whether a stimulus is A or B is irrelevant) and hence the parameters for the RNNs can be shared (siamese network). Additional features about stimuli and the rating can be added via concatenation

14.5.1 Model Architecture

The task of preference ranking is *asymmetric* in the sense that if the two stimuli to be compared are swapped, then the comparison result is the opposite. This has two consequences: (a) parameters for sequence analysis of both stimuli can be shared which is called a *siamese model* (Bromley et al., 1994) and reduces the degrees of freedom of the model, making learning more efficient, and (b) the outputs from sequence analysis of each stimulus can simply be subtracted and the difference be subjected to a final decision layer.

In our model and as shown in Fig. 14.4, we use two layers of bidirectional RNN (LSTMs Hochreiter & Schmidhuber, 1997 or GRUs (Cho et al., 2014)) to model the feature sequence of each stimulus and concatenate the outputs of the forward and backward pass. We can also concatenate additional stimulus-level features into the representation at this time, e.g., measures of signal quality such as ITU-T P.563 (Malfait et al., 2006) (cmp. Sect. 14.4.2), or meta information about the speaker or the audio recording (such as gender or bitrate, cmp. Sect. 14.4.1).

Given that our final decision is based on the quality *difference* alone, not the overall quality, we subtract both stimuli's vectors.[9] We then pass the difference to one hidden layer and a final binary softmax layer that models the preference decision. We can

[9] We found, in initial experimentation, that this performs much better than concatenating the outputs of each speaker.

also optionally include additional meta features of the rating (such as identity, age, or dialectal region of the rater). These can easily be concatenated in before the hidden layer, in order to model the relative preferences of individual raters or rater groups. As we found that preferences differ, this could be useful information.

14.5.2 Data and Evaluation

The original purpose of the rating collection reported in Sect. 14.3 was to create a ranking and effort was put into maximizing the efficiency of human annotation by focusing the human effort on 'difficult' pairs using *active sampling* of stimulus pairs that focus human annotation effort on 'similarly good' speakers. As a result, the stimulus pairs that were rated by participants are much more similar in their quality than randomly selected stimulus pairs would be.

In addition, inconsistency in the data set is high, as are pairs of stimuli that have been rated multiple times. Above, we have computed the minimum feedback arc set, i.e., the subset of ratings that lead to a fully consistent ranking (Eades et al., 1993). We found the proportion of conflicting arcs to be 29%. This can act as an indicator of the proportion of ratings that are inconsistent (where potentially different raters have different preferences, or simply cannot reliably tell the difference). In addition, we here compute an oracle correctness for all pairs that have been rated more than once, by checking for each rating, if it is the majority rating for this pair (deciding randomly to resolve draws). We find that such an *oracle classifier* reaches a correctness of only 65% for those pairs that have been rated more than once. Pairs that were rated just once *potentially* are easier to classify, which makes it possible to beat this oracle.

For evaluations, we report multiple settings below. The settings are meant to counterbalance the difficulties introduced by the data elicitation technique and to test different aspects of listener preference classification:

naïve we sample randomly among the evaluation instances from the corpus of human-rated pairs; as the corpus focuses on difficult pairs, we cannot expect a spectacular performance;

easy based on the median ranking derived in Sect. 14.4, we sample instances with 'large' ranking differences (distance on the ranking scale >0.25 or >0.5), in order to test if our classifiers fare better with stronger preference differences (and hence easier to identify differences in speech quality).

Given that stimuli were presented in random order, the data set is balanced in terms of which stimulus outperforms the other. Thus, we focus on accuracy as the only evaluation metric.

14.5.3 *Features and Conditions*

Using a sliding window, we derive a multitude of local features from the audio stream that might capture aspects of speech quality. All features use a frame shift of 10 ms. In particular, we measure Mel-frequency cepstral coefficients (MFCCs, 12 + 1 energy) to capture voice and recording characteristics, f_0 (measured using `Snack`'s `esps` implementation) as a first measure of speech melody, and fundamental frequency variation (FFV) features (Laskowski, Heldner, & Edlund, 2008) as these are more robust (and might contain more valuable information) than single f_0. Using Praat (Boersma, 2002), we compute jitter (PPQ5), shimmer (APQ5), and harmonics-to-noise ratio (Boersma, 1993). We do not perform z-scale normalization on the feature streams.

The Spoken Wikipedia Corpus also contains phonetic alignments that were computed using the MAUS tool (Schiel, 2004). The alignments allow us to assign phone annotations to every frame. With this information, the model is informed explicitly that different phones have different phonetic characteristics (as expressed in the MFCCs) and can condition its learning of speech quality on these characteristics. In other words: the model can learn to focus on a phone's quality aspects (e.g., nasalization) without needing to learn to differentiate phones.

One frame of features for every 10 ms may overwhelm the model with very large amounts of parameters, reducing training efficiency as well as effectiveness. In order to keep training tractable, we subsample the feature frames with various values (see *seq. step size* in Table 14.2). When we do so, we use mean aggregation for numeric values (ignoring missing values for pitch and HNR).

14.5.4 *Experiments and Results*

We separate out about 1/10th of the 5440 ratings as the test data: the **naïve** test set contains 400 ratings, and we sample among ratings with 'large' differences 100 ratings each for the >0.25 and >0.5 **easy** test sets.

We implemented our network in dynet (Neubig, 2017). In the experiments reported below, we train for 50 epochs using AdamTrainer and no dropout. We concatenate the various audio features that are computed for every frame. We use embeddings to characterize the phonetic labels.

14.5.4.1 Meta Parameter Optimization

As originally reported in Baumann (2018), we have performed an optimization to find good sizes for various meta parameters of the model:

- To reduce the length of the sequence that need to be learned by the LSTMs (and to avoid the problem of vanishing gradients through long sequences), we subsample

Table 14.2 Meta parameters considered in grid search (best values are shown in boldface)

Meta parameter	Values
Sequence step size	**5**, 10, 15
Phone embed size	8, **16**, 24
Sequence state size	24, 32, **48**, 64
Hidden layer size	2, **3**, 4 × sequence state size

the audio features by mean-aggregating values over a number of frames (5, 10, or 15).

- To represent the discrete phonetic labels, we use embeddings of varying sizes (8, 16, or 24), in order to allow the model to cluster similar phones.
- The sequential LSTM state size determines how many dimensions can be considered during the sequence analysis and we experiment with various sizes (24, 32, 48, or 64).
- The output from concatenation of both forward and backward LSTMs doubles the size of the next layer's input. For the hidden layer size, we hence consider scaling factors (2, 3, or 4) over the size of the sequential state size.

We performed a grid search over the possible meta parameter values as summarized in Table 14.2 and focusing on the naïve data set. We found an optimum for sequence step size of 5 (i.e., one feature frame for every 50 ms of speech), phone embedding with 16 dimensions, sequence state size of 48, and hidden layer size of $3 \times 48 = 144$ (sequence state size of 32 and $4 \times 32 = 128$ was a close contender).

At these settings, our model yields an accuracy of 67.25% on the naïve test set, 93% on the **easy-0.25** test set and 97% on the **easy-0.5** test set. The accuracy on the naïve test set is close to what we estimated as the upper limit for the harder part of our training data.

14.5.4.2 Ablation Study on Phonemic Alignments

We hypothesized above that our performance gain over previous work may be largely due to the model being able to perform prosodically meaningful aggregations and could, for example, relate prosodic parameters to the phones spoken. To test this hypothesis, we perform an ablation study and remove the phoneme embeddings from the input features. We perform this experiment with the other meta parameters set to their optima as found in the previous subsection. As shown in Table 14.3, we find performance to drop substantially when the phone identity feature is removed. We believe this is because the model is unable to make maximum sense of features such as MFCCs given speech quality is obviously just a secondary feature, far behind phone identities. If the model is not informed about the phonetic identities, it needs to resolve whether input has good quality, whereas the full model only needs to resolve the quality of a feature given the particular speech sound.

Table 14.3 Accuracy (in percent) of full and reduced feature set (without phone alignment)

Setting	Accuracy		
	naïve	easy-0.25	easy-0.5
Full mode	67.25	93	97
Without phone alignment	58.75	73	80

14.6 Conclusions and Future Work

We have presented a method for creating crowdsourced speaker likability rankings from pairwise comparisons. The material that we base our ratings on is freely available and we likewise publish the ratings and the software to derive rankings from those ratings under the same terms. Unlike Gallardo (2016) which uses Bradley–Terry–Luce models, our method does not require a complete comparison of all pairs, and works on a small subset (in our case: 7% of possible comparisons) jointly provided by many participants.

One advantage of the Spoken Wikipedia corpus is the availability of much more data from each speaker beyond the short stimuli that are used in the ranking experiment. Thus, more complex characteristics of a speaker, such as accentuation and other prosodic idiosyncrasies (which listeners presumably would be able to judge in one sentence), can be derived from up to an hour of (closely transcribed and aligned) speech. In fact, we found in Sect. 14.4.2 that extracting the 50% pitch range as an estimate of liveliness significantly correlates with likability, at least if liveliness is extracted from the full speech, rather than just the one sentence used in human ratings, potentially because this circumvents effects from faulty fundamental frequency extraction.

We have also presented a neural architecture for determining which of two speech stimuli is rated as the better of the two in noisy human annotations. Our model yields good performance most likely because the RNN provides for complex aggregations of the (conventional) feature sequences. Our model's aggregations are able account for sequential information, in particular it is able to relate acoustic features to the phones spoken, unlike more coarse-grained aggregation functions as have been used before.

In Burkhardt et al. (2011), the authors train classifiers to differentiate whether a stimulus is better/worse than average and reach a classification accuracy of 67.6%. Their setup is comparable to our decisions for stimuli that are relatively far apart on the rating scale, in which case the neural aggregation and classification yields a classification accuracy of 93–97%. We believe this to be caused by the better temporal modeling of our approach and the use of phonetic identities during aggregation.

Despite the relatively good results, our method is still basic in terms of the neural architecture employed. In particular, our method does not yet employ an attention mechanism that could help to better weigh the speech quality encoding. Given that all speakers in our corpus speak (more or less) the same content, we envision that our model would profit greatly if the comparison between both stimuli could attend to particular differences rather than only the comparison of the final BiLSTM output vectors. An attention model would also help the analysis of *why* a speaker is rated as better than another, as it would indicate the relative importance of parts of the stimuli in the comparison. Another venue, at least for comparisons on shared text would be connectionist temporal classification to temporally relate the feature streams before comparison for a better notion of timing differences between the stimuli. Finally, it might be worthwhile to pre-train the intermediate representations of the model.

In the end, our model could weigh slight mis-pronunciations against voice quality or prosodic phrasing, and we intend to use analysis techniques to ultimately understand the relative weights of these aspects in comparisons.

We have limited our study to one identical stimulus sentence in order to exclude contextual differences, and to one stimulus per speaker. We plan to extend the study to other stimulus pairs where the sentences (or sentence fragments) are spoken by different speakers across the Spoken Wikipedia. In this way, we hope to get a better judgement of the speakers, based on more than (on average) 4.7 s of their speech.

Acknowledgments We thank our listeners/raters as well as the volunteers of the Spoken Wikipedia for donating their time and voice. This work has been partially supported by a Daimler and Benz Foundation PostDoc grant.

References

Baumann, T. (2018). Learning to determine who is the better speaker. In *Proceedings of Speech Prosody*.

Baumann, T., Köhn, A., & Hennig, F. (2018). The Spoken Wikipedia corpus collection: Harvesting, alignment and an application to hyper listening. In *Language resources and evaluation. Special Issue representing significant contributions of LREC 2016*.

Boersma, P. (1993). Accurate short-term analysis of the fundamental frequency and the harmonics-to-noise ratio of a sampled sound. *Proceedings of the Institute of Phonetic Sciences*, *17*, 97–110.

Boersma, P. (2002). Praat, a system for doing phonetics by computer. *Glot International*, *5*(9/10), 341–345.

Bradley, R. A., & Terry, M. E. (1952). Rank analysis of incomplete block designs: I. The method of paired comparisons. *Biometrika*, *39*(3/4), 324–345.

Bromley, J., Guyon, I., LeCun, Y., Säckinger, E., & Shah, R. (1994). Signature verification using a" siamese" time delay neural network. In *Advances in neural information processing systems* (pp. 737–744).

Burkhardt, F., Schuller, B., Weiss, B., & Weninger, F. (2011). Would you buy a car from me? On the likability of telephone voices. In *Proceedings of Interspeech. ISCA*.

Cho, K., van Merrienboer, B., Gulcehre, C., Bahdanau, D., Bougares, F., Schwenk, H., & Bengio, Y. (2014). Learning phrase representations using RNN encoder-decoder for statistical machine translation. In *Proceedings of the 2014 Conference on Empirical Methods in Natural Language Processing (EMNLP)* (pp. 1724–1734). Doha, Qatar: Association for Computational Linguistics. http://www.aclweb.org/anthology/D14-1179.

Dykema, J., Diloreto, K., Price, J. L., White, E., & Schaeffer, N. C. (2012). ACASI gender-of-interviewer voice effects on reports to questions about sensitive behaviors among young adults. *Public Opinion Quarterly*, *76*(2), 311–325.

Eades, P., Lin, X., & Smyth, W. F. (1993). A fast and effective heuristic for the feedback arc set problem. *Information Processing Letters*, *47*(6), 319–323.

Elo, A. E. (1978). The rating of chessplayers, past and present. Arco Pub.

Eyben, F., Wöllmer, M., & Schuller, B. (2010). Opensmile: The Munich versatile and fast open-source audio feature extractor. In *Proceedings of the 18th ACM International Conference on Multimedia* (pp. 1459–1462). ACM.

Gallardo, L. F. (2016). A paired-comparison listening test for collecting voice likability scores. In *Speech Communication; 12. ITG Symposium; Proceedings of VDE* (pp. 1–5).

Herbrich, R., Minka, T., & Graepel, T. (2007). TrueSkill^{TM}: A Bayesian skill rating system. In *Advances in neural information processing systems* (vol. 20, pp. 569-576). MIT Press.

Hochreiter, S., & Schmidhuber, J. (1997). Long short-term memory. *Neural Computation*, *9*(8), 1735–1780.

Kraft, S., & Zölzer, U. (2014). BeaqleJS: HTML5 and JavaScript based framework for the subjective evaluation of audio quality. In *Linux Audio Conference*.

Langville, A. N., & Meyer, C. D. (2012). *Who's # 1 ?. The Science of Rating and Ranking*. Princeton University Press.

Laskowski, K., Heldner, M., & Edlund, J. (2008). The fundamental frequency variation spectrum. In *Proceedings of FONETIK 2008*.

Malfait, L., Berger, J., & Kastner, M. (2006). P.563—The ITU-T standard for single-ended speech quality assessment. *IEEE Transactions on Audio, Speech and Language Processing*, *14*(6), 1924–1934.

Neubig, G. et al. (2017). DyNet: The dynamic neural network toolkit. In arXiv preprint arXiv:1701.03980.

Sakaguchi, K., Post, M., & Van Durme, B. (2014). Efficient elicitation of annotations for human evaluation of machine translation. In *Proceedings of the Ninth Workshop on Statistical Machine Translation* (pp. 1–11). Baltimore, Maryland, USA: ACL.

Schiel, F. (2004). MAUS goes iterative. In *Proceedings of the LREC*.

Chapter 15
Multidimensional Mapping of Voice Attractiveness and Listener's Preference: Optimization and Estimation from Audio Signal

Yasunari Obuchi

Abstract In this chapter, a new framework of listener-dependent quantification of voice attractiveness is introduced. The probabilistic model of paired comparison results is extended to the multidimensional merit space, in which the intrinsic attractiveness of voices and the preference of listeners are both expressed as vectors. The attractiveness for a specific listener is then obtained by calculating the inner product of two vectors. The mapping from the paired comparison results to the multidimensional merit space is formulated as the maximization problem of the likelihood function. After the optimal mapping is obtained, we also discuss the possibility of predicting the attractiveness from the acoustic features. Machine learning approach is introduced to analyze the real data of Japanese greeting phrase "irasshaimase," and the effectiveness is confirmed by the higher prediction accuracy.

Keywords Paired comparison · Mapping · Optimization · Listener's preference Multidimensional · Acoustic feature · Machine learning

15.1 Introduction

Most people believe that there are attractive voices and unattractive voices. However, they also believe that there are voices that are attractive and unattractive depending on who is listening. This chapter deals with such objectivity and subjectivity of voice attractiveness. We start the discussion by establishing a framework of voice attractiveness quantification based on the probabilistic analysis of experimental results. Once the quantification framework is given, we then try to predict the attractiveness of a new voice from its acoustic characteristics.

In this chapter, we focus on the social attractiveness, especially in a commercial context. For example, if you make a commercial video for your product with some narration, its attractiveness has strong influence on your business. In the pre-Internet

Y. Obuchi (✉)
School of Media Science, Tokyo University of Technology, 1404-1 Katakura, Hachioji, Tokyo 192-0982, Japan
e-mail: obuchiysnr@stf.teu.ac.jp

B. Weiss et al. (eds.), *Voice Attractiveness*, Prosody, Phonology and Phonetics,
https://doi.org/10.1007/978-981-15-6627-1_15

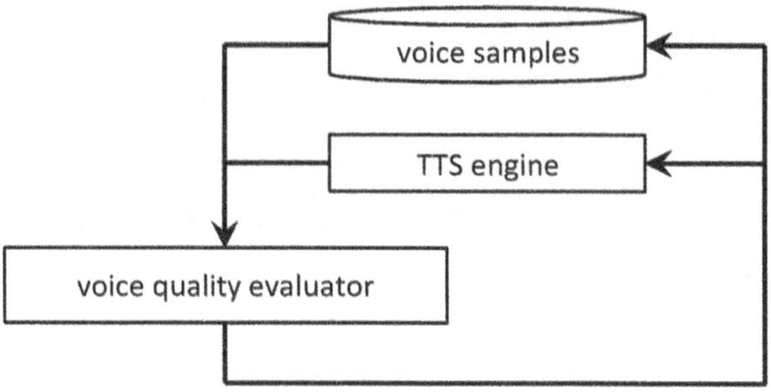

Fig. 15.1 Schematic diagram of voice selection process

era, voice attractiveness was evaluated as a scalar, typically the average of evaluation by a mass audience. However, in the Internet era, in which the contents can be customized on delivery, it is important to evaluate voice attractiveness for each customer.

The voice selection process in commercial applications is illustrated in Fig. 15.1. If the system uses recorded voice samples, the quality of each sample is evaluated, and the sample with the highest score is selected. If the system uses text-to-speech (TTS) software, the evaluation score is fed back to the TTS engine, and the system parameters are adjusted. In both cases, the voice quality evaluator plays the central role.

Voice quality evaluation can be realized by collecting a mass of human judgments. Typical evaluation methods include the mean opinion score (MOS) test and preference test. The MOS test is designed to give an absolute score for each voice sample, whereas the preference test focuses on relative quality of two or more voice samples. Both tests are based on subjective judgment of human listeners, and it requires days or weeks of evaluation process in the development cycle. If we can replace the human-based evaluator with the computer-based automatic evaluator, the development cycle would be accelerated dramatically.

Although we have limited insight into the physical features representing voice attractiveness, an automatic evaluator can be developed using the machine learning framework. If we have plenty of data with correct attractiveness label, machine learning algorithms such as support vector machine can provide a model which connects voice signals and their attractiveness.

Figure 15.2 shows the way to train the model from a large database. Before starting the training process by a machine learning tool, we have to prepare appropriate labels of voice attractiveness. We know from our daily experiences that the definition of attractiveness is ambiguous, and the evaluation results of human listeners are often inconsistent. Therefore, the first important problem is how to prepare correct attractiveness labels. For this problem, we start with the paired comparison test (Shah et al., 2014). Since it is difficult to give a concrete definition of the attractiveness scale, it is easier for a typical listener to answer the question "which voice is more attractive, A or B?" than the question "how attractive is this voice in the scale of one to five?".

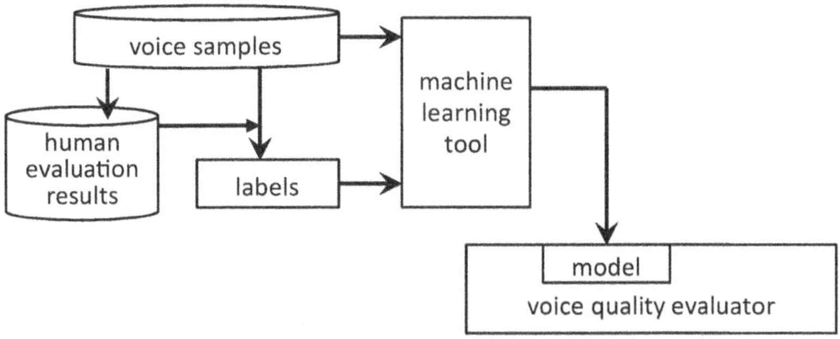

Fig. 15.2 Schematic diagram of voice quality evaluator training

The drawback of paired comparison test is that it includes more inconsistency than the absolute evaluation such as MOS test. In addition to the inconsistency between listeners, sometimes a listener gives inconsistent evaluation results within his/her own comparisons. A typical example is that A is better than B, B is better than C, and C is better than A. Such kind of discussion leads us to the probabilistic model of paired comparison test, in which the intrinsic attractiveness of voices works as the parameter of winning probability.

In this chapter, a new probabilistic model of voice attractiveness is introduced, in which the intrinsic attractiveness of voices and the intrinsic preference of listeners are handled in a unified form. In this model, the voice attractiveness is represented as a multidimensional vector in the "merit space." The preference of listener is described as a direction in the merit space, and the evaluation of a voice by a listener is expressed as the inner product of those vectors. The process to obtain the optimal mapping from the paired comparison results will also be discussed.

After the optimal mapping is obtained, we move on to the discussion of merit vector estimation from the acoustic features of a new voice. If the estimation scheme is established, we can predict the comparison results for the new voice, and the effectiveness of the whole framework can be confirmed by the correctness of prediction.

15.2 Analysis of Paired Comparison

In this section, we discuss how to model the paired comparison test of voice attractiveness. In the development process of text-to-speech (TTS) systems, voice quality assessment by MOS (Ribeiro, Florêncio, Zhang, & Seltzer, 2011) and paired comparison (Zen, Tokuda, Masuko, Kobayashi, & Kitamura, 2004; Junichi, Onishi, Masuko, & Kobayashi, 2005) test are both used. Although the MOS test has the advantage that the results can be used directly as the absolute attractiveness value, it imposes a heavy burden on listeners and the results are often unreliable. That is the reason why we decided to use the paired comparison test. Below we introduce various models of paired comparison result interpretation.

15.2.1 Universal Attractiveness Model

A straightforward interpretation of paired comparison test is that each voice has own attractiveness and the listener compares two attractiveness values to make a decision. In this paper, such interpretation is referred to as the *universal attractiveness model* because each voice sample is assumed to have one attractiveness value which is applicable to everyone.

The easiest case is that all comparison results are completely consistent. If so, we can obtain at least the order of the voices. Any mapping rule can be acceptable if it satisfies the order. However, such results are rarely obtained, and we need a mapping rule to handle inconsistent results that are unavoidable. From that viewpoint, various analysis methods were applied in various fields, including an example in sports competition (Cattelan, Varin, & Firth, 2013).

A typical approach of probabilistic modeling is based on minimization of the log likelihood function. Assuming that the probability of the voice i's winning against the voice j only depends on the difference of their attractiveness values, the total log likelihood function L can be written as follows.

$$L = \sum_i \sum_j w_{ij} \log f(d_{ij}) \tag{15.1}$$

$$d_{ij} = a_i - a_j \tag{15.2}$$

where a_i and a_j are the attractiveness of the voice i and j, $f(d_{ij})$ is the probability that the voice i wins against the voice j, and w_{ij} is the number of voice i's winning against the voice j.

Historically, there have been two major models of $f(d_{ij})$. In the Bradley–Terry model (Bradley & Terry, 1952), two voices behave as though competing for the shared resource and the probability represents the ratio of one's resource over the other.

$$f(d_{ij}) = \frac{e^{a_i}}{e^{a_i} + e^{a_j}} = \frac{1}{1 + e^{-d_{ij}}} \tag{15.3}$$

In the Thurstone-Mosteller Case V model (Mosteller, 1951), the observation probability of each voice is assumed to be a Gaussian whose mean corresponds to its own attractiveness. The probability that the voice i wins against the voice j is equal to the probability that the observation of voice i is larger than the observation of voice j, which is calculated by

$$f(d_{ij}) = \frac{1}{2}(1 + \mathrm{erf}(d_{ij})) \tag{15.4}$$

where erf represents the error function.

In both cases, a scaling factor can be multiplied to d_{ij}. A larger scaling factor induces less frequent "upset," in which a less attractive voice wins against a more attractive voice. However, it was omitted in the definition of $f(d_{ij})$ because the attractiveness a_i itself has the freedom of scale.

If the definition of $f(d_{ij})$ is fixed, it is easy to obtain the optimal set of $\{a_i\}$ for a given result of paired comparison. We start with randomly selected initial values of $\{a_i\}$, and update them on the direction of gradient ascent of L.

$$\frac{\partial L}{\partial a_i} = \sum_{j \neq i} f'(d_{ij}) \left(\frac{w_{ij}}{f(d_{ij})} - \frac{w_{ji}}{1 - f(d_{ij})} \right) \tag{15.5}$$

15.2.2 Personalized Attractiveness Model

In paired comparison test, two listeners may have opposing opinions for a given pair. In the universal attractiveness model, such event is interpreted simply as an occurrence of less likely event. However, in our social experience, we would react to such situations by saying "tastes differ." However, we also feel that there are some voices which many people tend to like. These two aspects of voice attractiveness—listener's preference and voice likability—can be modeled by defining voice attractiveness as a function of voice-originated character and listener-originated character. Such relationship can be visualized by mapping voices onto a multidimensional merit space, in which the voices are given as points and the preferences are given as directions.

There have been some studies proposing multidimensional extension of Bradley–Terry model. Fujimoto, Hino, and Murata (2009) proposed a mixture model and applied it as a visualization method to the movie rating task. The idea of calculating the inner product between the voice-originated vector and listener-originated vector is an extension of Fujimoto's model. Another example if the work of Causeur and Husson (2005), in which a two-dimensional model representing ranking and relevance axes is proposed and applied to the consumer's preference of cornflakes.

In the proposed personalized attractiveness model, the likelihood function has the same form as the universal attractiveness model, but an additional parameter of listener's index k is introduced.

$$L = \sum_i \sum_j \sum_k w_{ijk} \log f(d_{ijk}) \tag{15.6}$$

where w_{ijk} is the number of times the listener k prefers the voice i to the voice j. The fact that d_{ijk} includes the listener index k means that the attractiveness a_{ik} depends on the lister k. A simple model to define the listener-dependent attractiveness is

$$d_{ijk} = a_{ik} - a_{jk} \tag{15.7}$$

$$a_{ik} = \mathbf{p}_k \cdot \mathbf{m}_i \tag{15.8}$$

where $\mathbf{p}_k(|\mathbf{p}_k| = 1)$ is the preference vector for the listener k and $\mathbf{m}_i$ is the merit vector intrinsic for the voice i.

The process to obtain the optimal set of preference vectors and merit vectors is similar to the case of universal attractiveness model. We start with randomly selected initial values of $\{\mathbf{p}_k\}$ and $\{\mathbf{m}_i\}$, and update them on the direction of gradient ascent of L.

In the 2-dimensional case, we can describe the preference and merit vectors as $\mathbf{p}_k = (\cos\theta_k, \sin\theta_k)^T$ and $\mathbf{m}_i = (\xi_i, \eta_i)^T$, where θ_k, ξ_i, and η_i are the parameters to be adjusted. The parameter θ represents the preference angle. Two parameters ξ an η are interchangeable, and represent the elements of merit vector. The differentiation of L in terms of those parameters are calculated as follows.

$$\frac{\partial L}{\partial \xi_i} = \sum_{j \neq i} \sum_{k} f'(d_{ijk}) \left(\frac{w_{ijk}}{f(d_{ijk})} - \frac{w_{jik}}{1 - f(d_{ijk})} \right) \cos\theta_k \tag{15.9}$$

$$\frac{\partial L}{\partial \eta_i} = \sum_{j \neq i} \sum_{k} f'(d_{ijk}) \left(\frac{w_{ijk}}{f(d_{ijk})} - \frac{w_{jik}}{1 - f(d_{ijk})} \right) \sin\theta_k \tag{15.10}$$

$$\frac{\partial L}{\partial \theta_k} = \sum_{i} \sum_{j \neq i} f'(d_{ijk}) \left(\frac{w_{ijk}}{f(d_{ijk})} - \frac{w_{jik}}{1 - f(d_{ijk})} \right) r_{ijk} \tag{15.11}$$

$$d_{ijk} = (\xi_i - \xi_j)\cos\theta_k + (\eta_i - \eta_j)\sin\theta_k \tag{15.12}$$

$$r_{ijk} = (\eta_i - \eta_j)\cos\theta_k - (\xi_i - \xi_j)\sin\theta_k \tag{15.13}$$

Additional restriction is applied to constrain the vectors in the unit square.

$$0 \le \xi_i \le 1 \tag{15.14}$$

$$0 \le \eta_i \le 1 \tag{15.15}$$

$$0 \le \theta_k \le \pi/2 \tag{15.16}$$

Using above equations, the optimization procedure can be described by the pseudocode shown in Fig. 15.3. Since the quality of solution strongly depends on the initial values, the procedure is repeated using various initial values, and the best combination is selected as the final solution.

The process described above can be extended to the higher dimensional cases easily. In the N-dimensional space, we assume

$$\begin{aligned}
\mathbf{m}_i &= [\xi_{1i}, \xi_{2i}, \ldots, \xi_{Ni}] \\
\mathbf{p}_k &= [\sin\theta_{1k}\sin\theta_{2k}\cdots\sin\theta_{N-1,k}\cos\theta_{Nk}, \\
&\quad \sin\theta_{1k}\sin\theta_{2k}\cdots\sin\theta_{Nk}, \\
&\quad \sin\theta_{1k}\sin\theta_{2k}\cdots\sin\theta_{N-2,k}\cos\theta_{N-1,k}, \\
&\quad \vdots \\
&\quad \sin\theta_{1k}\cos\theta_{2k},
\end{aligned} \tag{15.17}$$

Fig. 15.3 Pseudocode for log likelihood maximization in the two-dimensional merit space

set small step value of s
repeat
  initialize $\{\xi_i\}$, $\{\eta_i\}$, $\{\theta_k\}$ randomly
  repeat
    for all i, k **do**
      calculate $\frac{\partial L}{\partial \xi_i}$, $\frac{\partial L}{\partial \xi_i}$, $\frac{\partial L}{\partial \xi_i}$ using (15.9),(15.10),(15.11)
    end for
    for all i, k **do**
      $\xi_i \leftarrow \xi_i + \frac{\partial L}{\partial \xi_i} s$
      $\xi_i \leftarrow \max(\min(\xi_i, 1), 0)$
      $\eta_i \leftarrow \eta_i + \frac{\partial L}{\partial \eta_i} s$
      $\eta_i \leftarrow \max(\min(\eta_i, 1), 0)$
      $\theta_k \leftarrow \theta_k + \frac{\partial L}{\partial \theta_k} s$
      $\theta_k \leftarrow \max(\min(\theta_k, \frac{\pi}{2}), 0)$
    end for
  until converge
  calculate L using (15.6) and store $\{\xi_i\}$, $\{\eta_i\}$, $\{\theta_k\}$, L
until N times
return $\{\xi_i\}$, $\{\eta_i\}$, $\{\theta_k\}$ that yielded the largest L

$$\cos\theta_{1k}] \tag{15.18}$$

and obtain the update rule, which is a straightforward extension of Eqs. (15.9)–(15.13).

Finally, we selected the Thurstone-Mosteller Case V model (15.4) for f, and the derivative is given by

$$f'(d_{ijk}) = \frac{1}{\sqrt{\pi}} e^{-d_{ijk}^2} \tag{15.19}$$

In fact, the factor $1/\sqrt{\pi}$ can be omitted because the step s has the freedom of arbitrary scale.

15.3 Estimating Merits from Acoustic Features

There have been many studies that tried to connect the subjective nature of voices and their physical characteristics. The largest field is emotion recognition from speech. Various acoustic features and machine learning techniques were applied to predict the emotional state of speaking person, and the achievements were compared in challenges (Ringeval et al., 2017; Schuller et al., 2017). Early researches focused on the prosodic features such as F0 and loudness (Tato, Santos, Kompe, & Pardo, 2002), but the cepstral features were also found to be effective (Sato & Obuchi, 2007). In recent years, it is common to prepare many features and apply machine learning algorithms to find the best feature set. The success of those studies encouraged us to connect the voice attractiveness in the multidimensional merit space and the acoustic features using the machine learning framework.

We start the analysis by preparing a redundant set of acoustic features. Those features are extracted using **OpenSMILE** (Eyben, Weninger, Groß, & Schuller, 2013) and **Julius** (Lee & Kawahara, 2009).

OpenSMILE is a multi-purpose feature extractor from audio signal. It divides the input voice into 25 ms overlapping frames with 10 ms frame interval, and extract various low-level descriptors (LLDs) including energy, pitch, and spectral centroid. Those LLDs are accumulated from all frames, and then various interframe features (functionals) are extracted from each type of LLD. As shown in Tables 15.1, we prepared 14 LLDs related to energy, pitch, and spectral features, and 23 functionals related to extremes, regression, and segment. The total number of features extracted by OpenSMILE is 322.

Julius is an open-source speech recognition engine. We assume that the transcription of voice is given, and Julius is used as the forced-alignment tool. The features provided by Julius include the acoustic model score, total utterance length, and the length of the final phoneme (mostly vowels in Japanese). The first feature indicates how typically the utterance was pronounced, because it represents the distance between the utterance and the standard acoustic model. The second feature indicates how fast the utterance was pronounced. The third feature represents the hesitation, which is frequently observed in Japanese conversation.

Starting with 325 baseline features (322 from OpenSMILE and 3 from Julius), we try to reduce the number of features using the backward stepwise selection (BSS) framework. For any set of features, candidate subsets are made by removing single feature, and each subset is evaluated by cross validation. After evaluating all subsets, the subset with the highest score survives as the set for next step. The same procedure is repeated until only one feature remains. We also tried forward stepwise selection (FSS) in which a null set is prepared as the baseline, and candidate features are added

Table 15.1 List of LLDs and functionals. Linreg stands for linear regression, qreg for quadratic regression, and seglen for segment length

LLDs		Functionals		
Energy/Pitch	Spectral	Extremes	Regression	Segment
RMS energy	Max position	Max	Linreg slope	Number of seg
Log energy	Min position	Min	Linreg offset	Seglen mean
F0	Centroid	Range	Linreg linear error	Seglen max
Voicing prob.	Entropy	Max position	Linreg quadratic error	Seglen min
	Variance	Min position	Qreg coef 1	Seglen std. dev.
	Skewness	Mean	Qreg coef 2	
	Kurtosis	Max−mean	Qreg coef 3	
	Slope	Mean−min	Qreg linear error	
	Harmonicity		Qreg quadratic error	
	Sharpness		Qreg contour centroid	

step by step. However, FSS achieves much worse results than BSS, so the detailed investigation was done with BSS only.

Prediction of the multidimensional merit values are realized by the regression algorithm called SMOreg (Shevade, Keerthi, Bhattacharyya, & Murthy, 2000, which is an extension of support vector machine algorithm. We use **WEKA** (Hall et al., 2009), which provides various machine learning algorithms including SMOreg.

15.4 Experimental Results

In this section, two important issues are examined by the experiments using a real database. The first issue is how efficient mappings of voices can be obtained by the optimization process of personalized attractiveness model. The second issue is how accurately those mappings can be reproduced from the unknown voice using acoustic features.

15.4.1 Recordings and Comparisons

For the experiments, we recorded voices of Japanese greeting "irasshaimase (welcome)" uttered by 115 university students. They were recorded using Panasonic RR-XS355 digital voice recorders with 44.1 kHz sampling rate, stereo recording and 16-bit quantization condition. Since "irasshaimase" is the phrase given by the shop clerk every time a customer comes in, it is uttered very frequently in commercial situations and its impression is very important for the business. The recording was done in a typical classroom situation, in which voluntary students with no payment were asked to say "irasshaimase" one by one. No instruction was given as for the speaking style. Silence was not kept during the recording and the recorded voices include some environmental noises.

In the feature extraction process, OpenSMILE version 2.3.0 rc1 and Julius version 4.2 grammar kit were used. OpenSMILE used the recorded data as their original format, and Julius used the converted version to 16 kHz monaural sampling. The original Japanese acoustic model delivered with the Julius main program was used.

Eighteen listeners participated in comparison experiments. Since we used a browser-based comparison system equipped with anonymous login function, gender and age distribution of the listeners are not available. However, we assume that the majority are in their twenties and there are more male listeners than female listeners. Each listener was given 38 or 39 sets of triplet voices, and asked to choose the most attractive one. The sets were made randomly. We used triplet comparison instead of paired comparison simply because we can obtain more comparison results with smaller number of trials, although we understand that it is controversial whether triplet comparison provides as reliable results as paired comparison. A single triplet comparison result was interpreted as two paired comparison results. If voice A was

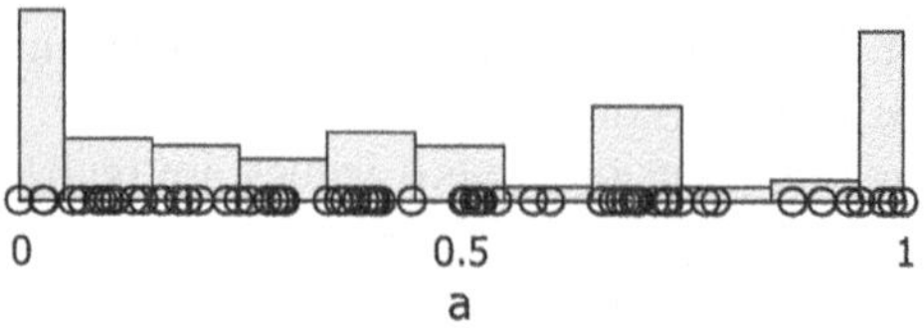

Fig. 15.4 Optimal mapping of 1-dimensional merit space. The variable a is the 1-dimensional merit value. Since there are many circles that are completely overlapped, bars were added to show the histogram. There are 26 voice mapped to a = 0 and 22 voices mapped to a = 1

chosen from the triplet {A, B, C}, it was interpreted that A won in the comparisons {A, B} and {A, C}. Finally, we collected 1,388 paired comparison results (76 or 78 randomly chosen comparisons for each listener) over 115 voice samples.

15.4.2 Mapping to Multidimensional Merit Space

First, we confirmed the effectiveness of mapping to 1-dimensional merit space, which corresponds to the universal attractiveness model. The likelihood function L of Eq. (15.1) was minimized in terms of 115 scaler values $\{a_1, a_2, \ldots, a_{115}\}$ using Eq. (15.5).

Figure 15.4 shows the obtained mapping. Using the attractiveness values shown in Fig. 15.4, we can calculate which voice deserves a win for each comparison. Accordingly, the human judgments are categorized into anticipated or surprising. The mapping efficiency is defined by the ratio of anticipated judgments.[1]

A common metric for such efficiency is called "Kendall rank correlation coefficient," or "Kendall's τ" in short. To deal with the incomplete comparison with ties, we modify "Kendall's τ_b" as

$$\tau_b = \frac{N_C - N_D}{\sqrt{N_C + N_D + N_T}\sqrt{N_C + N_D}} \qquad (15.20)$$

where N_C is the number of concordant (anticipated) pairs, N_D is the number of discordant (surprising) pairs, and N_T is the number of tied pairs in which two voices have the same attractiveness. τ_b becomes 1 if all comparisons are concordant and -1 if all comparisons are discordant. In the case of Fig. 15.4, τ_b was 0.622.

Next, the same procedure was applied to the higher dimensional cases. The procedure in 2-d was described in Fig. 15.3. In the cases with higher dimension, it was extended in a natural manner. In each case, we repeated the update 80 times with random initialization, but they converged to several mappings only.

[1] It is similar to the athletes' ranking. If the high-ranked player always wins, the ranking is efficient. If there are many upsets in which the low-ranked player wins, the ranking is not efficient.

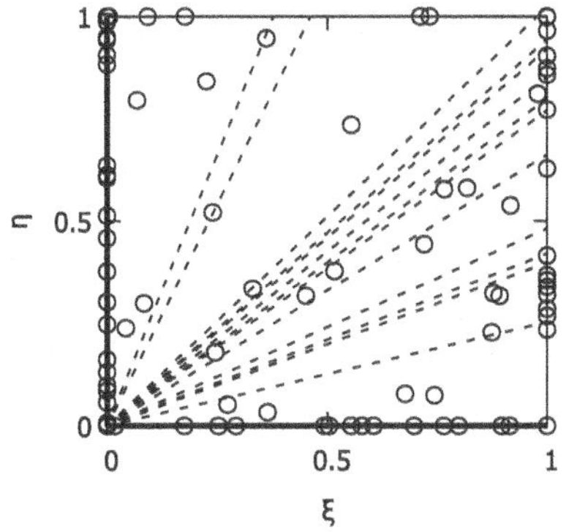

Fig. 15.5 Optimal mapping of 2-dimensional merit space. The voices are represented by circles. The listeners are represented by dashed lines. There are two listeners with $\theta_k = 0$ (x-axis) and three listeners with $\theta_k = \pi/2$ (y-axis)

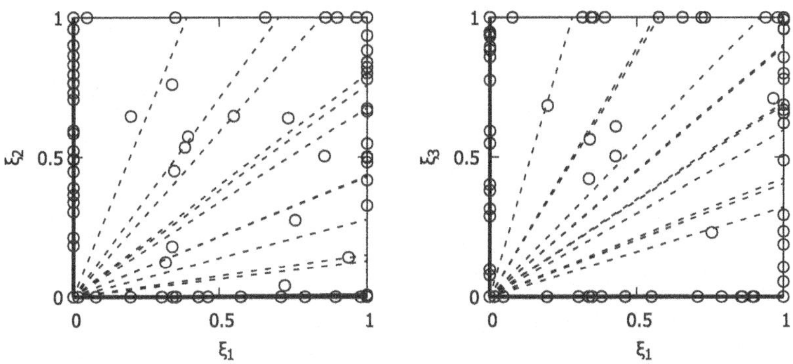

Fig. 15.6 Optimal mapping of 3-dimensional merit space. The left plot shows the first and second dimension, and the right plot shows the first and third dimension. There is one listener with $\theta_{1k} = 0$, three listeners with $\theta_{1k} = \pi/2$, four listeners with $\theta_{2k} = 0$, and two listeners with $\theta_{2k} = \pi/2$

Figures 15.5 and 15.6 are the optimal mapping in 2-d and 3-d cases. It can be seen that many voices have either 0 or 1 as an element of **m**, meaning that the goodness or badness in terms of specific viewpoint is judged unanimously. The voices located on the right-top corner are perfectly attractive voices for everyone. The voices on the left-bottom are perfectly unattractive for everyone. There are 16 perfectly attractive and 15 perfectly unattractive voices in 2-d mapping, and 8 perfectly attractive and 7 unattractive voices in 3-d mapping.

Table 15.2 shows τ_b values in the mapping in various dimensions up to eight. Since the larger number of free parameters have more power to solve inconsistency of comparison results, it is natural that τ_b increases as the larger dimension is introduced. However, it can be noticed that τ_b seems to saturate at around $D = 5$.

Table 15.2 Values of τ_b calculated from the optimal map

Dim	1	2	3	4	5	6	7	8
N_C	1068	1143	1200	1237	1306	1262	1309	1310
N_D	232	186	159	132	79	121	71	76
N_T	88	59	29	19	3	5	8	2
τ_b	0.622	0.705	0.758	0.802	0.885	0.824	0.895	0.890

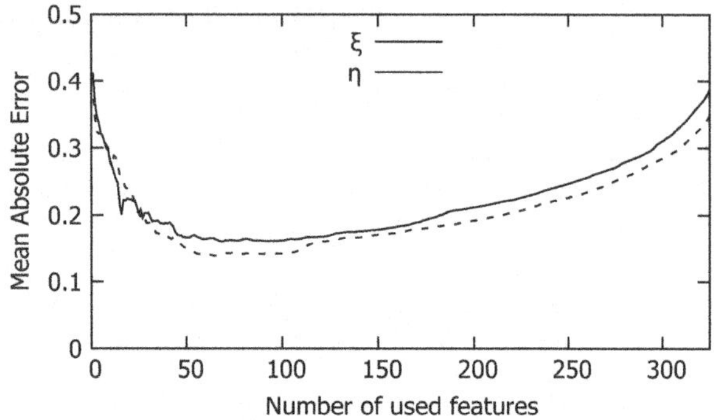

Fig. 15.7 Results of backward stepwise selection (BSS). The experiments started with 325 features (far right), and proceeded to the left

15.4.3 Merit Estimation from Acoustic Features

After confirming the effectiveness of multidimensional mapping, our concern shifted to the relationship between the merit space and the acoustic features. In particular, the most interesting question would be whether we can predict $\mathbf{m}_i$ from the voice itself. If $\mathbf{m}_i$ is predictable, we can predict the voice attractiveness at least for the known listeners whose preference vector $\mathbf{p}_k$ is given.

Since the size of our database is not large enough for two-stage (optimal mapping and merit estimation) fully open condition experiments, we conducted evaluation experiments under a semi-open condition. The optimal mapping in multidimensional merit space was obtained using all data. However, after the merit values for all voices are fixed, the predictability of those merit values from the acoustic features was evaluated using WEKA version 3.6.13 under an open condition using tenfold cross validation. As mentioned before, we started the experiments using all of the 325 features. SMOreg estimator with the second-order polynomial kernel was used. BSS was carried out using the criteria of mean absolute error between real and predicted values.

Figure 15.7 shows how BSS reduced the mean square error when the number of used features changed in the 2-dimensional case. In the case of ξ, the error drops from 0.39 with all features to 0.16 with 68 carefully selected features. The error of η drops from 0.35 with all features to 0.14 with 66 features. Using the predicted values of ξ and η in cross-validation for all 115 voices, we obtained the estimation map shown in Fig. 15.8.

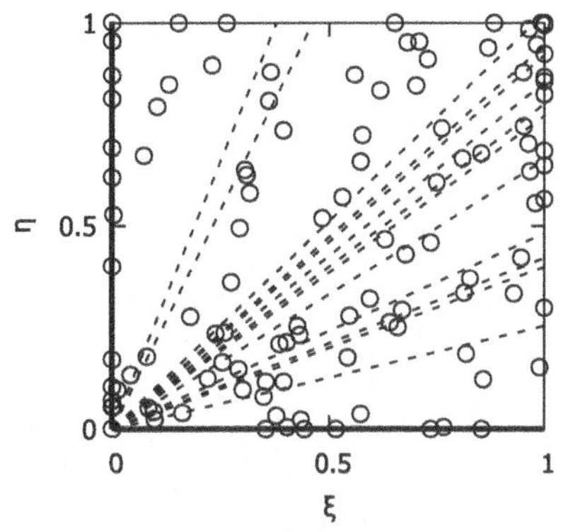

Fig. 15.8 Optimal mapping of 2-dimensional merit space. The voices are represented by circles. The listeners are represented by dashed lines. There are two listeners with $\theta_k = 0$ (x-axis) and three listeners with $\theta_k = \pi/2$ (y-axis)

Table 15.3 Values of τ_b calculated from the estimated mapping

Dim	1	2	3	4	5	6	7	8
N_C	1057	1127	1175	1206	1221	1194	1243	1252
N_D	320	260	211	182	167	194	144	136
N_T	11	1	2	0	0	0	1	0
τ_b	0.533	0.625	0.695	0.738	0.759	0.720	0.792	0.804

It is straightforward to predict paired comparison results from the mapping of Fig. 15.8. The efficiency of prediction is quantified by τ_b. Among 1,388 comparisons, we obtained 1,127 concordant pairs, 260 discordant pairs, and 1 tie prediction. The value of τ_b was calculated as 0.625, which is slightly better than τ_b obtained with 1-dimensional optimal mapping.

After all, we carried out experiments in various dimensions, and obtained τ_b values of estimated attractiveness as shown in Table 15.3. Since the mapping itself becomes more powerful as the dimension increases, the value of τ_b also increases as the higer dimension is introduced. The tendency that the efficiency saturates at around $D = 5$ does not change.

15.5 Conclusions

In this chapter, a multidimensional mapping scheme of voice attractiveness was proposed. Intrinsic attractiveness of voice samples are represented as vectors in the merit space, and listener-dependent preferences are represented as directions in the same space. The attractiveness of a voice for a listener is calculated as the inner product of these two vectors. This mapping scheme provides a better-fit model for the comparison results to which the universal attractiveness model assigned small likelihood values.

The effectiveness of the proposed model was confirmed by the experiments using real voices and their attractiveness judgments. The multidimensional mapping scheme achieves the higher likelihood for the Thurstone-Mosteller model, and better prediction of comparisons.

We also tried to predict the merit values of a new voice sample from its acoustic features. If we can do so, we can predict the comparison result at least for the known listener. A set of machine learning-based experiments confirmed the feasibility of such prediction. If we use four or higher dimension merit space, more than 1,200 of 1,388 paired comparisons can be predicted correctly.

The proposed scheme can be applied to select attractive voice samples for commercial applications. Moreover, it can also be applied to the development process of TTS systems. Since it is easier to synthesize various voices than to prepare a large set of recorded voices, a TTS-based system can speak with the tailor-made voice for the customer.

Although the results presented in this chapter are promising, there are three important problems to solve. First, the experiments presented in this chapter are not fully open. In a sense, the optimization process of multidimensional mapping and feature selection are optimized using the evaluation data. If such optimization tends to overfit the training data, we would have less accurate results with completely new data, especially in the higher dimension cases. To avoid that problem, we need more data and more experiments under the fully open condition.

The second problem is that the paired comparison data were collected with only small number of listeners. Due to the limited data size, it remains an open question whether the model trained in a certain listener group is transferable to another listener group. In addition to the data size problem, anonymousness of the listeners made it impossible to analyze the age and gender dependency of the preference.

The third problem is that all results in this chapter were obtained for just one phrase "irasshaimase." Although it is a very important phrase in the commercial context, we may have something different if we use different phrases. However, the methodology to deal with the merit space and acoustic features is applicable to any phrase, and that is the most important achievement of the work described in this chapter.

References

Bradley, R. A., & Terry, M. E. (1952). Rank analysis of incomplete block designs: I. The method of paired comparisons. *Biometrika, 39*(3–4), 324–345.

Cattelan, M., Varin, C., & Firth, D. (2013). Dynamic Bradley-Terry modelling of sports tournaments. *Journal of the Royal Statistical Society: Series C (Applied Statistics), 62*(1), 135–150.

Causeur, D., & Husson, F. (2005). A 2-dimensional extension of the Bradley-Terry model for paired comparisons. *Journal of Statistical Planning and Inference, 135*, 245–259.

Eyben, F., Weninger, F., Groß, F., & Schuller, B. (2013). Recent developments in openSMILE, the Munich open-source multimedia feature extractor. In *Proceedings of ACM Multimedia (MM), Barcelona, Spain* (pp. 835–838).

Fujimoto, Y., Hino, H., & Murata, N. (2009). Item-user preference mapping with mixture models—Data visualization for item preference. In *Proceedings of International Conference on Knowledge Discovery and Information Retrieval* (pp. 105–111).

Hall, M., Frank, E., Holmes, G., Pfahringer, B., Reutemann, P., & Witten, I. H. (2009). The WEKA data mining software: an update. *SIGKDD Explorations 11*(1), 10–18.

Junichi, Y., Onishi, K., Masuko, T., & Kobayashi, T. (2005). Acoustic modeling of speaking styles and emotional expressions in HMM-based speech synthesis. *IEICE Transaction on Information and Systems 88*(3), 502–509.

Lee, A., & Kawahara, T. (2009). Recent development of open-source speech recognition engine Julius. In *Proceedings of APSIPA Annual Summit and Conference, Sapporo, Japan* (pp. 1–7).

Mosteller, F. (1951). Remarks on the method of paired comparisons: I. The least squares solution assuming equal standard deviations and equal correlations. *Psychometrika, 16*(1), 3–9.

Ribeiro, F., Florêncio, D., Zhang, C., & Seltzer, M. (2011). CROWDMOS: An approach for crowd-sourcing mean opinion score studies. In *Proceedings of IEEE International Conference on Acoustics, Speech, and Signal Processing, Prague, Czech Republic* (pp. 1–7).

Ringeval, F. et al. (2017). AVEC2017 Real-life depression, and affect recognition workshop and challenge. In *Proceedings of 7th Annual Workshop on Audio/Visual Emotion Challenge, Mountain View, CA, USA* (pp. 3–9).

Schuller, B. et al. (2017). The Interspeech 2017 computational paralinguistics challenge: addressee, cold & snoring. In *Proceedings of INTERSPEECH 2017, Stockholm, Sweden* (pp. 3442–3446).

Sato, N., & Obuchi, Y. (2007). Emotion recognition using mel-frequency cepstral coefficients. *Journal of Naturan Language Processing 14*(4), 83–96.

Shah, N. B., Balakrishnan, S., Bradley, J., Parekh, A., Ramchandran, K. & Wainwright, M. (2014). When is it better to compare than to score? CoRR abs/1406.6618.

Shevade, S. K., Keerthi, S. S., Bhattacharyya, C., & Murthy, K. (2000). Improvements to the SMO algorithm for SVM regression. *IEEE Transaction on Neural Networks 11*(5), 1188–1193.

Tato, R., Santos, R., Kompe, R., & Pardo, J. (2002). Emotional space improves emotion recognition. In *Proceedings of 7th International Conference on Spoken Language Processing (ICSPL2002), Denver, USA* (pp. 2029–2032).

Zen, H., Tokuda, K., Masuko, T., Kobayashi, T., & Kitamura, T. (2004). Hidden semi-Markov model based speech synthesis. In *Proceedings of Interspeech 2004, Jeju Island, Korea* (pp. 1393–1396).

Part V
Technological Applications

Chapter 16
Trust in Vocal Human–Robot Interaction: Implications for Robot Voice Design

Ilaria Torre and Laurence White

Abstract Trust is fundamental for successful human interactions. As robots become increasingly active in human society, it is essential to determine what characteristics of robots influence trust in human–robot interaction, in order to design robots with which people feel comfortable interacting. Many interactions are vocal by nature, yet the vocal correlates of trust behaviours have received relatively little attention to date. Here we examine the existing evidence about dimensions of vocal variation that influence trust: voice naturalness, gender, accent, prosody and interaction context. Furthermore, we argue that robot voices should be designed with specific robot appearance, function and task performance in mind, to avoid inducing unrealistic expectations of robot performance in human users.

Keywords Speech · Robots · Voice design · Human-robot interaction · Trustworthiness · Speech prosody

16.1 Introduction

Trust is an essential foundation for human societies. Numerous approaches have been taken towards understanding the means by which it is negotiated. For background, the reader is referred to texts in biology (Bateson, 2000), evolutionary theory (Harcourt, 1991), sociology (Luhmann, 1979), economics (Berg, Dickhaut, & McCabe, 1995) and neuroscience (Bzdok et al., 2011). Here, it will suffice to say that trust relates both to attribution—when someone makes a decision to trust someone else—and to states and traits, when someone acts, in the short term or over the long term, in a trustworthy manner.

I. Torre (✉)
Department of Electronic and Electrical Engineering, Trinity College Dublin, Dublin , Ireland
e-mail: torrei@tcd.ie; ilariat@kth.se

L. White
School of Education, Communication and Language Sciences, Newcastle University, Newcastle, UK
e-mail: laurence.white@ncl.ac.uk

B. Weiss et al. (eds.), *Voice Attractiveness*, Prosody, Phonology and Phonetics,
https://doi.org/10.1007/978-981-15-6627-1_16

Human social evolution has made us very sensitive to cues that may provide information about state or trait trustworthiness in others (e.g. Jones & George, 1998), to the point that a short extract of someone's speech (McAleer, Todorov, & Belin, 2014), or a short exposure to someone's face (Willis, 2006) are enough to make us form a consistent impression of that person's trustworthiness. As robots increasingly become part of our daily lives, it is important to understand what makes people trust robots and, conversely, how we can design robots to appear trustworthy, in order to facilitate human–robot interaction (HRI) and collaboration. While much effort is put into designing robots to look trustworthy and appropriate for their task (e.g. Saldien, Goris, Yilmazyildiz, Werner, & Lefeber, 2008; DiSalvo, Gemperle, Forlizzi, & Kiesler, 2002; Lütkebohle et al., 2010; Oh et al., 2006), less consideration is given to designing voices for these robots (e.g. Sandygulova & O'Hare, 2015). As we argue in this chapter, voice is a very powerful cue used in judgements of trustworthiness, and it should be carefully considered—in conjunction with the robot's appearance/function and with the users' expectations—when designing a robot.

With regard to vocal attractiveness more generally, this chapter considers how characteristics of a robot's voice that contribute to an impression of trustworthiness reinforce, and are reinforced by, features of vocal attractiveness. Dion, Berscheid, & Walster (1972) used the phrase 'what is beautiful is good' to refer to the fact that attractiveness is strongly perceived as related to other positive traits, including that of trustworthiness. Indeed, attractiveness and trustworthiness loaded on the same factor in McAleer et al. (2014)'s study of vocal features. Additional evidence of the close link between attractiveness and trustworthiness comes from neuroscience (Bzdok et al., 2011) and neurobiology Theodoridou, Rowe, Penton-Voak and Rogers (2009), Bzdok et al. (2011), for example, concluded that specific brain regions, such as the amygdala, might selectively reinforce sensory information with high social importance, such as information concerning potential relationships (e.g.: 'Is this person attractive? I might date them in the future.'; 'Is this person trustworthy? I might collaborate with them in the future'.). Here, we examine what characteristics of a robot's voice, by analogy with human voices, contribute to an impression of trustworthiness as a socially relevant cue for human–robot interaction.

16.2 Trust in Voices

Most human communication is predicated on some degree of mutual trust between interlocutors. When we ask a stranger for directions, we trust that they will give us the correct information, to the best of their knowledge (cf, Cooperative Principle, Grice, 1975). Moreover, like the fabled 'boy who cried wolf', untrustworthy communicators tend to be downgraded as interlocutors once their deceitfulness has been exposed.

As the spoken channel is typically our main mode of communication, we have naturally developed vocal means to signal our trustworthiness and to detect it in others. Indeed, the natural tendency to trust speech is mediated by heuristics that give us indicators about when the speaker might not be trustworthy. Not being able to determine a speaker's background can contribute to this impression, as can vocal identifiers of social affiliations disfavoured by the perceiver, or prosodic indicators of aggression or dominance. Conversely, positive evidence for trustworthiness can be inferred from many vocal features, such as accent (e.g. Lev-Ari & Keysar, 2010), prosodic cues (e.g. Miller, Maruyama, Beaber, & Valone, 1976), or emotional expressions (e.g. Schug et al., 2010).

Regarding accents, the literature suggests that foreign accents tend to be trusted less than native accents (Lev-Ari & Keysar, 2010), and that, within a language, 'prestigious' and 'standard' accents are trusted more than regional accents (Giles, 1970). For example, in the context of the UK, Standard Southern British English (SSBE) is generally evaluated as more trustworthy than, for example, typical London or Birmingham accents (Bishop, Coupland, & Garrett, 2005). Furthermore, experimental evidence suggests that such first impressions of trustworthiness might persist over time, despite being mediated by experience of a speaker's actual behaviour (Torre, White, & Gosli, 2016).

Results are less conclusive, indeed sometimes contradictory, regarding the direction of influence of various prosodic features on trust attributions. For example, O'Connor and Barclay (2017) found that people had greater trust in higher pitched male and female voices (based on average fundamental frequency, f0). By contrast, Villar, Arciuli, and Paterson (2013), amongst other studies, have found that participants raise their vocal pitch when lying, and Apple (1979) showed that speakers with a high f0 and slow speech rate were rated as 'less truthful'. A fast speaking rate was found to be a feature of charismatic and persuasive speakers (Jiang & Pell, 2017; Chaiken, 1979), but has also been found to detract from charisma in speech (Niebuhr, Brem, Novák-Tót, & Voße, 2016). Finally, higher pitch and slower speech rate predicted greater trusting behaviour in an economic game (Torre et al., 2016). Such variable results might be due to the fact that the studies employed different methods, such as questionnaires or behavioural measures, and looked at different aspects of trust, such as deception, economic trust, voting behaviour, charisma, and so on. They might also reflect quantitative variation in the prosodic features examined in the different studies: it is unlikely that the relationships between trust attributions and, for example, speech rate or pitch range are strictly linear. Additionally, rather than individual vocal characteristics, it is more likely to be a combination of features that determine the perceiver's assessment of trustworthiness, along with how vocal features interact with physical appearance, interaction context and the perceiver's emotional state.

Voice is a powerful medium through which a diversity of speaker-specific indexical information is transmitted and interpreted, and robot voices are likely to be subjected to similar appraisals. Thus, the design of robot voices should be influenced by the nature of the attributions appropriate to the purposes of particular human–robot interactions.

16.3 Trust in Robot Voices

People tend to attribute personality traits to computers and robots as if they were human agents (Nass & Lee, 2001; Nass, Moon, Fogg, Reeves, & Dryer, 1995; Walters, Syrdal, Dautenhahn, Te Boekhorst, & Koay, 2008), and to respond to robots as if they had a personality (Lee, Peng, Jin, & Yan, 2006). Given also that people attribute traits to human speakers based on subtle speech characteristics (e.g. Torre et al., 2016; McAleer et al., 2014), there is reason to assume that voice information will be used to attribute traits—e.g. of trustworthiness—to robots as well. Thus, voice selection should be an integral part of the overall robot's design. Issues to take into consideration are numerous and diverse, the following being just a selection. Should large robots have lower pitched voices than small robots, congruent with anthropomorphic expectations about larger larynxes? Should human-like robots have natural human voices? Should robot voices have regional accents? If so, should these be chosen to reflect the accent of the person with whom they are interacting or, for example, to reflect a stereotyped association between particular voice styles and the specific functions that the robot will perform? The latter approach risks reinforcing stereotypes, but ignoring any considerations of voice-function congruency could be problematic for the naturalness of the interaction.

It seems, however, that relatively little attention is currently paid to how the selection of robot voices in HRI research might affect our interaction with robots. For example, McGinn and Torre (2019) conducted an informal survey of researchers whose paper at the Human–Robot Interaction 2018 conference featured a speaking robot. Specifically, they asked if they chose the voice of their robot and, if so, why. Of the 18 responses received, six had used the Nao robot built-in voice, seven had used a voice generated with a Text-To-Speech system, either because it was freely available or because it was the voice that the robot came with, three pre-recorded the voice using actors, and two simply described what the voice sounded like (e.g. 'androgynous, child-like voice'). In addition, six of these authors specified that they had adjusted the robot voice in terms of pitch or speech rate to increase intelligibility or to elicit the perception of a particular voice age. Only one author mentioned the accent that the voice had, and only one author mentioned looking for a voice that would specifically suit the task that the robot had to carry out in the experiment. About the reasons for their choice, two authors specified that 'it was the only good one' and 'because it was open source'. While 11 mentioned the gender of the robot's voice, only a minority considered other voice characteristics such as prosody or accent, or the context in which the interaction would take place. However, as we show in this chapter, all of these features influence human perception of robots, and should not be neglected.

Studies experimentally manipulating a robot voice in order to measure its effect on users' perceptions and behaviours are relatively scarce, but here we review work in which a robot's voice was manipulated, or where vocal characteristics were considered in the analyses. As trust is related to other positive traits—a typical 'halo effect' Dion et al. (1972)—and as studies examining the effect of robot voices on trust are limited, we evaluate voice-based research in human–robot interaction in general, considering the implications for trustworthiness in particular.

16.3.1 Voice Naturalness

One key aspect of voice that is often taken into account when designing robots is naturalness. While current efforts in the speech technology community are dedicated to creating the most natural-sounding artificial voices, it might not be the case that people actually prefer interacting with a robot or other artificial agent with a perfect natural-sounding voice (Hinds, 2004). For example, Sims et al. (2009) showed that being able to speak with either a synthetic or a natural voice was enough for a robot to be treated as a competent agent: people gave more commands to a robot that had a voice, whether synthetic or natural, and fewer to a robot that communicated with beeps. They hypothesised that people assumed that speechless robots would not understand language, and thus they did not speak either. Within the speaking robot condition, however, participants gave more commands to the synthetic-voiced robot than the natural one: Sims et al., (2009) suggested that participants might have thought that a robot with a human voice was more competent and therefore needed fewer commands. Taking a different perspective, Mitchell et al. (2011) argued that incongruence in the human likeness of a character's face and voice can elicit feelings of eeriness. In contrast, Tamagawa, Watson, Kuo, MacDonald, and Broadbent (2011) argued that, for the sake of clarity and familiarity, people would prefer such an 'incongruent' robot. In Eyssel, Kuchenbrandt, Hegel, and de Ruite (2012), participants were shown a video of a Flobi robot saying: 'it's quarter past three' and were asked to rate the robot in terms of anthropomorphism, likeability, psychological closeness and intentions. The robot had either a natural or a synthetic voice. Interestingly, voice had an effect only on participants' ratings of likeability, with people rating the natural voice higher. On the other hand, in Theodoridou et al. (2009), people implicitly trusted robots with synthetic voices more than those with natural voices when they were behaving trustworthily, but found the opposite effect when the robots were behaving untrustworthily. This also points to the importance of interaction context for robot voice design (Sect. 16.3.5).

More generally, Hegel (2012) did not find strong evidence that the human likeness of a robot's appearance influenced the perception of its social capabilities. If the same were true for the human likeness of robot *voices*, this would argue that voice naturalness might not be critical for creating feelings of trust. However, another factor to take into account when considering naturalness is listening effort: thus, listening to synthetic voices can increase cognitive load relative to natural voices (Simantiraki, Cooke, & King, 2018; Francis & Nusbaum, 2009). In turn, high cognitive load hinders strategic thinking and can lead to trust misplacement, for example, to trusting untrustworthy individuals (Duffy & Smith, 2014; Samson & Kostyszyn, 2015). This suggests that—especially if the robot is meant to sustain an extended vocal communication with a person—it should be given a natural—or high-quality synthetic,—voice, notwithstanding any contradictions with the robot's mechanical looks.

16.3.2 Voice Gender

Talking specifically about trust, research on human–human interaction has not found consistent differences in trust judgments towards men or women (e.g. Nass & Brave, 2005; Chaudhuri, 2007; Boenin & Serra, 2009; Slonim & Guillen, 2010). Given that people's mental models of humanoid social robots are generally similar to human models (Lee et al., 2005; Kiesler & Goetz, 2002), it would be reasonable to expect a lack of overall difference, when it comes to trusting a 'female' or 'male' robot. Indeed, Crowell, Scheutz, Schermerhorn, and Villano (2009) failed to find any difference in how people reacted to a mechanical robot that had either a female or a male voice. Thus, in terms of voice design, the straightforward expectation would be that robots that are designed to look more feminine or masculine should have a voice corresponding to their apparent gender.

The problem of voice gender selection may be further simplified by the fact that many robots are not perceived as having a clearly defined gender. For example, in a recent study (partially described in Theodoridou et al. (2009)), we used a Nao robot with two different natural female voices, with participants interacting with both. At the end of the experiment, a random sample of the 120 participants was asked what gender they thought the robots had. Of the 66 randomly sampled participants, 23 said they thought the robot was always female, 17 always male, 20 did not associate any gender, and 6 associated a different gender to the two robots they played with. This suggests that even a natural female voice does not consistently convey information about the gender of the robot with that voice. Similarly, the majority of participants in Walters et al. (2008) who interacted with a robot that had either a pre-recorded male voice, a pre-recorded female voice, or a synthesised voice, gave either a male or a neutral name to the robot, even when the robot had a female voice.

Thus, it seems that the gender of a robot voice does not necessarily influence whether people will perceive the robot to have the same gender. However, describing a study involving 9–11-year-old children, Sandygulova and O'Hare (2015) suggested that children assigned a gender to a Nao robot on the basis of the voice alone. This was a synthetic male or female voice. However, participants heard all the possible voices in succession with the same robot, and so a contrast effect may have contributed to the gender attribution being based on voice in this case.

While there is no evidence that voice gender influences a positive human–robot interaction, it is possible that it could interact with presumed gender-specific knowledge (e.g. Powers et al., 2005). As discussed later (Sect. 16.3.5), the context in which the interaction takes place might be more important for trustworthy voice design than voice gender as an isolated feature.

16.3.3 Voice Accent

Everyone has an accent. The term 'accent' refers to systematic patterns of realisation of the sounds of a language—phonetic and phonological—that people belonging to certain geographically or socially defined groups tend to have in common (Lippi-Green, 1997). Accents thus provide immediate information about whether or not two interlocutors belong to similar social and/or regional groups, information that we tend to implicitly use in judgements of trustworthiness (e.g. Kinzler et al., 2009). Specifically, in-group membership elicits favourable first impressions, including with robots (Kuchenbrandt, Eyssel, Bobinger, & Neufeld, 2013). Given that every speaker has an accent, and that these accents affect the way we interpret interpersonal communication, should speaking machines have purpose-specific human accents?

There is, unsurprisingly, evidence of straightforward accent preferences in interactions with robots. For example, children based in Ireland showed a preference towards male and female UK English over US English in a Nao robot (Sandygulova & O'Hare, 2015). We can also contribute some survey data regarding overall preferences for robot accents. All the participants of various UK-based studies run over 3 years were asked what accent they would like a robot to have. The question was open-ended, so we re-coded the answers to fit in broad categories (e.g. 'West Country' and 'South West' would both be coded as 'South West'; labels such as 'English', 'British', 'RP' would be coded as 'SSBE'). Figure 16.1 shows these standardised answers from all 503 participants who answered this question. As the figure shows, the majority of respondents answered with 'SSBE', followed by 'Neutral' accent (which in the UK is also likely to mean the non-regional SSBE), followed by 'Irish'. All of the respondents were native British English speakers, with the following self-reported regional identities: southwest England (58%), southeast England (22%), Midlands (8%), Wales (5%), East Anglia (3%), with participants from northeast England, northwest England and Scotland comprising almost all of the remaining 3–4%. As shown in Fig. 16.1, very few people reported a preference for a robot to have a machine-like voice. There were also relatively few preferences for a regional accent reflecting one's own origins: 58% of respondents were from the southwest but only about 5% of all respondents said they would like the robot to have a southwestern accent (which here we use to encompass Bristol, Cornwall, Devon, Plymouth or general South-West).

Preferences for robot accents may well also be influenced by the nature of the interaction, however. For example, research from Andrist (2015) on the Arabic language showed an interaction between accent and behaviour in human–robot interaction (see Sect. 16.3.5): participants believed that robots with the same regional accent as theirs were more credible—when the robots were knowledgeable—than those with a standard accent, whereas robots with standard accents were perceived to be the more credible when the robots had little knowledge. Similar interactions between accents and behaviour are, of course, likely with other languages. For example, Tamagawa et al. (2011) ran two experiments comparing synthesised British, American, and New Zealand English accents. In the first experiment, participants from New

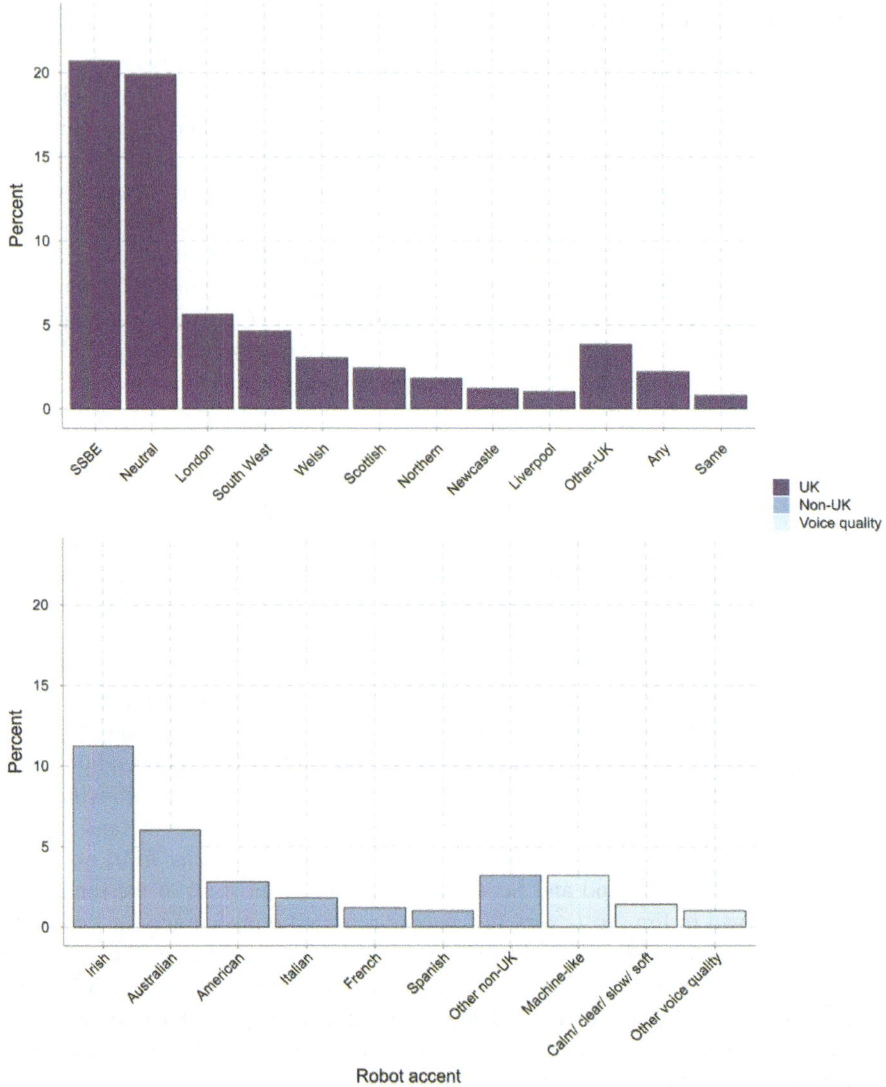

Fig. 16.1 Preference for a robot's accent from a survey of 503 native British English speakers. The question allowed a free response and the bars indicate the overall proportion of all responses that fitted within each accent category (see text for how accent responses were categorised). 'Any' means 'any accent'; 'Same' means 'the same accent as me'

Zealand explicitly rated the disembodied UK accent more positively than the US one, while their own New Zealand accent was not rated significantly differently to either of the other accents. In the second experiment, participants were told by a healthcare robot, in one of the three accents, how to take blood pressure measurements on themselves. After the interaction, they completed a questionnaire, and reported more positive emotions towards the New Zealand-accented robot than the US-accented robot, and thought that the New Zealand robot performed better (no other pairwise comparison was statistically significant). These results also point to the differential effect that accents might have in different interaction contexts (e.g. disembodied voice versus speaking robot).

16.3.4 Voice Prosody

Pitch gives information about a speaker's size, being inversely proportional to the body mass (Ohala, 1983). Thus, it might be straightforward to think that bigger robots should have lower pitched voices than smaller robots. However, as we saw in the discussion on voice gender (Sect. 16.3.2), intuitive assumptions regarding appropriate voices do not necessarily apply in practice, and experimental work is required. In Niculescu, van Dijk, Nijholt, and See (2011), after interacting with a robot with either a high-pitched or a low-pitched female voice, participants' questionnaire responses indicated an overall preference for the higher pitched voice. In another study, Yilmazyildiz et al. (2012) asked participants which of two voices, with lower or higher pitch, was more suitable for a NAO—a child-like humanoid robot—or a Probo—a green furry elephant-like robot: participants preferred the higher pitched voice for NAO and the lower pitched voice for Probo.

Vocal prosody is also a feature that often manifests convergence. Linguistic convergence—sometimes also called adaptation, entrainment or synchrony, although we prefer the specificity of 'convergence'—is a phenomenon by which two speakers tend to unconsciously imitate each other's speech characteristics as interactions proceed (Beňuš, 2014). According to Communication Accommodation Theory (Giles, Coupland, & Coupland, 1991), convergence is a signal of openness and positive attitude—including trust—towards the interlocutor. For example, Manson, Bryant, Gervais, and Kline (2013) showed that people who converged in terms of their speech rate trusted each other more in a Prisoner's Dilemma task. Looking at linguistic convergence more generally, Scissors, Gill, Geraghty, and Gergle (2009) found that some types of linguistic similarity positively correlate with behavioural trust in text-based interaction, while others negatively correlate with it: for example, trusting individuals exhibited convergence in the use of words linked to positive emotions, while deceiving individuals exhibited convergence in the use of negative emotions words.

The well-documented occurrence of convergence phenomenon in human–human interaction led researchers to examine it in human-agent interaction as well. In a computer game where participants followed the advice of an owl-shaped avatar that was either converging or diverging from the participants' own prosody, Beňuš et al. (2018) found that female participants followed the advice of the diverging avatar more often than the converging one, while there was no effect for male participants. Also contrary to some expectations, Strupka, Niebuhr, and Fischer (2016) found that participants tended to diverge prosodically from Keepon robots whose prosody was manipulated. On the other hand, Sadoughi et al. (2017) found that children who played a game with a converging social robot had higher levels of engagement at the end of the interactions than children who played the game with a non-converging robot. The apparent differences in convergence behaviour might be due to several factors, notably whether one is concerned with factors promoting convergence by human speakers towards the vocal features of agents or with the impact of convergence by agents on human behaviours and attitudes. Additionally, there may be an influence of age differences in participants, adults in Benuš et al. (2018) and Strupka et al. (2016), compared with children in Sadoughi et al. (2017): potentially, for example, children may have fewer implicit socio-cognitive biases towards artificial agents. More generally, discrepancies between studies may arise because of intrinsic differences in the artificial voices used. Human speech has been shown to converge with artificial voices in terms of phonetics and prosody when the artificial voice is of high quality, but less so when it is of low quality (Gessinger, Raveh, Le Maguer, Möbius, & Steiner, 2017; Gessinger et al., 2018). Differences could also be due to appearance contrasts between artificial agents: for example, in the studies reported above, there was an owl-shaped avatar in Benuš et al. (2018), a small, rudimentarily humanoid robot in Strupka et al. (2016) and a life-size humanoid robot in Sadoughi et al. (2017). Interactions between robot appearance and convergence behaviours cannot be ruled out.

Prosody conveys important information on the emotional state of the speaker (e.g. Bänziger, & Scherer, 2005; Auberge & Cathiard, 2003). In this regard, it is known that displaying a positive emotion generally leads to attributions of other positive traits—including trustworthiness—a typical ‘halo’ effect (Lau, 1982; Penton-Voak, Pound, Little, & Perrett, 2006; Schug et al., 2010). Indeed, voice-based Embodied Conversational Agents that were smiling were trusted more than those with a neutral facial expression (Elkins, 2013). Smiling in the face also led to trusting avatars and robots more (Krumhuber et al., 2007; Mathur & Reichling, 2016). Thus, a robot expressing positive affect in its prosody could similarly increase the human user’s feeling that it can be trusted. The situation-congruent expression of affect might increase trust even when it is not displaying a positive emotion. For example, portraying stress and urgency through the voice increased performance in a joint human–robot collaborative task (Scheutz et al., 2006).

Apart from signalling a speaker’s mood or emotional state, prosodic cues also contribute to an individual’s vocal profile, that is, what makes a voice unique. Arguably, distinct-looking robots should have different-sounding voices, in order to: (a) contribute to the impression that they are individual agents; (b) be congruent with their physical appearance; (c) elicit personality attributions congruent with the primary functions. In a recent study (partially described in Theodoridou, Rowe, Penton-Voak, and Rogers, (2009), people played a trust game with robots having different voices. We obtained a natural recording of two female SSBE speakers, which we then resynthesised to sound robotic, thus generating four voices altogether: Speaker 1 natural, Speaker 1 synthetic, Speaker 2 natural, Speaker 2 synthetic. As mentioned earlier (Sect. 16.3.1), much of the variance in trust was explained by the voice naturalness variable: specifically, people trusted robots with synthetic voices more than those with natural voices when they were behaving trustworthily, but the opposite when the robots were behaving untrustworthily (Theodoridou et al., 2009). However, people also demonstrated greater implicit trust to one of the two speaker voices over the other, both in natural–natural and synthetic–synthetic comparisons. This is consistent with previous studies showing that very fine speech characteristics, which are independent from higher level features such as accent, affect impression formation (e.g. Gobl & Chasaide, 2003; Trouvain et al., 2006). It also suggests that people's preference for certain individual voices might apply when these voices are embodied in a robot. Thus, idiolectal characteristics, such as those conveyed by prosody, seem to contribute to trusting behaviours as well.

Overall, it seems simplistic to relate trustworthiness judgments purely to isolated vocal features—such as gender, naturalness or pitch—and a holistic view of voice might be better suited for promoting positive interactions, rather than only considering specific individual vocal features.

16.3.5 Voice Context and Expectations

As discussed earlier, some studies have shown that people perceive robots differently depending on the context in which the interaction takes place (Sims et al., 2009; Andrist, 2015). Thus, the nature of the specific human–robot interaction may affect the optimal characteristics of the robot (see also Theodoridou et al., 2009). For example, Wang, Arndt, Singh, Biernat, and Liu (2013) found that, in a favourable context, such as a satisfactory customer/employee call centre interaction, customers with an American English background tended to suppress their negative prejudices towards employees with an Indian English accent. On the other hand, when the interaction was not satisfactory, customers tended not to suppress their accent prejudice (Wang et al., 2013). Similarly, Bresnahan, Ohashi, Nebashi, Liu, and Shearman (2002) examined accent perception as a function of the message that the accented speaker was delivering. They recorded two non-native speakers of American English, one very intelligible and one not very intelligible, and one native speaker, reading passages in a 'friend' and 'teaching assistant' condition. Participants were undergraduate students of various ethnic origins, but mostly white Americans. They found that the 'friend' context was judged as more attractive and dynamic than the 'teaching assistant' context, in all accent conditions. Also, participants with and a strong ethnic identity regarded the native accent as higher in status, dynamism and attractiveness, while the opposite was found for participants with a weak ethnic identity, who attributed higher status and attractiveness to the not very intelligible foreign accent, as compared to the native one. Thus, not only the interaction context, but also the specific background context of the human interlocutor is likely to influence the interaction success.

In HRI, Salem, Ziadee, and Sakr (2013) found that participants' perception—in terms of politeness, competency, extraversion, perceived warmth and shared reality—

of a receptionist robot differed according to the context of the interaction, which was either goal-oriented or open-ended. By contrast, the variation in the robot's politeness level did not influence participants' perception. Additionally, in the aforementioned study by Sims et al. (2009), participants watched videos of a robot in different scenarios (robot damaged, robot in danger, robot requiring more information, robot has located target, robot has completed task). They found that, for example, participants gave more commands to the robot in the videos where the robot needed assistance, and concluded that a robot's voice should be chosen based on task context. In particular, this would allow for the transmission of pragmatic information which may increase the operation success. For example, in a search and rescue operation, a synthetic voice for a robot might be the appropriate choice, because—while it conveys to the person being rescued that the robot is able to help and may be capable of understanding human speech—the fact that the robot voice is not fully human-like could suggest to its human teammates that their input in the operation is still necessary.

As reviewed above, a robot's voice, along with its appearance, will have an influence on the first impressions of that robot's trustworthiness. Given the role of interaction context, however, these first impressions should be validated over long-term interactions with that robot. In fact, several experiments on trusting behaviour in human–machine interaction showed that incongruency between first impressions of trustworthiness and experience of a speaker's actual trustworthiness can drastically reduce trust (Theodoridou et al., 2009). Thus, if a robot's voice gives the impression that the robot will function well, people might have more negative reactions in the case that the robot's performance does not live up to expectations. If it is expected that a robot will operate with some degree of error, perhaps its design (appearance, voice) should reflect the fact that its performance will not always be flawless, so as not to set the users' expectations too high from the beginning (Van den Brule, Dotsch, Bijlstra, Wigboldus, & Haselager, 2014). For example, Hegel (2012) found that people attributed higher social capabilities, including honesty, to robots that looked more sophisticated. Whether robots can deliver on their promise of sophisticated performance is a different matter, however, and over-reliance on a robot according to positive first impressions could have major negative consequences (Robinette, Li, Allen, Howard, & Wagner, 2016; Hancock et al., 2011; Salem, Lakatos, Amirabdollahian, & Dautenhahn, 2015).

Emotional expression might also elicit different trusting behaviours depending on the interaction context. Van Kleef, De Dreu, and Manstead, (2010), in the 'Emotions as Social Information' (EASI) model, suggest that emotions are used to make sense of ambiguous situations, and that their effect depends on the situation in which the interaction takes place, being specifically mediated by its cooperative or competitive nature. Thus, displaying a positive emotion, such as happiness, in a cooperative context will reinforce the parties' belief that everyone is gaining, and will elicit more cooperative behaviours. By contrast, displaying a negative emotion, such as anger, in a cooperative context will hinder future cooperative behaviours. Accordingly, Antos (2011) found that, in a negotiation game, participants tended to select as partners those computer agents which displayed emotions congruent with their actions. Those agents were also perceived as more trustworthy than agents whose emotional expression and action strategy did not match, even though the strategy itself was the same. In summary, emotional expression is helpful only if it is congruent with behaviour.

16.4 Conclusion

This chapter offers an overview of some of the aspects to consider when designing a trustworthy voice to be used in human–robot interaction. Given that many studies in HRI employing a speaking robot have not carefully considered their robot's voice, the present work aims to be a starting point for subsequent research involving speaking robots.

In particular, we summarised work on the effect that voice naturalness, gender, accent, and prosody can have on trust attributions in human–robot interactions, along with the interactions of such vocal features with the characteristics and demands of the specific human–robot encounter. Naturalness, accent, and prosody seem to be the features with the highest likelihood of shaping trusting behaviour, while voice gender appears secondary. Moreover, carefully controlling for context might be more important than, for example, manipulations of naturalness in the voice: specifically, successful interactions over time may be hindered by inaccurate user expectations arising from mismatches between robot's voice features and robot's competence and performance.

It is possible that voice has been a secondary concern in human–robot interaction research so far because vocal interactions have often been scripted, or generated by an imperfect dialogue system, meaning that other aspects of the interaction, such as the robot's movements or attention, might have been prioritised. However, recent advances in the field of natural language and speech processing (such as WaveNet) mean that fluent autonomous human–robot conversations are getting closer to being commonplace. It is time to consider more carefully what the robot's input into these conversations should actually sound like.

Acknowledgments The first author is funded by the European Union's Horizon 2020 research and innovation programme under the Marie Skłodowska–Curie grant agreement No. 713567. She is also funded by the ADAPT Centre for Digital Content Technology, which is funded under the SFI Research Centres Programme (Grant 13/RC/2016) and is co-funded by the European Regional Development Fund. We are grateful to all the HRI conference authors who generously took time to reply to the survey questions.

References

Andrist, S., Ziadee, M., Boukaram, H., Mutlu, B., & Sakr, M. (2015). Effects of culture on the credibility of robot speech. In *Proceedings of the Tenth Annual ACM/IEEE International Conference on Human-Robot Interaction—HRI'15* (pp. 157–164). ACM. ACM Press.

Antos, D., De Melo, C., Gratch, J., & Grosz, B. J. (2011). The influence of emotion expression on perceptions of trustworthiness in negotiation. In *Proceedings of the 25th AAAI Conference on Artificial Intelligence*.

Apple, W., Streeter, L. A., & Krauss, R. M. (1979). Effects of pitch and speech rate on personal attributions. *Journal of Personality and Social Psychology, 37*(5), 715–727.

Aubergé, V., & Cathiard, M. (2003). Can we hear the prosody of smile? *Speech Communication, 40*(1–2), 87–97.

Bänziger, T., & Scherer, K. R. (2005). The role of intonation in emotional expressions. *Speech Communication, 46*(3–4), 252–267.

Bateson, P. (2000). The biological evolution of cooperation and trust. In D. Gambetta (Ed.), *Trust: Making and breaking cooperative relations* (pp. 14–30). Oxford: Department of Sociology, University of.

Berg, J., Dickhaut, J., & McCabe, K. (1995). Trust, reciprocity, and social history. *Games and Economic Behavior, 10*(1), 122–142.

Benuš, V. (2014). Social aspects of entrainment in spoken interaction. *Cognitive Computation, 6*(4), 802–813.

Benuš, V., Trnka, M., Kuric, E., Marták, L., Gravano, A., Hirschberg, J., & Levitan, R. (2018). Prosodic entrainment and trust in human-computer interaction. In *Proceedings of the Ninth International Conference on Speech Prosody 2018*. Poznan, Poland: ISCA.

Bishop, H., Coupland, N., & Garrett, P. (2005). Conceptual accent evaluation: Thirty years of accent prejudice in the UK. *Acta Linguistica Hafniensia, 37*(1), 131–154.

Boenin, A., & Serra, D. (2009). Gender pairing bias in trustworthiness. *The Journal of Socio-Economics, 38*, 779–789.

Bresnahan, M. J., Ohashi, R., Nebashi, R., Liu, W. Y., & Shearman, S. M. (2002). Attitudinal and affective response toward accented English. *Language & Communication, 22*(2), 171–185.

Bzdok, D., Langner, R., Caspers, S., Kurth, F., Habel, U., Zilles, K., et al. (2011). ALE meta-analysis on facial judgments of trustworthiness and attractiveness. *Brain Structure and Function, 215*(3–4), 209–223.

Chaiken, S. (1979). Communicator physical attractiveness and persuasion. *Journal of Personality and Social Psychology, 37*(8), 1387.

Chaudhuri, A., & Gangadharan, L. (2007). An experimental analysis of trust and trustworthiness. *Southern Economic Journal*, 959–985.

Crowell, C. R., Scheutz, M., Schermerhorn, P., & Villano, M. (2009). Gendered voice and robot entities: Perceptions and reactions of male and female subjects. In *IEEE/RSJ International Conference on Intelligent Robots and Systems, 2009. IROS 2009* (pp. 3735–3741). IEEE.

DiSalvo, C. F., Gemperle, F., Forlizzi, J., & Kiesler, S. (2002). All robots are not created equal: the design and perception of humanoid robot heads. In *Proceedings of the 4th Conference on Designing Interactive Systems: Processes, Practices, Methods, and Techniques* (pp. 321–326). ACM.

Dion, K., Berscheid, E., & Walster, E. (1972). What is beautiful is good. *Journal of Personality and Social Psychology, 24*(3), 285.

Duffy, S., & Smith, J. (2014). Cognitive load in the multi-player prisoner's dilemma game: Are there brains in games? *Journal of Behavioral and Experimental Economics, 51*, 47–56.

Elkins, A. C., & Derrick, D. C. (2013). The sound of trust: Voice as a measurement of trust during interactions with embodied conversational agents. *Group Decision and Negotiation, 22*(5), 897–913.

Eyssel, F.; Kuchenbrandt, D.; Hegel, F. & de Ruiter, L. (2012). Activating elicited agent knowledge: How robot and user features shape the perception of social robots. In *IEEE International Workshop on Robot and Human Interactive Communication, 2012. ROMAN 2012* (pp. 851–857). IEEE.

Francis, A. L., & Nusbaum, H. C. (2009). Effects of intelligibility on working memory demand for speech perception. *Attention, Perception, & Psychophysics, 71*(6), 1360–1374.

Gessinger, I., Raveh, E., Le Maguer, S., Möbius, B., & Steiner, I. (2017). Shadowing synthesized speech—Segmental analysis of phonetic convergence. *Proceedings of Interspeech, 2017*, 3797–3801.

Gessinger, I., Schweitzer, A., Andreeva, B., Raveh, E., Möbius, B., & Steiner, I. (2018). Convergence of pitch accents in a shadowing task. In *Proceedings of the 9th International Conference on Speech Prosody 2018* (pp. 225–229).

Giles, H. (1970). Evaluative reactions to accents. *Educational Review, 22*(3), 211–227.

Giles, H., Coupland, N., & Coupland, J. (1991). Accommodation theory: Communication, context, and consequence. In H. Giles, N. Coupland, & J. Coupland (Eds.), *Contexts of accommodation: Developments in applied sociolinguistics* (pp. 1–68). Press: Cambridge University.

Gobl, C., & Ní Chasaide, A. (2003). The role of voice quality in communicating emotion, mood and attitude. *Speech Communication, 40*, 189–212.

Grice, H. P. (1975). Logic and conversation. In P. Cole & J. L. Morgan (Eds.), *Syntax and semantics: Speech acts* (vol. 3, pp. 41–58). New York: Academic Press.

Hancock, P. A., Billings, D. R., Schaefer, K. E., Chen, J. Y. C., De Visser, E. J., & Parasuraman, R. (2011). A meta-analysis of factors affecting trust in humanrobot interaction. *Human Factors: The Journal of the Human Factors and Ergonomics Society, 53*(5), 517–527.

Harcourt, A. H. (1991). Help, cooperation and trust in animals. In R. A. Hinde& J. Groebel (Eds.), *Cooperation and Prosocial Behaviour* (pp. 15–26). Cambridge University Press.

Hegel, F. (2012). Effects of a robot's aesthetic design on the attribution of social capabilities. In *IEEE RO-MAN: The 21st IEEE International Symposium on Robot and Human Interactive Communication* (pp. 469–475). IEEE.

Hinds, P. J., Roberts, T. L., & Jones, H. (2004). Whose job is it anyway? A study of human-robot interaction in a collaborative task. *Human-Computer Interaction, 19*(1), 151–181.

Jiang, X., & Pell, M. D. (2017). The sound of confidence and doubt. *Speech Communication, 88*, 106–126.

Jones, G. R., & George, J. M. (1998). The experience and evolution of trust: Implications for cooperation and teamwork. *Academy of Management Review, 23*(3), 531–546.

Kiesler, S., & Goetz, J. (2002). Mental models of robotic assistants. In *Proceedings of the CHI (2002) Conference on Human Factors in Computer Systems*. New York: ACM.

Kinzler, K. D., Shutts, K., DeJesus, J. M., & Spelke, E. S. (2009). Accent trumps race in guiding children's social preferences. *Social Cognition, 27*(4), 623.

Krumhuber, E., Manstead, A. S. R., Cosker, D., Marshall, D., Rosin, P. L., & Kappas, A. (2007). Facial dynamics as indicators of trustworthiness and cooperative behavior. *Emotion, 7*(4), 730–735.

Kuchenbrandt, D., Eyssel, F., Bobinger, S., & Neufeld, M. (2013). When a robot's group membership matters. *International Journal of Social Robotics, 5*(3), 409–417.

Lau, S. (1982). The effect of smiling on person perception. *The Journal of Social Psychology, 117*(1), 63–67.

Lee, K. M., Peng, W., Jin, S., & Yan, C. (2006). Can robots manifest personality? An empirical test of personality recognition, social responses, and social presence in human-robot interaction. *Journal of Communication, 56*(4), 754–772.

Lee, S., Lau, I. Y., Kiesler, S., & Chiu, C. (2005). Human mental models of humanoid robots. In *Proceedings of the 2005 IEEE International Conference on Robotics and Automation* (pp. 2767–2772). IEEE.

Lev-Ari, S., & Keysar, B. (2010). Why don't we believe non-native speakers? The influence of accent on credibility. *Journal of Experimental Social Psychology, 46*(6), 1093–1096.

Lippi-Green, R. (1997). *English with an accent: Language, ideology, and discrimination in the United States*. Psychology Press.

Luhmann, N. (1979). *Trust and power*. Wiley.

Lütkebohle, I., Hegel, F., Schulz, S., Hackel, M., Wrede, B., Wachsmuth, S., & Sagerer, G. (2010). The bielefeld anthropomorphic robot head "Flobi". In *2010 IEEE International Conference on Robotics and Automation (ICRA)* (pp. 3384–3391). IEEE.

Manson, J. H., Bryant, G. A., Gervais, M. M., & Kline, M. A. (2013). Convergence of speech rate in conversation predicts cooperation. *Evolution and Human Behavior*, *34*(6), 419–426.

Mathur, M. B., & Reichling, D. B. (2016). Navigating a social world with robot partners: A quantitative cartography of the Uncanny Valley. *Cognition*, *146*, 22–32.

McAleer, P., Todorov, A., & Belin, P. (2014). How do you say Hello? Personality impressions from brief novel voices. *PLoS ONE*, *9*(3), e90779.

McGinn, C., & Torre, I. (2019). Can you tell the robot by the voice? An exploratory study on the role of voice in the perception of robots. In *Proceedings of the 14th Annual ACM/IEEE International Conference on Human-Robot Interaction—HRI'19*. ACM.

Miller, N., Maruyama, G., Beaber, R. J., & Valone, K. (1976). Speed of speech and persuasion. *Journal of Personality and Social Psychology*, *34*(4), 615.

Mitchell, W. J., Szerszen, K. A., Lu, A. S., Schermerhorn, P. W., Scheutz, M., & MacDorman, K. F. (2011). A mismatch in the human realism of face and voice produces an uncanny valley. In *i-Perception 2.1* (pp. 10–12).

Nass, C. I., & Brave, S. (2005). *Wired for speech: How voice activates and advances the human-computer relationship*. Cambridge, MA: MIT Press.

Nass, C. I., & Lee, K. M. (2001). Does computer-synthesized speech manifest personality? Experimental tests of recognition, similarity-attraction, and consistencyattraction. *Journal of Experimental Psychology: Applied*, *7*(3), 171–181.

Nass, C. I., Moon, Y., Fogg, B. J., Reeves, B., & Dryer, C. (1995). Can computer personalities be human personalities? *International Journal of Human-Computer Studies*, *43*(2), 223–239.

Niculescu, A., van Dijk, B., Nijholt, A., & See, S. L. (2011). The influence of voice pitch on the evaluation of a social robot receptionist. In *2011 International Conference on User Science and Engineering (i-USEr)* (pp. 18–23). Shah Alam, Selangor, Malaysia: IEEE.

Niebuhr, O., Brem, A., Novák-Tót, E., & Voße, J. (2016). Charisma in business speeches—A contrastive acoustic-prosodic analysis of Steve Jobs and Mark Zuckerberg. In *Proceedings of the 8th International Conference on Speech Prosody. Boston, MA, USA*.

O'Connor, J. J. M.& Barclay, P. (2017). The influence of voice pitch on perceptions of trustworthiness across social contexts. In *Evolution and human behavior*.

Oh, J. -H., Hanson, D., Kim, W. -S., Han, Y., Kim, J. -Y., & Park, I. -W. (2006). Design of android type humanoid robot Albert HUBO. In *2006 IEEE/RSJ International Conference on Intelligent Robots and Systems* (pp. 1428–1433). IEEE.

Ohala, J. J. (1983). Cross-language use of pitch: An ethological view. *Phonetica*, *40*, 1–18.

Penton-Voak, I. S., Pound, N., Little, A. C., & Perrett, D. I. (2006). Personality judgments from natural and composite facial images: More evidence for a "Kernel Of Truth" in social perception. *Social Cognition*, *24*(5), 607–640.

Powers, A., Kramer, A. D. I., Lim, S., Kuo, J., Lee, S.-l., & Kiesler, S. (2005). Eliciting information from people with a gendered humanoid robot. In *2005 IEEE International Workshop on Robot and Human Interactive Communication, ROMAN 2005* (pp. 158–163). IEEE.

Robinette, P., Li, W., Allen, R., Howard, A. M., & Wagner, A. R. (2016). Overtrust of robots in emergency evacuation scenarios. In *Proceedings of the 11th Annual ACM/IEEE International Conference on Human-Robot Interaction—HRI '16* (pp. 101–108). IEEE Press.

Sadoughi, N., Pereira, A., Jain, R., Leite, I., & Lehman, J. F. (2017). Creating prosodic synchrony for a robot co-player in a speech-controlled game for children. In *Proceedings of the 12th Annual ACM/IEEE International Conference on Human-Robot Interaction—HRI '17* (pp. 91–99). ACM.

Saldien, J., Goris, K., Yilmazyildiz, S., Werner, V., & Lefeber, D. (2008). On the design of the huggable robot probo. *Journal of Physical Agents*, *2*(2), 3–11.

Salem, M., Ziadee, M., & Sakr, M. (2013). Effects of politeness and interaction context on perception and experience of HRI. In *International Conference on Social Robotics* (pp. 531–541). Springer.

Salem, M., Lakatos, G., Amirabdollahian, F., & Dautenhahn, K. (2015). Would you trust a (faulty) robot?: Effects of error, task type and personality on human robot cooperation and trust. In *Proceedings of the 10th Annual ACM/IEEE International Conference on Human-Robot Interaction—HRI '15* (pp. 141–148). ACM.

Samson, K., & Kostyszyn, P. (2015). Effects of cognitive load on trusting behavior—An experiment using the trust game. *PloS ONE*, *10*(5), e0127680.

Sandygulova, A., & O'Hare, G. M. P. (2015). Children's perception of synthesized voice: Robot's gender, age and accent. In A. Tapus, E. André, J.-C. Martin, F. Ferland & M. Ammi (Eds.), *Social robotics* (pp. 594–602). Springer International Publishing.

Scheutz, M., Schermerhorn, P. W., & Kramer, J. (2006). The utility of affect expression in natural language interactions in joint human-robot tasks. In *Proceedings of the First Annual ACM/IEEE International Conference on Human-Robot Interaction—HRI '06* (pp. 226–233).

Schug, J., Matsumoto, D., Horita, Y., Yamagishi, T., & Bonnet, K. (2010). Emotional expressivity as a signal of cooperation. *Evolution and Human Behavior*, *31*(2), 87–94.

Scissors, L. E., Gill, A. J., Geraghty, K., & Gergle, D. (2009). In CMC we trust: The role of similarity. In *Proceedings of the SIGCHI Conference on Human Factors in Computing Systems* (pp. 527–536). ACM.

Simantiraki, O., Cooke, M., & King, S. (2018). Impact of different speech types on listening effort. In *Proceedings of Interspeech 2018* (pp. 2267–2271). https://doi.org/10.21437/Interspeech.2018-1358.

Sims, V. K., Chin, M. G., Lum, H. C., Upham-Ellis, L., Ballion, T., & Lagattuta, N. C. (2009). Robots' auditory cues are subject to anthropomorphism. In *Proceedings of the Human Factors and Ergonomics Society Annual Meeting* (Vol. 53, pp. 1418–1421). San Antonio, Texas, USA: SAGE Publications.

Slonim, R., & Guillen, P. (2010). Gender selection discrimination: Evidence from a trust game. *Journal of Economic Behavior & Organization*, *76*(2), 385–405.

Strupka, E., Niebuhr, O., & Fischer, K. (2016). Influence of robot gender and speaker gender on prosodic entrainment in HRI. In *2016 IEEE International Workshop on Robot and Human Interactive Communication, ROMAN 2016*. IEEE.

Tamagawa, R., Watson, C. I., Kuo, I. H., MacDonald, B. A., & Broadbent, E. (2011). The effects of synthesized voice accents on user perceptions of robots. *International Journal of Social Robotics*, *3*(3), 253–262.

Theodoridou, A., Rowe, A. C., Penton-Voak, I. S., & Rogers, P. J. (2009). Oxytocin and social perception: Oxytocin increases perceived facial trustworthiness and attractiveness. *Hormones and Behavior*, *56*(1), 128–132.

Torre, I., White, L., & Goslin, J. (2016). Behavioural mediation of prosodic cues to implicit judgements of trustworthiness. In *Proceedings of the 8th International Conference on Speech Prosody 2016*. Boston, MA, USA: ISCA.

Torre, I., Goslin, J., White, L., & Zanatto, D. (2018). Trust in artificial voices: A "congruency effect" of first impressions and behavioural experience." In *Proceedings of APAScience '18: Technology, Mind, and Society (TechMindSociety'18)*. Washington, DC, USA.

Trouvain, J., Schmidt, S., Schröder, M., Schmitz, M., & Barry, W. J. (2006). Modelling personality features by changing prosody in synthetic speech. In *Proceedings of the 3rd International Conference on Speech Prosody, Dresden, Germany*.

Van Kleef, G. A., De Dreu, C. K. W., & Manstead, A. S. R. (2010). An interpersonal approach to emotion in social decision making: The emotions as social information model. *Advances in Experimental Social Psychology*, *42*, 45–96.

Van den Brule, R., Dotsch, R., Bijlstra, G., Wigboldus, D. H. J., & Haselager, P. (2014). Do robot performance and behavioral style affect human trust? *International Journal of Social Robotics*, *6*(4), 519–531.

Villar, G., Arciuli, J., & Paterson, H. (2013). Vocal pitch production during lying: Beliefs about deception matter. *Psychiatry, Psychology and Law*, *20*(1), 123–132.

Walters, M. L., Syrdal, D. S., Dautenhahn, K., Te Boekhorst, R., & Koay, K. L. (2008). Avoiding the uncanny valley: Robot appearance, personality and consistency of behavior in an attention-seeking home scenario for a robot companion. *Autonomous Robots*, *24*(2), 159–178.

Wang, Z., Arndt, A. D., Singh, S. N., Biernat, M., & Liu, F. (2013). "You Lost Me at Hello": How and when accent-based biases are expressed and suppressed. *International Journal of Research in Marketing*, *30*, 185–196.

Willis, J., & Todorov, A. (2006). First impressions: Making up your mind after a 100-ms exposure to a face. *Psychological Science*, *17*(7), 592–598.

Yilmazyildiz, S., Patsis, G., Verhelst, W., Henderickx, D., Soetens, E., Athanasopoulos, G., Sahli, H., Vanderborght, B., & Lefeber, D. (2012). Voice style study for human-friendly robots: Influence of the physical appearance. In: *Proceedings of the 5th International Workshop on Human-Friendly Robotics*.

Chapter 17
Exploring Verbal Uncanny Valley Effects with Vague Language in Computer Speech

Leigh Clark, Abdulmalik Ofemile, and Benjamin R. Cowan

Abstract Interactions with speech interfaces are growing, helped by the advent of intelligent personal assistants like Amazon Alexa and Google Assistant. This software is utilised in hardware such as smart home devices (e.g. Amazon Echo and Google Home), smartphones and vehicles. Given the unprecedented level of spoken interactions with machines, it is important we understand what is considered appropriate, desirable and attractive computer speech. Previous research has suggested that the overuse of humanlike voices in limited-communication devices can induce uncanny valley effects—a perceptual tension arising from mismatched stimuli causing incongruence between users' expectations of a system and its actual capabilities. This chapter explores the possibility of verbal uncanny valley effects in computer speech by utilising the interpersonal linguistic strategies of politeness, relational work and vague language. This work highlights that using these strategies can create perceptual tension and negative experiences due to the conflicting stimuli of computer speech and 'humanlike' language. This tension can be somewhat moderated with more humanlike than robotic voices, though not alleviated completely. Considerations for the design of computer speech and subsequent future research directions are discussed.

Keywords Speech interface · Voice interface · Intelligent personal assistant · Uncanny valley · Humanlike · Politeness · Vague language

L. Clark (✉) · B. R. Cowan
School of Information, & Communication Studies, University College Dublin, Dublin, Ireland
e-mail: leigh.clark@ucd.ie; l.m.h.clark@swansea.ac.uk

B. R. Cowan
e-mail: benjamin.cowan@ucd.ie

L. Clark
Computational Foundry, Swansea University, Swansea, UK

A. Ofemile
English Department, FCT College of Education, Zuba, Abuja, Nigeria
e-mail: abdulmalikkuka@gmail.com

B. Weiss et al. (eds.), *Voice Attractiveness*, Prosody, Phonology and Phonetics,
https://doi.org/10.1007/978-981-15-6627-1_17

17.1 Introduction

As a mode of interaction, speech can affect peoples' perceptions of others in terms of identity, personality, power and attractiveness (Cameron, 2001; Coulthard, 2013; Goffman, 2005; Zuckerman & Driver, 1988). Speech can impact these perceptions in both the language used and the voice quality used to produce it; the latter defined here as 'those characteristics which are present more or less all the time that a person is talking' (Abercrombie, 1967, p. 91 in Laver, 1980, p. 1). As with human–human interaction (HHI), this impact on perceptions can be seen in human–computer interaction (HCI), where speech has become a more prominent mode of interaction. This prominence has been accelerated with the advent of intelligent personal assistants (IPAs) such as Amazon Alexa and Google Assistant featuring in home-based smart speakers like Amazon Echo and Google Home, as well as in mobile devices and vehicles. These are in addition to longer standing speech-based technologies like interactive voice response (IVR) and navigation systems. Although we are beginning to understand more about how people use and communicate with these types of devices (Cowan et al., 2017; Luger & Sellen, 2016; Porcheron, Fischer, Reeves, & Sharples, 2018; Porcheron, Fischer, & Sharples, 2017), less is known about the psychological and behavioural effects of speech interface design choices on users (Clark, Cabral, & Cowan, 2018).

While we are aware that design choices in speech-based HCI can affect user experience (UX) and interaction behaviour, we are still lacking theoretical understandings and subsequent design considerations supporting them (Clark et al., 2019b). Consequently, it is not always clear what linguistic or voice styles may be appropriate, desirable or even attractive to users in HCI. Mimicking aspects of humanness in speech interfaces, for example, may not always be an appropriate design objective and can result in systems being perceived as creepy or even deceitful (Aylett, Cowan, & Clark, 2019). Recent research (Moore, 2017a) has argued that humanlike voices are not always appropriate for non-human artefacts, as they may heighten peoples' expectations of what artefacts are capable of, in contrast to more robotic voices. This heightened perception of humanness can result in a gap between users' perceptions of a system's abilities or *partner models* and the reality of its limitations observed through interaction (Cowan et al., 2017). As well as the quality of a system's voice, there are also less explored questions as to what are considered appropriate styles of language for computer speech, and how humanlike or 'machinelike' they are expected to be Clark (2018), Clark et al., (2019a).

This chapter explores the concepts of three interpersonal linguistic strategies—politeness, relational work and vague language (VL)—as a lens to examine the possibility of *verbal uncanny valley effects* that exist in users' perceptions towards both voice and language in computer speech. This may underpin some of the user behaviour and perceptions of appropriateness, desirability and attractiveness directed towards speech interfaces in previous research, as well as peoples' expectations and partner models of their computer interlocutors. It is hoped that these discussions may drive theoretical understandings of our interactions with speech interfaces, which may in turn encourage design considerations in the field.

17.2 Uncanny Valley

The *uncanny valley* hypothesis suggests that non-human artefacts approaching close to human likeness, but retaining obvious differences from human norms, can induce negative responses from people due to one or more obvious differences from expected human appearance or behaviour (Mori, 1970; Mori, MacDorman, & Kageki, 2012). These responses may be referred to as concepts like eeriness, revulsion, or a sense of unease, signifying perceptions of undesirable or unattractive characteristics. Disfluencies between appearance and motion, for instance, may be more disliked than entities displaying more congruent features—contrasting an android that is humanlike in appearance yet displaying robotic movements with an all human and all robot alternative (Carr, Hofree, Sheldon, Saygin, & Winkielman, 2017).

While empirical evidence for the uncanny valley is somewhat scarce, a review of uncanny valley research papers highlighted support for two perceptual mismatch hypotheses (Kätsyri, Förger, Mäkäräinen, & Takala, 2015). The first of these hypotheses suggests that uncanny valley effects arise due to mismatches between the human likeness of different sensory cues (e.g. obviously non-human eyes on a fully humanlike face). The second hypothesis posits that the effects occur because of a higher sensitivity towards exaggerated features on more humanlike characters that differ from expected humanlike norms (e.g. 'grossly enlarged eyes, Kätsyri et al., 2015, p. 7). Similar explanations for uncanny valley effects are discussed by Moore (2012). In developing a Bayesian explanation for the uncanny valley effect, Moore points to conflicting cues creating a perceptual distortion and subsequent perceptual tension at category boundaries. These categories refer to stimuli that are discriminately perceived as being different from one another. Stimuli perceived to be at the boundaries of these categories may incur more perceptual distortion than those stimuli perceived to be prototypical examples of those categories.

Whereas most uncanny valley research has focused on the visual, there are an increasing number of works that include audio as an additional modality of interest in exploring perceptual mismatches. Grimshaw (2009) discusses the concept of an audio uncanny valley, with the view that further theoretical understandings may be useful for sound design in horror-based computer games in creating perceptions of fear and apprehension. The author provides examples of features that may induce uncanny valley effects, including uncertainty about the location of sound sources and exaggerated articulation of the mouth whilst speaking. Mitchell et al. (2011) and Meah and Moore (2014) explored the concepts of misaligned voice and face cues (or mismatched stimuli) in robots and humans. Both experiments showed that mismatches in voice and face (e.g. robotic voice and human face or human voice and robotic face) result in higher ratings of perceived eeriness than matched stimuli.

These experiments give credence to the uncanny valley existing in audio as well as visual stimuli, although the focus in the above work is on multimodal cues and the audio is primarily centred on the voice quality. With the increasing number of speech interfaces, users are exposed to unprecedented levels of primarily speech-based interactions with machines. However, there remain important design considerations on what is considered appropriate speech output by speech interfaces. Moore (2017a), for example, highlights the proliferation of humanlike rather than more robotic sounding voices in computer speech is not always an appropriate design choice. Using humanlike voices can create mismatches between users' expectations of a machine's capabilities and the reality of what it can achieve through speech. This may result in unsuccessful engagement with speech-based, non-human artefacts. Less is understood as to what may be considered appropriate language in spoken interactions with machines—perceptual mismatches may also occur on a linguistic as well as a voice level, potentially resulting in unwanted negative effects to UX (Clark, 2018). The subsequent sections of this chapter reflect on recent research into the use of interpersonal linguistic strategies in spoken computer instructions and discuss the possible boundaries of appropriate language use (as opposed to solely the appropriate humanlike synthesis choices) in light of uncanny valley theories and mismatched stimuli (Clark, Bachour, Ofemile, Adolphs, & Rodden, 2014; Clark, Ofemile, Adolphs, & Rodden, 2016).

17.3 Politeness and Relational Work

The concept of politeness is often discussed in terms of Brown and Levinson's (1987) work that associates politeness with the concept of *face*—the social self-image that we present to others during interaction (Goffman, 1955). This self-image is dependent on sociocultural and contextual factors and dynamically progresses between and within interactions. Face theory discusses it being in speakers' own interests to avoid damaging the face of oneself or the face of others during interaction. Conducting this is known as *facework*.

In Brown and Levinson's (1987) research, facework can be accomplished using politeness strategies. *Positive face* refers to desires of being liked and approved. Positive politeness strategies include showing group membership between partners, paying attention to the wants and desires of others, and presenting approval. *Negative face* refers mainly to the desire not to be imposed upon by others. Negative politeness strategies often focus on minimising this potential imposition. This can be accomplished by being indirect rather than direct, for example, when issuing instructions or making requests that may create an imbalance of power.

Relational work seeks to expand Brown and Levinson's (1987) politeness theory to include the whole polite–impolite spectrum (Locher, 2004, 2006; Locher & Watts, 2005, 2008). This includes all work by individuals for the 'construction, maintenance, reproduction and transformation of interpersonal relationships among those engaged in social practice' (Locher & Watts, 2008, p. 96). As with facework and politeness described above, relational work is similarly discursive and on-going (Locher & Watts, 2005, 2008; Watts, 2003).

17.3.1 Politeness in Machines

While there are disagreements in politeness and relational work, the politeness strategies discussed in this chapter focus on the polite end of the relational work spectrum and discuss a combination of positive and negative politeness strategies discussed in Brown and Levinson's (1987) theory. In some previous research, politeness strategies have been explored in both the HCI and human–robot interaction (HRI) communities, although the visual modality and/or the use of embodiment was as prominent as speech. For example, Wang et al. (2008) employed politeness strategies in a Wizard-of-Oz experiment providing tutorial feedback to students. The tutorial interface contained visual features—in the form of text and an animated robotic character that produces gestures—and text-to-speech (TTS) synthesis that would appear to come from the robotic character. In comparing polite and direct feedback, the authors note that students receiving the polite tutorial feedback learned better than those receiving the direct feedback. Furthermore, politeness appeared to be especially effective for students who displayed a preference for indirect help or were judged to have less ability to complete the task.

In an HRI-based experiment, positive attitudinal results were observed. Torrey, Fussell and Kiesler (2013) conducted a study in which participants observed videos of human and robot helpers giving advice to a person learning to make cupcakes. In creating the communication conditions, the authors used combinations of hedges and discourse markers. Hedges (e.g. *sort of*, *I guess*) are described by the authors as a negative politeness strategy mitigating the force of messages and reducing threats to a listener's autonomy. The authors acknowledge that descriptions of discourse markers (e.g. *like*, *you know*) have no standard definition,[1] though for the purposes of their study they are described in similar terms hedges in being used to 'soften commands' (Torrey et al., 2013, p. 277). Four communication conditions were created: direct (no hedges/discourse markers), hedges with discourse markers, hedges without discourse markers and discourse markers without hedges. Results of the experiment showed that hedges and discourse markers as individual strategies improved perceptions towards helpers in terms of considerateness, likeability and the helper being controlling compared to the direct condition. However, the combination of the two strategies did not show significant differences compared to the individual strategies. While positive improvements in perceptions towards both human and robot helpers were

[1]Discourse markers may also be referred to, amongst other terms, as *discourse particles, pragmatic particles* and *pragmatic expressions*. Their purposes can include switching topics, marking boundaries between segments of talk, helping to conduct linguistic repair and being used as hedging devices (Jucker & Ziv, 1998).

observed, participants only observed videos of interactions with helpers, rather than interact with any themselves.

In a similar study, Strait, Canning and Scheutz (2014) analysed both observations and actual interactions with robots providing advice in a drawing task. The authors created an experiment comparing three different interaction modalities: remote third-person (observations of interactions), remote first person (one-to-one with a robot via a laptop) and co-located first person (one-to-one with robot in the same room). As with the experiment by Torrey, Fussell, and Kiesler (2013), two communication conditions were presented. The indirect condition used a combination of positive politeness strategies (e.g. giving praise, being inclusive) and negative politeness strategies (e.g. being indirect, using discourse markers), whereas the direct condition referred to the absence of these strategies in the robot helper's speech. A further condition was included in the robot's appearance, which compared one robot with a more humanlike appearance and another with a more typical robotic appearance. The results of the experiment showed politeness strategies in the indirect speech condition improve ratings of likeability and reduced ratings of perceived aggression when compared to the direct speech condition. Improved ratings for considerateness were also observed in indirect speech, but only in the remote third-person interaction modality. The findings showed that previous results from observations of interaction of robots do not necessarily transfer to actual interaction.

17.3.2 Politeness in Non-embodied Computer Speech

The above studies highlight the mixed user responses towards different types of machines and interaction modalities using politeness strategies, focusing in particular on interactions with partners who are embodied or are represented visually. Many modern speech interface technologies like Google Assistant can include a minimal amount of visual output, depending on the device being used but do not necessarily include embodied features.

With this in mind, two further studies explored the use of politeness strategies in HCI, in which participants were tasked with constructing models under the instruction of a speech interface (Clark et al., 2014, 2016). In both studies, VL was used to create indirectness as a form of overall negative politeness strategy.[2] VL refers to language that is deliberately imprecise and can achieve a wide range of functional and interpersonal goals (Channell, 1994). For example, lexical hedges like *just* and *partly* can be used as a tension-management device to play down the perceived significance of research during academic conferences (Trappes-Lomax, 2007). Furthermore, vague nouns such as *thing* and *whatsit* can be used to replace a typical noun if speakers

[2]These were adaptors, e.g. *more or less, somewhat* (reduce assertiveness, minimise imposition); discourse markers, e.g. *so, now* (structure talk, mitigate assertive impact of utterance); minimisers, e.g. *just, basically* (structure talk, reduce perceived difficulty, mitigate utterance impact) and vague nouns, e.g. *thing, bit* (improve language efficiency) (Clark et al., 2016).

and listeners have both established what the vague nouns are referring to (Channel, 1994). While not all VL has functions in being polite, this is the primary purpose of which it used in the speech interface studies—the indirectness and imprecision of VL can contribute to lessening the perception of speakers being too authoritative (McCarthy & Carter, 2006) and help create an informal and less direct atmosphere during interaction.

In the first speech interface study using VL, two communication conditions were developed—a vague condition containing politeness strategies and a non-vague condition excluding these politeness strategies (Clark et al., 2014). Participants were tasked with building Lego models under the verbal instructions of a computer interface, the speech of which was produced by the TTS voice Cepstral Lawrence.[3] During this study, participants interacted with an interface on a MacBook Pro 10.2. This was a minimalistic interface using an HTML file linked to a library of pre-recorded speech files. The interface allowed participants to proceed to the next instruction or repeat a current instruction, with the pace being dictated by the participants. Results of this study showed that the non-vague interface was rated as significantly more direct and authoritative than the vague interface. However, post-task interviews revealed participants perceived the vague interface to be inappropriate in terms of its language choice. This was partly a result of the quality of the voice. People's expectations of a relatively robotic voice were matched more with the non-vague interface than the vague interface, with the latter discussed as being insincere and its language more appropriately suited to a more natural (i.e. humanlike) sounding voice.

A follow-up experiment explored vague communication conditions across three different voices (Clark et al., 2016). Two of these were TTS-synthesised voices—Cepstral Lawrence as per the previous experiment—and CereProc Giles.[4] The third voice was provided by a professional voice actor who was deemed to sound similar in age and accent to the two synthesised voices. Participants followed verbal instructions to build models using two of the three voices in two separate tasks. These tasks used the same style of interface as the first experiment. Results showed that the voice actor was perceived as significantly more likeable, more humanlike and less annoying than the two synthesised voices. Furthermore, it was perceived as more coherent than Giles, and both the voice actor and Lawrence were rated as allowing more task completion than Giles. Analysis of post-task interview data also revealed that VL in both synthesised voices was perceived negatively. Participants cited it as inappropriate and often commented on the jarring nature between the quality of the voice and the language being used. However, while the voice actor was seen as a more appropriate fit for VL, results were not wholly convincing. Despite the increased naturalness and humanlikeness, participants still highlighted the disparity between the more machinelike nature of the voice and the humanlike nature of the language. Even with a human voice, there were comments discussing it as 'just a machine' that is not capable of executing VL or politeness strategies, unlike other people, due to their inherent interpersonal and social linguistics purposes. This suggests that the

---

[3]https://www.cepstral.com.

[4]https://www.cereproc.com.

medium of speech delivery, in this case a machine, can also impact on perceptions of appropriateness and attractiveness.

17.4 Implications for Verbal Uncanny Valley Effects

In terms of what may be considered appropriate computer and human speech, the experiments discussed above raise the possibility of category boundaries existing on a linguistic level—verbal uncanny valley effects. While participants could not always explicitly identify individual lexical items that caused negative reactions towards the interfaces, they were able to identify a general disparity between the language being used and the interface that provided the language. Although this was not the case for all participants, there was a general trend towards describing the vague conditions in both experiments as humanlike language, whereas in Clark et al. (2014), the non-vague condition was cited as being appropriately machinelike.

In the sense of the latter, the use of direct and non-vague language was seen to match people's expectations of appropriate language use with a robotic synthesised voice. This is an example of matched speech-based stimuli, whereby categories of preconceived 'machine likeness' are aligned. Subsequently, there is little discussion about feelings of the uncanny arising, which are focused more on misaligned stimuli (Mitchell et al., 2011; Moore, 2012a). This also draws similarities with Moore's (2012a) discussion of appropriate voices in non-human artefacts. With non-vague and direct instructions provided by a robotic voice, appropriateness is seemingly determined as it matches people's expectations of what their interaction partner is capable of. These expectations and beliefs of what a communicative partner can produce may be referred to as peoples' partner models (e.g. Cowan, Branigan, Obregón, Bugis, & Beale, 2015). Previous research with infrequent users of IPAs has suggested that speech qualities such as regional accents can signal the communicative attributions people make towards artificial assistants (Cowan et al., 2017). Similarly, this may operate with the quality of a system's voice, the language it uses, and how these two relate to one another. A robotic voice may relate more to signals of using direct than indirect language that is absent in relational work, vague language or politeness strategies. In terms of users' expectations, these linguistic concepts may not be seen as residing in the category of appropriate computer speech.

This can be observed in the vague conditions of the two experiments (Clark et al., 2014, 2016). In the synthesised voices, in particular, the combination of a robotic sounding voice with language that is used to undertake social goals creates a mismatch in stimuli. Subsequently, uncanny valley effects can be observed, especially in participants' descriptions of their interactions with the interfaces. In the second experiment (Clark et al., 2016), however, using a pre-recorded human voice appeared to cause less perceived stimuli mismatch in the vague conditions than the synthesised voices. This may indicate that perceived categories of appropriate computer and human speech can be blurred somewhat with the introduction of more humanlike voices—a human voice can signal a perceptual cue of being capable of producing more humanlike language, even in a computer interface. However, the mismatch is not alleviated completely. Other cues, such as the medium and/or context of interaction (laptop interface providing task-based instructions), may alter what is perceived as appropriate speech even with a human voice.

17.4.1 Identifying Appropriateness in Computer Speech

Indeed, the combination of socially driven linguistic cues and computer speech output may create a *habitability gap* (Moore, 2017b), whereby there is a gap between a users' model of a system and the reality of the actual system (Hone & Graham, 2000). Users' models of computer speech may not include the use of interpersonal linguistic strategies and subsequently the presentation of actual computer speech that includes these creates feelings of unease or *perceptual tension* (Moore, 2012).

The mismatching of cues and accompanying perceptual tension in spoken interactions with computers and other machines appears strongly linked to perceptions of what is considered appropriate communication. In addition to a possible habitability gap, it may also be the case that perceived inappropriateness of politeness, relational work or vague language in computer speech is aligned with the socially driven nature of these concepts. Relational work and politeness strategies, for example, are primarily focused on establishing and maintaining interpersonal relationships with other people (Locher & Watts, 2008; Brown & Levinson, 1987). It is debatable as to what extent this can be accomplished in HCI, how achievable this is as a design goal, and how much users would desire this feature in a speech-based device. The social rules that underpin much HHI do not automatically transfer to HCI and the latter may be markedly diminished in comparison. Moore (2017b, p. 8) highlights a similar possible phenomenon—that there may be a 'fundamental limit' to the linguistic interactions between humans and machines due to them being '*unequal* partners'. The very nature of humans and machines means there are inherent differences in capabilities, and this is likely present in the partner models users create in speech-based HCI. When these partner models clash with experiences, this may lead to negative user experiences and perceptions of inappropriate, undesirable or unattractive speech interface partners.

The social rules underpinning HCI and HHI also do not automatically align. Relational work and politeness strategies are primarily focused on interpersonal relationships. Brown and Levinson's (1987) theory on politeness in particular is strongly associated with the process of facework during interaction. However, the maintenance of face during interaction with machines is different than with other people—machines do not have a face as such to protect and, in turn, users do not have another self-image they have to consider during interaction. There may be elements of corporate rather than individual self-images present during interaction, and users can still be imposed upon by machines. However, this remains markedly different from interaction with other people. Indeed, recent research observed that, while descriptions of conversations with people often discuss social and interpersonal wants and needs, interactions with machines are described in very functional and tool-like terms (Clark et al., 2019a). This may be due to a lack of familiarity and experience from which to draw upon. However, spoken interactions with machines lack many of the conversational complexities seen in human communication and are often limited to isolated question–answer pairs (Porcheron et al., 2018).

17.5 Future Work and Considerations for Computer Speech

This chapter has presented the possible existence of verbal uncanny valley effects—that perceptual tension and negative user experiences and attitudes can emerge in spoken interactions with computers when using linguistic strategies that are inherently social and interpersonal. This effect appears to be intensified with more robotic voices and lessened, though not entirely, with more humanlike voices. This differs from previous discussions of an auditory uncanny valley (e.g. Grimshaw, 2009; Meah & Moore, 2014) in that it focuses on both language and voice quality, and the relationship between them. Verbal uncanny valley effects suggest there may be category memberships that exist with styles of language that focus on relational work—i.e. that other people are members of this category, whereas computers do not become automatic members by virtue of employing the same strategies. Doing so may create an impression of machines encroaching upon the verbal space of people. This is similar to Moore's (2017b) discussion of there being a fundamental limit to spoken interaction between humans and machines. Moore (2015) mentions that endowing machines with features like humanlike voices can create the mismatched stimuli that lead to perceptual tension, and this may also hold true for certain linguistic styles. With similar considerations, it appears that reducing perceptual tension with verbal uncanny valley effects may depend partly on the relationship between voice and language. If using a very robotic voice, interpersonal linguistic strategies may not be appropriate and may be subsequently undesirable and unattractive. Conversely, if wanting to employ these strategies, a more humanlike voice would be more appropriate. However, there remains the possibility that no matter what voice is used, certain interpersonal language may be evaluated negatively regardless due to fundamental and embedded differences in user expectation between humans and computers as interlocutors.

It is likely that this is not always the case—this argument stops short of saying all types of interpersonal linguistic strategies are off-limits. However, there are design choices around voice and language to consider for computers using speech. There are also other choices to consider. The discussions of politeness strategies and VL in this chapter tend to focus on task-based scenarios in HCI. While this is arguably where most speech-based HCI still currently remains at a linguistic level, it may be the case that instruction-giving or advice-giving computers in task-based scenarios are not appropriate vessels for interpersonal language. If the aim of an interaction between speaking computers and humans is fundamentally an interpersonal one (e.g. social talk Gilmartin, Cowan, Vogel, & Campbell, 2017) or in healthcare dialogues

(Bickmore et al., 2018), then these linguistic styles may be more appropriate. Similarly, the role in which both computer and human play in any given interaction may also influence evaluations of speech—an instruction-giver may be treated differently to a machine that operates more on a peer-level or as a caregiver, due to varying levels of power and exactly what linguistic possibilities these roles may afford. Similarly, human-controlled speech synthesis output, such as the use of a vocal synthesiser to create the ability to speak, may be evaluated differently to speech synthesis output that is controlled by a machine. Furthermore, the direction of interaction may have an effect. Previous experiments often focus on speech output only from a system, whereas two-way dialogue may induce different evaluations. Previous research has shown that politeness can be reciprocated back and forth in an interaction with an in-car help system (Large, Clark, Quandt, Burnett, & Skrypchuk, 2017), though the work does not provide insight into people's actual evaluations of the system.

However, while these ideas are rooted in evidence from previous research, there is still the need to test them further. As noted in Sect. 17.2, the evidence for the uncanny valley alone is scarce, with Moore's (2012) Bayesian approach offering a rare quantitative verification of its existence. Future research endeavours can explore the concept of a verbal uncanny valley and its effects further in both quantitative and qualitative means, although any notions of a valley in terms of the shape are arguably less important than the effects caused by underlying concepts of fundamental communicative limits. Comparisons with actual human stimuli as well as computers may also prove beneficial. Indeed, quantifying what constitutes 'humanlike' or 'machinelike' communication is a complex process. Given the increasing prevalence of computer speech, what is perceived as 'machinelike' may well change over the years as familiarity with these devices increases. Longitudinal studies may also uncover further evidence on the effects of prolonged interaction with devices and the extent to which this may affect any verbal uncanny valley effects.

17.6 Summary and Conclusion

Determining what is considered appropriate speech in HCI remains a challenge. Moore (2017a) offers examples of how to determine appropriateness in the voices of non-human artefacts and avoid uncanny valley effects—robotic rather than humanlike in less sophisticated systems may be better at matching users' expectations of a system with reality. Language use, however, is arguably a more complex affair. This chapter discusses three concepts of interpersonal linguistic strategies (politeness, relational work and VL) to explore what may be considered appropriate language use in speech-based HCI. In linking previous experiments on these strategies with research on the uncanny valley, we find that the social rules that underpin human interaction do not automatically transfer to HCI. The concept of face—the social self-image presented to others—is mostly non-existent on the part of the system during interaction. The need to conduct facework, i.e. protecting this self-image, is then diminished. While users can still be imposed upon by an interface, using strategies like politeness and VL may not always be appropriate and may be undesirable. The combination of computer speech and interpersonal language gives rise to perceptual mismatch at the category boundaries between human and computer speech, creating potential for negative user evaluations of systems. Consequently, this raises the potential of verbal uncanny valley effects, whereby the use of very 'humanlike' language creates feelings of perceptual tension in HCI. While a humanlike voice can act as a moderator for these effects, it does not alleviate perceptual tension completely. Future research should explore the empirical testing of the verbal uncanny valley and its effects, identify what linguistic concepts are seen to reside in the category of appropriate and inappropriate computer speech, and understand what further phenomena (like voice) may influence its evaluation by users.

Acknowledgments This research was funded by a New Horizons grant from the Irish Research Council entitled "The COG-SIS Project: Cognitive effects of Speech Interface Synthesis" (Grant R17339).

References

Abercrombie, D. (1967). *Elements of general phonetics* (Vol. 203). Edinburgh: Edinburgh University Press.

Aylett, M. P., Cowan, B. R., & Clark, L. (2019). Siri, echo and performance: You have to suffer darling. In *Proceedings of the 2019 CHI Conference on Human Factors in Computing Systems*. ACM.

Bickmore, T. W., Trinh, H., Olafsson, S., O'Leary, T. K., Asadi, R., Rickles, N. M., & Cruz, R. (2018). Patient and consumer safety risks when using conversational assistants for medical information: An observational study of Siri, Alexa, and Google Assistant. *Journal of Medical Internet Research, 20*(9). https://doi.org/10.2196/11510.

Brown, P., & Levinson, S. C. (1987). *Politeness: Some universals in language usage*. Cambridge University Press.

Cameron, D. (2001). *Working with spoken discourse*. SAGE.

Carr, E. W., Hofree, G., Sheldon, K., Saygin, A. P., & Winkielman, P. (2017). Is that a human? Categorization (dis)fluency drives evaluations of agents ambiguous on human-likeness. *Journal of Experimental Psychology: Human Perception and Performance, 43*(4), 651–666. https://doi.org/10.1037/xhp0000304.

Channell, J. (1994). *Vague language*. Oxford University Press.

Clark, L. (2018). Social boundaries of appropriate speech in HCI: A politeness perspective. In *Proceedings of British HCI*.

Clark, L., Cabral, J. & Cowan, B. R. (2018). The CogSIS project: Examining the cognitive effects of speech interface synthesis. In *Proceedings of British HCI*.

Clark, L., Doyle, P., Garaialde, D., Gilmartin, E., Schlögl, S., Edlund, J., ... & Cowan, B. R. (2019a). The state of speech in HCI: Trends, themes and challenges. *Interacting with Computers, 31*(4), 349–371. https://doi.org/10.1093/iwc/iwz016.

Clark, L., Pantidi, N., Cooney, O., Doyle, P., Garaialde, D., Edwards, J., ... & Cowan, B.R. (2019b, May). What makes a good conversation? challenges in designing truly conversational agents. In *Proceedings of the 2019 CHI Conference on Human Factors in Computing Systems* (pp. 1–12). https://doi.org/10.1145/3290605.3300705.

Clark, L. M. H., Bachour, K., Ofemile, A., Adolphs, S., & Rodden, T. (2014). Potential of imprecision: Exploring vague language in agent instructors (pp. 339–344). ACM Press. https://doi.org/10.1145/2658861.2658895

Clark, L., Ofemile, A., Adolphs, S., & Rodden, T. (2016). A multimodal approach to assessing user experiences with agent helpers. *ACM Transactions on Interactive Intelligent Systems, 6*(4), 29:1–29:31. https://doi.org/10.1145/2983926.

Coulthard, M. (2013). *Advances in spoken discourse analysis*. Routledge.

Cowan, B. R., Branigan, H. P., Obregón, M., Bugis, E., & Beale, R. (2015). Voice anthropomorphism, interlocutor modelling and alignment effects on syntactic choices in human – computer dialogue. *International Journal of Human-Computer Studies, 83,* 27–42. https://doi.org/10.1016/j.ijhcs.2015.05.008.

Cowan, B. R., Pantidi, N., Coyle, D., Morrissey, K., Clarke, P., Al-Shehri, S., ... Bandeira, N. (2017). *'What can I help you with?': Infrequent users' experiences of intelligent personal assistants* (pp. 1–12). ACM Press. https://doi.org/10.1145/3098279.3098539.

Gilmartin, E., Cowan, B. R., Vogel, C., & Campbell, N. (2017). Exploring multiparty casual talk for social human-machine dialogue. In *International Conference on Speech and Computer* (pp. 370–378). Springer.

Goffman, E. (1955). On face-work. *Psychiatry, 18*(3), 213–231. https://doi.org/10.1080/00332747.1955.11023008.

Goffman, E. (2005). *Interaction ritual: Essays in face to face behavior*. AldineTransaction.

Grimshaw, M. (2009). The audio Uncanny Valley: Sound, fear and the horror game. *Audio Mostly,* 21–26.

Hone, K. S., & Graham, R. (2000). Towards a tool for the subjective assessment of speech system interfaces (SASSI). *Natural Language Engineering, 6*(3–4), 287–303.

Jucker, A. H., & Ziv, Y. (1998). *Discourse markers: Descriptions and theory*. John Benjamins Publishing.

Kätsyri, J., Förger, K., Mäkäräinen, M., & Takala, T. (2015). A review of empirical evidence on different uncanny valley hypotheses: Support for perceptual mismatch as one road to the valley of eeriness. *Frontiers in Psychology, 6*. https://doi.org/10.3389/fpsyg.2015.00390.

Large, D. R., Clark, L., Quandt, A., Burnett, G., & Skrypchuk, L. (2017). Steering the conversation: A linguistic exploration of natural language interactions with a digital assistant during simulated driving. *Applied Ergonomics, 63,* 53–61. https://doi.org/10.1016/j.apergo.2017.04.003.

Laver, J. (1980). *The phonetic description of voice quality* (Cambridge Studies in Linguistics). Cambridge: Cambridge University Press.

Locher, M. A. (2004). *Power and politeness in action: Disagreements in oral communication*. Walter de Gruyter.

Locher, M. A. (2006). *Polite behavior within relational work: The discursive approach to politeness*. Walter de Gruyter.

Locher, M. A., & Watts, R. J. (2005). Politeness theory and relational work. *Journal of Politeness Research. Language, Behaviour, Culture, 1*(1). https://doi.org/10.1515/jplr.2005.1.1.9

Locher, M. A., & Watts, R. J. (2008). *Relational work and impoliteness: Negotiating norms of linguistic behaviour*. Mouton de Gruyter.

Luger, E., & Sellen, A. (2016). 'Like having a really bad PA': The gulf between user expectation and experience of conversational agents. In *Proceedings of the 2016 CHI Conference on Human Factors in Computing Systems* (pp. 5286–5297). New York, NY, USA: ACM. https://doi.org/10.1145/2858036.2858288.

McCarthy, M., & Carter, R. (2006). This that and the other: Multi-word clusters in spoken English as visible patterns of interaction. *Explorations in Corpus Linguistics*, 7.

Meah, L. F. S., & Moore, R. K. (2014). The uncanny valley: A focus on misaligned cues. In M. Beetz, B. Johnston, & M.-A. Williams (Eds.), *Social robotics* (pp. 256–265). Springer International Publishing.

Mitchell, W. J., Szerszen, K. A., Lu, A. S., Schermerhorn, P. W., Scheutz, M., & MacDorman, K. F. (2011). A mismatch in the human realism of face and voice produces an Uncanny Valley. *I-Perception, 2*(1), 10–12. https://doi.org/10.1068/i0415.

Moore, R. K. (2012). A Bayesian explanation of the 'Uncanny Valley' effect and related psychological phenomena. *Scientific Reports, 2*(1). https://doi.org/10.1038/srep00864.

Moore, R. K. (2015). *From talking and listening robots to intelligent communicative machines.* In Robots that talk and listen: de Gruyter.

Moore, R. K. (2017a). Appropriate voices for artefacts: Some key insights. In *1st International Workshop on Vocal Interactivity in-and-between Humans, Animals and Robots.*

Moore, R. K. (2017b). Is spoken language all-or-nothing? Implications for future speech-based human-machine interaction. In *Dialogues with Social Robots* (pp. 281–291). Springer, Singapore. https://doi.org/10.1007/978-981-10-2585-3_22.

Mori, M. (1970). The uncanny valley. *Energy, 7*(4), 33–35.

Mori, M., MacDorman, K. F., & Kageki, N. (2012). The uncanny valley [from the field]. *IEEE Robotics and Automation Magazine, 19*(2), 98–100.

Porcheron, M., Fischer, J. E., Reeves, S., & Sharples, S. (2018). Voice interfaces in everyday life. In *Proceedings of the 2018 CHI Conference on Human Factors in Computing Systems* (p. 640). ACM.

Porcheron, M., Fischer, J. E., & Sharples, S. (2017). *'Do animals have accents?': Talking with agents in multi-party conversation* (pp. 207–219). ACM Press. https://doi.org/10.1145/2998181.2998298.

Strait, M., Canning, C., & Scheutz, M. (2014). *Let me tell you! Investigating the effects of robot communication strategies in advice-giving situations based on robot appearance, interaction modality and distance* (pp. 479–486). ACM Press. https://doi.org/10.1145/2559636.2559670.

Torrey, C., Fussell, S. R., & Kiesler, S. (2013). *How a robot should give advice* (pp. 275–282). IEEE. https://doi.org/10.1109/HRI.2013.6483599

Trappes-Lomax, H. (2007). Vague language as a means of self-protective avoidance: Tension management in conference talks. In *Vague language explored* (pp. 117–137). Springer.

Wang, N., Johnson, W. L., Mayer, R. E., Rizzo, P., Shaw, E., & Collins, H. (2008). The politeness effect: Pedagogical agents and learning outcomes. *International Journal of Human-Computer Studies, 66*(2), 98–112. https://doi.org/10.1016/j.ijhcs.2007.09.003.

Watts, R. J. (2003). *Politeness.* Cambridge University Press.

Zuckerman, M., & Driver, R. E. (1988). What sounds beautiful is good: The vocal attractiveness stereotype. *Journal of Nonverbal Behavior, 13*(2), 67–82. https://doi.org/10.1007/BF00990791.

The manufacturer's authorised representative in the EU is Springer Nature Customer Service Centre GmbH, Europaplatz 3, 69115 Heidelberg, Germany. If you have any concerns regarding our products, please contact ProductSafety@springernature.com

Printed and bound by CPI Group (UK) Ltd, Croydon, CR0 4YY
30/09/2026
02225330-0010